Decision Analytics

Decision Analytics

Mathematical Models and Algorithms for Sequential Decision-Making

Brian T. Denton

University of Michigan, Ann Arbor, MI, USA

IEEE Press
WILEY

Published by John Wiley & Sons, Inc., Hoboken, New Jersey.

Library of Congress Cataloging-in-Publication Data has been applied for:

Print ISBN: 9781394345786
ePDF ISBN: 9781394345809
ePub ISBN: 9781394345793
oBook ISBN: 9781394345816

Cover Design: Wiley
Cover Image: Erhan Dayi/Shutterstock

To my wife, Laura, for her love and support.

Contents

About the Author

Brian Denton is the Stephen M. Pollock Professor of Industrial and Operations Engineering (IOE) at the University of Michigan. His research interests are data-driven decision-making and optimization under uncertainty with applications to healthcare delivery, public health, scheduling, supply chain management, and other industrial systems. He received his PhD from McMaster University in Canada. He served as department chair of IOE from 2018–2023. He was a visiting professor and fellow at Oxford University during the 2023 academic year. Before joining the University of Michigan, he worked at IBM, Mayo Clinic, and North Carolina State University. His honors and awards include the INFORMS George Kimball Medal, the National Science Foundation Career Award, the INFORMS Daniel H. Wagner Prize, and the Institute of Industrial Engineers Outstanding Publication Award. He has co-authored over 140 journal articles, conference proceedings, book chapters, and patents. He is an elected Fellow of INFORMS and IISE and past President of INFORMS.

Acknowledgments

I have many people to thank for their generous help in completing this book. First and foremost, I thank my wife, Laura, for her patience and support during the many long hours I spent working to complete it.

I am grateful to the many students in my dynamic programming classes at the University of Michigan for the feedback they provided on early versions of this material, which became the foundation of this book and helped shape my perspective on how to teach these topics.

Many people provided feedback on chapter drafts, including David Kaufman, Jack Fistrum, Allison Grimsted, and Lingzhi Hao. I also thank the Wiley editors and anonymous reviewers who provided feedback on the book proposal and the initial draft and whose suggestions helped improve the manuscript.

Finally, I am grateful to the University of Michigan for the opportunity to take a sabbatical and to Agni Orfanoudaki for sponsoring my visiting fellowship at Exeter College and Oxford University, where much of this book was written.

About the Companion Website

This book is accompanied by a companion website.

www.wiley.com/go/denton/decisionanalytics1

To access the website materials, click on the link under Related Resources on the book's page on wiley.com. You can also find this by searching www.wiley.com for the ISBN [9781394345786].

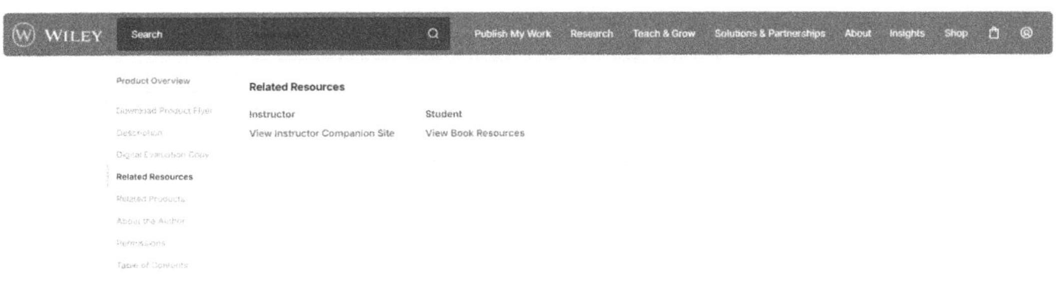

The website includes:

- Practice Exercises and Solution Manual

1

Introduction to Decision Analytics

1.1 Introduction

Sequential decision-making is a fundamental part of the human experience. The ability to think and act rationally over time to achieve an objective can be broken down into the following steps: observe the available data, process it logically, make a decision based on the processing, and repeat this process. This can be summarized succinctly by the question: *What shall I do next?* Sequential decision-making arises in many contexts, including personal decisions, business operations, energy systems, finance, healthcare delivery, transportation systems, sports, and many other situations where a *decision maker* needs to select among multiple choices to decide what to do next.

In some situations, the "What next?" question is easy and determined by others' preferences or expectations (e.g., customer expectations, employer policies). In other cases, the outcome of alternative decisions may be inconsequential, in which case the decision maker may be satisfied with several decisions and happy to let chance dictate what happens next (e.g., flip a coin). However, decision-makers quickly become overwhelmed when the stakes are high and complicating factors are present. For instance, decision-making is challenging when allocating scarce or expensive resources, when there are numerous alternatives, and when there is uncertainty about the future. Complexity is high, especially when present decisions have implications for future decisions.

People have been making decisions since the dawn of humanity, collecting data through personal experience or learning from others (e.g., parents, peers, teachers), and using their acquired knowledge to inform their decisions. However, it is only in the last century that computers and algorithms have significantly aided decision-makers. The first half of the 20th century witnessed significant advances across key areas essential to informed decision-making, including computer science, economics, mathematics, psychology, and statistics. The origins of *operations research*, a multidisciplinary field that blends these key areas, are often associated with military decision-making during World War II, when scientists were recruited to apply approaches from multiple fields to analyze data and recommend decisions on resource allocation, logistics planning, risk mitigation, and more. Gass and Assad [1] provide a brief history of the field for interested readers. Since its early days, operations research and the associated scientific and mathematical approaches to decision-making have continually evolved to address increasingly complex decision-making situations.

The progression of modern computing technology began in the 1950s with the creation of mainframes, followed by a steady stream of advances, including personal computers, laptops, and increasingly compact computing devices, such as tablets and smartphones. Over time, advances in computational speed, data storage efficiency, and cost have enabled continuous improvements in algorithms that automatically store and process enormous datasets, employing mathematical decision-making models that far surpass human capabilities in many tasks. However, this raw computing power is only as useful as the models and algorithms that transform data into decisions; thus, human decision-makers' judgment remains as important as ever in guiding the appropriate use of computing technology.

1.2 Mathematical Models for Decision-making

When faced with important and complex decisions, decision-makers might wish they had an *oracle* to consult. While oracles exist only in fictional settings, imagining their use for decision-making is a worthwhile thought experiment that reveals important considerations. First, the Oracle must have a way to characterize the decision-maker's situation and the decisions to be made. In other words, the Oracle must be aware of the context in which the decision is to be made. Next, there must be an agreed-upon collection of feasible decisions that can be selected in each situation. Finally, the decision maker's criterion for evaluating decision quality must be known to the Oracle. With these characteristics in place, a communication protocol with an Oracle can be defined as follows: (1) the decision maker describes their situation to the Oracle; (2) the Oracle recommends a decision from the collection of feasible decisions; (3) the decision maker acts on the recommendation, subsequently finding themselves in a new situation, faced with new decisions; (4) the decision maker describes their new situation to the Oracle, receives a recommended decision, and so on. This multi-step process may extend over many steps or even continue indefinitely if there is no defined endpoint for the decision-making process.

The notion of a decision maker working with an oracle is analogous to that of a decision maker working with computational models and algorithms, which require several of the same key elements. The most important element is the *state* of the *system*. The state encapsulates only the information needed to make an optimal decision at a given point in time, while the system is the environment in which decisions are made. For example, consider a global positioning system (GPS) that directs your travel from home to work over a network of roads, avoiding traffic delays. The state is the location of the decision-maker, and the system is the transportation network in a geographic region. In principle, additional information could be added to the state's definition, such as what the decision maker had for breakfast or their favorite movie, but such information is clearly not materially relevant to the task at hand. Recommendations from the oracle are instructions for decisions to take at discrete points in time (e.g., drive 1.2 km and then turn left onto State St., then drive 400 m and turn right onto Huron St., ...). Thus, states should define the minimal amount of information necessary to make the best possible decision (e.g., when to turn). Other key elements include the set of feasible decisions and a mathematical model that defines transitions between states over time. Finally, a quantitative measure of reward is necessary to define the decision-maker's criterion (e.g., minimizing distance traveled).

This book focuses on how to formulate and solve mathematical models to support decision-makers. Defining the system's state is the first challenge in formulating a model. Equally important, but often less difficult to define, is the set of possible decisions at each state. Next, the decision-maker's criterion must be defined quantitatively for objective comparison and ranking of alternative decisions using a *reward function* that yields feedback on the contribution decisions taken in each state made towards the criterion. The ultimate goal of the models discussed in this book is to find an optimal *policy*. The notion of a policy in sequential decision-making ties together the states, decisions, and rewards. Policies prescribe the decision to make for every possible state (the role of the Oracle above). Figure 1.1 illustrates the sequential nature of the decision-making process as a cycle with three key elements: states, decisions, and rewards. The order of steps is: (1) observe the current state of the system; (2) make a decision; (3) receive a reward that reflects the contribution of the state and decision towards the decision-maker's criterion. In sequential decision-making environments, these steps occur repeatedly until some goal state is reached, or until the end of a finite *decision horizon*, or, alternatively, indefinitely over an infinite horizon if no specific endpoint exists.

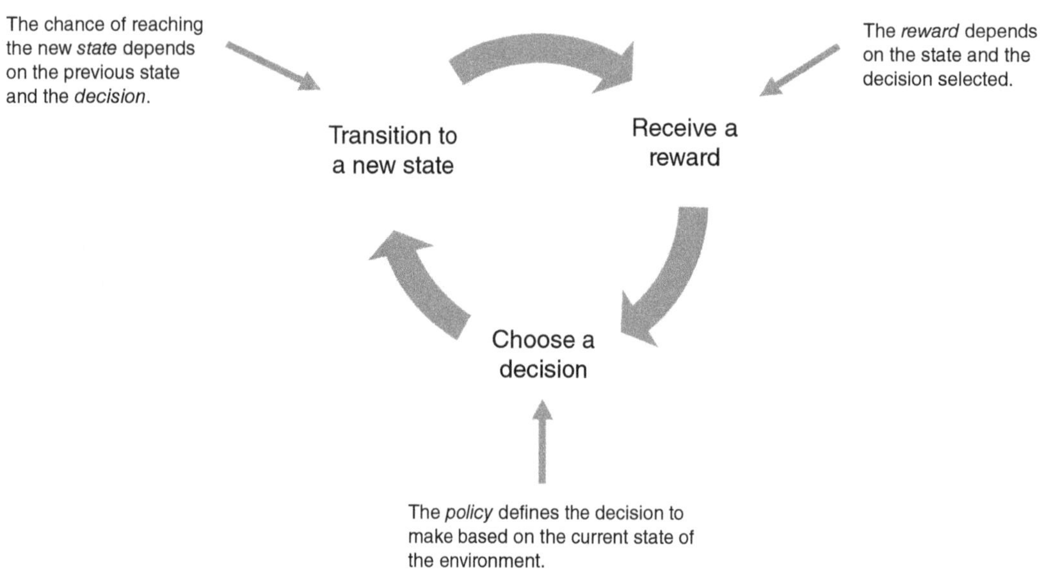

Figure 1.1 The key steps of the sequential decision-making process.

This book explores various types of mathematical decision-making models and the algorithms that can solve them. Many alternative approaches have been developed and recognized for their merit, each offering distinctive advantages in different contexts. The following chapters cover common features of models, providing an understanding of the principles at the heart of mathematical decision-making and their limitations. Unlike oracles (who actually see the future) or human decision-makers (who use their judgment), the approaches in this book provide a scientific, repeatable means to find an optimal policy that aligns with the decision-maker's criterion for success. Intertwined with the discussion of models is the discussion about the algorithms that solve them. The following chapters cover a range of alternative models and algorithms that can be applied in various settings. Many examples are provided as a basis for learning about which models are best suited for alternative settings that can arise due to factors such as uncertainty about the outcomes of decisions and future states, alternative criteria that decision makers may have, imperfect information about the state of the system, situations involving multiple decision makers, and many others. Before jumping into mathematical models and algorithms, this chapter introduces some intuitive concepts and examples of applications.

Finding the best policy involves identifying a policy that maximizes rewards (or minimizes penalties) for the decision-maker to achieve the specified criterion (e.g., maximizing profits, minimizing risk, minimizing travel time). In the GPS example, the criterion could be to achieve the shortest distance traveled, the shortest travel time, or find the most scenic route, for instance. The most straightforward algorithms use *total enumeration* to find the optimal policy. Letting π define the abstract notion of a policy (a mapping of states to decisions), and V the value of expected rewards associated with a particular policy, the following pseudocode serves to solidify the basic concept of finding an optimal policy.

For all policies π in the set of allowable policies, Π:
  $V = \text{Evaluate}(\pi)$
  If V is higher than the previous best policy:
    $\pi_{\text{best}} = \pi$
  end if
end.

Cycling through and comparing all policies will work, in theory; however, it is often computationally infeasible to enumerate all policies in practice. For example, in the GPS example, the reader can imagine the numerous possible routes from one location to another as the distance from the origin to the destination increases (e.g., compare the number of possible routes from your home to your nearest grocery store versus the number of possible routes from Burlington, VT, to San Francisco, CA). All but the most trivial examples become computationally intractable when enumerating all possible policies. Nevertheless, the total enumeration algorithm provides a simple conceptual framework for understanding the problem. With the essential elements of a sequential decision-making model (states, decisions, and rewards) in place, the remainder of this book explores alternative approaches to formulating, analyzing, and solving mathematical models. Before delving into how to formulate and solve the models, several examples are used to motivate practical applications, many of which will be covered in later chapters.

1.3 Examples of Practical Applications

The overarching goal of this book is to introduce mathematical models and algorithms for quantifying and optimizing sequential decisions. At this point, a working definition of the key elements of a sequential decision-making process has been established: states, decisions, rewards, and policies. As will become clear, the approaches covered in this book apply to many domains. For now, the following is a list of motivating examples where the approaches discussed in this book have been used.

1.3.1 Healthcare

The applications of sequential decision-making methods in healthcare have become increasingly prevalent in recent years, as the data required to create and test models has become more readily available in electronic health records. One example is cardiovascular disease, a leading cause of death in most developed countries. Fortunately, medications for controlling cholesterol and blood pressure, the major controllable risk factors associated with cardiovascular disease, have become available. However, decisions to initiate medication must weigh the benefits against its side effects. Mason et al. [2] developed a model to describe the stochastic (random) variation in cholesterol and blood pressure over time, using data from a cohort of high-risk patients who were followed for many years. The model aimed to optimize the timing and sequence of decisions regarding medication initiation as patients age and their cholesterol and blood pressure levels, and hence their risk, change. The states were defined by clinically relevant ranges of cholesterol and blood pressure (low, medium, high, and very high), and the decisions were whether to start one or more medications at annual time points corresponding to annual preventive care visits to a physician. An optimal policy recommends treatment for each state, defined by the combination of age, cholesterol, blood pressure, and a history of medications already initiated. Results showed that an optimal policy could significantly outperform published medical guidelines. Other healthcare examples include liver transplants [3], breast cancer screening [4], and HIV treatment [5].

1.3.2 Operations Management

There are numerous applications of sequential decision-making to business problems, including inventory control, machine maintenance, production planning, and logistics. Drawing on a particular example, Ferreira, Arruda, and Marujo [6] formulated a sequential decision-making model for inventory management in the context of humanitarian logistics, where many important items (e.g., food, blood supply, medications) have a limited shelf life, and at the same time, demand for items is uncertain and evolves according to a stochastic process. Ferreira et al. describe a model in which the states are defined by the levels of unexpired inventory, and decisions are made on how much of each item to order at the beginning of each week. The optimal policies in these problems prescribe ordering decisions for each state to minimize costs, including the cost of inventory holding, the purchasing and transportation cost for orders, the cost of disposal of expired items, and a shortage cost when items in inventory are insufficient to cover demand for a particular week.

1.3.3 Energy

Optimizing energy systems management is another important application area, with many examples of sequential decision-making models. Perera and Kamalaruban [6] reviewed examples of models of energy systems using reinforcement learning, a sequential decision-making approach discussed in Chapter 8 that seeks to learn the optimal policy through a process of *exploration* (experimenting with new decisions) and *exploitation* (using previous experience with decisions). Pereira and Pinto [7] formulated a multi-stage stochastic program, a topic covered in Chapter 5, for optimizing energy systems decisions in light of uncertainty in supply and demand. They applied their approach to the case of a hydroelectric power system with energy stored in a collection of water reservoirs. The sequential decisions in their model involve determining how much water to release to meet uncertain demand that is observed sequentially over time, taking into account additional uncertainty in future supply due to limited reservoir capacity. Their model aimed to minimize operating costs, the primary component of which is the amount of costly hydrothermal energy required to meet any shortfalls in the hydroelectric energy supply.

1.3.4 Games

Games have often been the subject of sequential decision-making models, serving as a valuable test bed for new methods. One such example is blackjack, a card game in which a *player* (decision-maker) seeks to beat the *dealer* by getting a higher total card sum without going over 21. Initially, the dealer deals two cards face up to the player, one face up and one face down to themselves. The player goes first in requesting an additional card (hit) or declining to take more cards (stick). The decision to hit or stick is sequential and continues until the player's sum of cards exceeds 21 or the player sticks. Next, the dealer draws cards until the dealer's sum exceeds or is equal to the player's sum (dealer wins) or the sum exceeds 21 (player wins). Importantly, the dealer wins in the event of a tie, and the dealer must continue to take cards while their score is below 17 (even if the sum already exceeds that of the player). Algorithms have been successfully applied to learn the player's optimal policy for blackjack, where the states are the player's sum and the dealer's card showing. The decisions are whether to hit or stick, with the latter ending the player's sequential decisions. Finally, the reward is 1 if the player wins and 0 otherwise. The optimal policy maximizes the long-term proportion of hands the player wins, which falls below 50% (so it is best to avoid betting on blackjack!).

1.3.5 Examples of Sequential Decisions in Everyday Life

Some of the above examples may seem far removed from typical situations encountered in real life. Thus, the following is a list of some everyday examples of sequential decision-making to provide further motivation and intuition about the types of problems this book addresses.

Navigation: Mapping software, which can be found on most smartphones, takes a location obtained from a GPS, which defines the current state and recommends decisions about the next steps to achieve some goal defined by a reward function. Rewards could be defined for alternative goals, such as finding the shortest distance or shortest travel time to reach a desired location, avoiding construction or toll roads.

Temperature Control: In buildings with heating or cooling systems, a thermostat monitors the temperature at the location and switches on heating or cooling to achieve the desired temperature. In this case, the state is the temperature, and the decision is whether to start cooling or heating, given the desired target temperature. Rewards may be quantified by measures such as the time the actual temperature is within a certain tolerance level of the desired temperature.

Autonomous Vehicles: Automobiles with automated driving functions must routinely make decisions about the speed and direction of the vehicle at each instant in time. The state definition encompasses the vehicle's physical location, as well as the presence of surrounding vehicles, pedestrians, or objects in the vehicle's path. A reward function defines the competing goals of safe and efficient travel to a desired location.

Sports: Most competitive sports involve sequential decisions made throughout a game or match. For example, in soccer, players with the ball must make a sequence of decisions about whether to keep the ball or pass the ball to another player based on changing conditions on the field that define the state, such as the remaining time, score of the game, proximity to the goal, the position of teammates, and the position of the opponents. The reward function is defined by the end goal of winning the game with a terminal reward of one (win) or zero (loss).

The examples above are the tip of the iceberg regarding the applications of models to sequential decision-making. These, and many other examples, will be discussed in future chapters. All of the examples were developed under the assumption that there is a human decision maker in the loop, as illustrated in Figure 1.1, and that the model and algorithm support the decision-maker. Even in learning-based approaches intended to act autonomously, it is assumed that the criteria defining the reward function are set by a human decision-maker and that the decisions are subject to their oversight.

1.4 Python and Computational Examples

This book includes dozens of computational examples to illustrate the application of the methods discussed. Python was chosen because it is readable and relatively easy to learn compared to compiled languages. It is also part of a very large open-source ecosystem and therefore widely available at no cost. The code for the Python examples is included inline in the text for convenience, so readers can review the implementation, the problems solved, and the results together. All code is provided verbatim, so an electronic version of the book provides the full code. Python version 3.12.5 was used for all examples. The reference in the bibliography provides a link to the website. Undoubtedly, some aspects of the code will eventually be deprecated. For this reason, revisions to the code can be found at https://github.com/btdenton/Decision-Analytics. All libraries used are open source and free, and some may require installation depending on the distribution. One exception is the commercial software package Gurobi version 13.0, which is used to solve the optimization models described in this book, starting with Chapter 5. Free access is provided by Gurobi for educational use and academic (non-commercial) research as of the book's publication. Readers who prefer fully open-source alternatives will find other optimization solvers referenced in the book.

1.5 Summary of Future Chapters

This book aims to provide an introduction to the various types of mathematical models and algorithms employed in sequential decision-making. It is assumed that the reader understands basic concepts in probability, statistics, and linear algebra at a level consistent with the typical requirements of a science, technology, engineering, and mathematics (STEM) undergraduate program. Advanced calculus, real analysis, and other advanced topics are avoided in favor of an accessible introductory treatment of the most important material. It is assumed the reader has some basic knowledge of Python—or a willingness to learn—because computational examples are used throughout the book. All the Python examples were developed with a bias toward ease of understanding over computational efficiency. Mathematical notation will be used extensively to formulate models and to describe solution methods. All notation is defined in a table in Appendix A.1 for easy reference.

Chapter 2 discusses *decision trees*, which visually represent a sequential decision-making process that involves multiple decisions and randomness in the model parameters. Key concepts include *conditional probabilities*, *expected values*, the *time value of money*, and various criteria for quantifying the decision maker's goal. The concepts of risk preferences and utility functions as measures of payoffs for decision-making are introduced. Finally, the idea of working backwards from the end of a decision tree to the beginning, known as *collapsing* the decision tree, is described. The chapter uses several examples, including a decision tree for evaluating the use of biomarkers for cancer detection and another for deciding what steps to take when launching a new product.

Chapter 3 introduces the powerful tool of *dynamic programming* and the *principle of optimality*. This key discovery led to the development of efficient methods for finding optimal policies for sequential decision-making in a wide range of applications. The reader is no doubt familiar with the notion that today's decisions influence tomorrow's outcomes and subsequent decisions, as this is fundamental to understanding time and cause-and-effect relationships. As discussed in Chapter 3, when optimizing sequential decisions, the converse is also true: the best decision today depends on what future (recourse) decisions will be possible in the future. Fundamental concepts such as the elements of a dynamic program and *optimality equations* are introduced, and their use is illustrated with examples, including shortest path problems, production planning, pattern recognition, and resource allocation.

Chapter 4 extends deterministic dynamic programs to the stochastic domain, where decisions must be made without perfect knowledge of the future. After a short review of key aspects of probability, Markov chains are introduced as the most common stochastic process to model uncertainty. Mathematical models and methods for solving Markov decision processes (MDPs) are discussed and supported by analysis for both finite- and infinite-horizon models. The most well-known algorithms are described, and the theoretical concepts required to demonstrate their correctness are presented. Examples of the algorithms' use include medicine, battery charging, and machine maintenance.

Chapter 5 introduces constrained optimization models, also known as *mathematical programming* models, which provide an alternative approach to modeling sequential decision-making. This chapter uses a small, two-dimensional example to illustrate key factors. The remainder of the chapter focuses on model formulation for multiple types of models, including linear, integer, and stochastic programs. Stochastic programs, a type of constrained optimization model used for sequential decision-making under uncertainty, are discussed, and evaluation metrics such as the expected value of perfect information (EVPI) and the value of the stochastic solution (VSS) are introduced. The commercial solver Gurobi is combined with Python to formulate and solve examples in the context of vaccine purchasing, scheduling truck arrivals to a depot, energy systems, and finance.

Chapter 6 covers *Monte Carlo Simulation*, a fundamental approach used to evaluate stochastic systems affected by uncertainty, which can be scaled up to extremely large models. The relevance of this chapter to analyzing and solving sequential decision-making models of various kinds, including MDPs and stochastic programs, is emphasized. Particular attention is paid to the use of Monte Carlo simulation for sensitivity analysis of sequential decision-making models. Examples from vehicle sales models, the repair of deteriorating systems, and financial options are used to illustrate the approach. The material covered in Chapter 6 provides an important foundation for future chapters that apply Monte Carlo simulation.

Chapter 7 covers sequential decision-making when the decision-maker has imperfect information about the system's state. First, models known as *hidden Markov models* (HMMs) are presented. HMMs model the underlying system behavior under the assumption that a decision-maker can, at best, infer the probability that the system is in a given state based on observations of signals that provide limited information about the system's state. Next, HMMS are combined with a reward function to formulate a partially observable Markov decision process (POMDP) for finding optimal policies, analogous to MDPs. However, in the case of POMDPs, the policies are defined over a continuous *belief vector* that represents the system's state as a probability distribution over the underlying unobservable states. Bayesian updating is presented as a means of probabilistic updating of belief vectors, and exact and approximate algorithms are covered. Examples, including optimization of decisions for active surveillance of cancer, are used to illustrate the use of POMDP models.

Chapter 8 introduces *reinforcement learning* (RL), a key topic in artificial intelligence that addresses situations in which mathematical models cannot be estimated in advance. RL methods, sometimes called *model-free* methods, seek to find optimal policies by repeatedly making decisions and observing the resulting outcomes to evaluate and improve the policy. The most well-known model, the *multi-armed bandit*, is used to illustrate key concepts, such as the trade-off between *exploitation* and *exploration* when choosing decisions to search for an optimal policy. Algorithms such as Monte Carlo policy improvement and Q-learning are presented, and examples from games, revenue management, and robotics are used to illustrate their application.

Chapter 9 discusses problems involving multiple decision makers, known as *multi-agent models*. The chapter begins with the case of a two-player game between a decision maker and an *adversary* that can influence the model parameters in response to the decision maker's choices. This framework is often used when there is uncertainty about model parameters to identify a policy that is robust to worst-case outcomes arising from inaccurate parameter estimation. Next, the chapter continues with examples of situations in which multiple agents collaborate to achieve a shared goal. Examples from search-and-rescue operations and robot teams are used to demonstrate the methods covered.

Finally, Chapter 10 introduces methods for addressing large-scale problems arising in practice, employing approximations that reduce the dimensionality of the resulting models. Chapter 10 is devoted to methods for large-scale models of complex systems for which approximations are necessary. The focus is on cases where the system's scale is too large to find an optimal policy. Instead, approximation methods that have been proven effective for finding near-optimal policies, such as value-function approximations, are presented. Examples such as the autonomous piloting of a lunar lander and the control of an autonomous automobile illustrate the concepts described.

1.6 Concluding Remarks

This book is the culmination of the author's experience teaching undergraduate and graduate courses on the topics covered herein. It draws on knowledge from excellent references that have developed over many decades. Each chapter concludes with remarks on various sources, attributing concepts to the original researchers as accurately as possible. Some particularly influential sources are included in the bibliography below, which provide more specialized and advanced treatment of the various approaches brought together in this book, such as decision analysis [8], dynamic programming [9], Markov decision processes [10], stochastic programming [11], simulation [12], reinforcement learning [13], and a unifying framework for large-scale models for sequential decision-making [14]. For readers interested in more motivating examples, [15] presents a variety of industry applications, with each chapter representing a different practical context, including healthcare, transportation, production, finance, and others.

1.7 Practice Exercises

1. What examples of sequential decision-making arise in your everyday life? What are the states, decisions, and rewards?

2. If you were tasked with finding an optimal policy for managing a financial portfolio, how would you define the states, decisions, rewards, and the time horizon? What kind of data would help you formulate a model?

3. Assume you are interested in writing a program to automate playing the game of *tic-tac-toe,* where your program will play against human decision makers. What are the states, decisions, and rewards you would use? Describe in words the optimal policy.

4. If you were designing an app for bicycle riders looking to choose a path for a ride for various purposes (e.g., from home to work, leisure, or exercise), what information would you request from the user to determine the criteria to be used for the reward function?

5. Consider the shortest path problem expressed as a network in Figure 1.2, where *nodes* are locations, *arcs* are options to choose from, and numbers assigned to arcs (the *arc weight*) are distances. How many policies exist for navigating from node A to node D? Which policy minimizes the distance traveled?

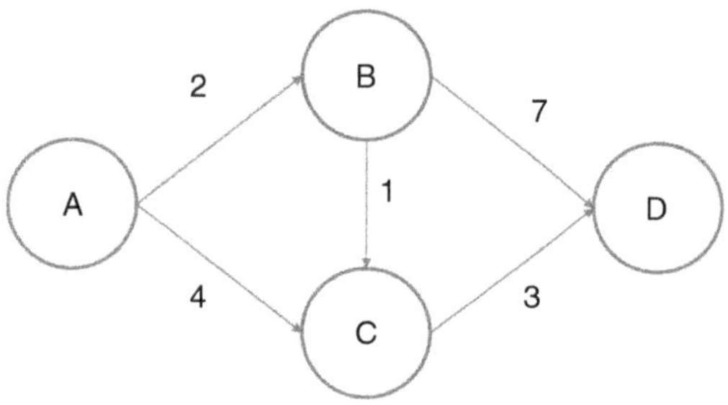

Figure 1.2 Example of a shortest path problem. Enumerate the policies to find the shortest path.

6. For the shortest path problem in Figure 1.2. Is it possible to change the arc weights such that all policies lead to the same distance traveled?

7. For the shortest path problem in Figure 1.2, assume that each time there is a choice between arcs, the decision maker flips an unbiased coin to decide which to take. What is the probability that the path length will be less than 7?

References

1 Gass, S. I. & Assad, A. A. (2011). History of operations research. In *INFORMS Tutorials in Operations Research* (pp. 1–14). INFORMS. https://doi.org/10.1287/educ.1110.0084.

2 Mason, J. E., Denton, B. T., Shah, N. D. & Smith, S. A. (2014). Optimizing the simultaneous management of blood pressure and cholesterol for type 2 diabetes patients. *European Journal of Operational Research, 233* (3), 727–738.

3 Alagoz, O., Maillart, L. M., Schaefer, A. J. & Roberts, M. S. (2004). The optimal timing of living-donor liver transplantation. *Management Science, 50* (10), 1420–1430.

4 Maillart, L. M., Simmons Ivy, J., Ransom, S. & Diehl, K. (2008). Assessing dynamic breast cancer screening policies. *Operations Research, 56* (6), 1411–1427.

5 Shechter, S. M., Bailey, M. D., Schaefer, A. J. & Roberts, M. S. (2008). The optimal time to initiate HIV therapy under ordered health states. *Operations Research, 56* (1), 20–33.

6 Perera, A. T. D. & Kamalaruban, P. (2021). Applications of reinforcement learning in energy systems. *Renewable and Sustainable Energy Reviews, 137*, 110618.

7 Pereira, M. & Pinto, L. (1991). Multi-stage stochastic optimization applied to energy planning. *Mathematical Programming, 52*, 359–375.

8 Raiffa, H. (1968). *Decision Analysis*. Random House.

9 Bellman, R. (1954).The theory of dynamic programming. *Bulletin of the American Mathematical Society, 60* (6), 503–515

10 Puterman, M. L. (1994). *Markov Decision Processes: Discrete Stochastic Dynamic Programming*. John Wiley & Sons. (Reprinted 2014.)

11 Birge, J. R. & Louveaux, F. (2011). *Introduction to Stochastic Programming*. Springer.

12 Law, A. M. & Kelton, W. D. (2007). *Simulation Modeling and Analysis* 4th ed. McGraw-Hill, New York, NY.

13 Sutton, R. S. & Barto, A. G. (2018). *Reinforcement Learning: An Introduction* 2nd ed. MIT Press.

14 Powell, W. B. (2022). *Sequential Decision Analytics and Modeling: Modeling with Python*. Now publishers, Boston–Delft. https://doi.org/10.1561/9781638280835.

15 Boucherie, R. J. & Van Dijk, N. M. (2017). *Markov Decision Processes in Practice*. Springer.

16 Alagoz, O., Hsu, H., Schaefer, A. J. & Roberts, M. S. (2010). Markov decision processes: A tool for sequential decision making under uncertainty. *Medical Decision Making, 30* (4), 474–483.

17 Ferreira, G. O., Arruda, E. F. & Marujo, L. G. (2018). Inventory management of perishable items in long-term humanitarian operations using Markov decision processes. *International Journal of Disaster Risk Reduction*, *31*, 460–469.

18 Gurobi Optimization, LLC (2026). *Gurobi Optimizer Reference Manual, Version 13.0.* https://docs.gurobi.com/. (Accessed February 16, 2026.)

19 Python Software Foundation (2026). *Python 3.11 Documentation.* https://docs.python.org/3.11/. (Accessed February 16, 2026.)

20 Steimle, L. N. & Denton, B. T. (2017). Markov decision processes for screening and treatment of chronic diseases. In Boucherie, R. J. & Van Dijk, N. M. (Eds.), *Markov Decision Processes in Practice* (pp. 189–222). Springer.

2

Decision Trees

2.1 Introduction

Decision trees are among the earliest approaches to decision-making under uncertainty. The basic idea of a decision tree is to draw, from left to right, a sequence of branches between *nodes* of three types: *decision nodes* (squares) where the decision-maker has a choice to make, *chance nodes* (circles) where the decision-maker observes the outcome of uncertain events, and *payoff nodes* (triangles) that define the termination of the sequential decision-making process, which culminates in a payoff (reward). Relating this to the notion of a policy in Chapter 1, the decision-maker aims to determine the decision at each decision node (state) in the tree that maximizes the expected future payoff.

Consider the example in Figure 2.1 in which a decision-maker (a patient in this case) chooses between two alternatives: whether to treat (T) a medical condition or not (NT). Depending on the choice, one of two branches is taken. Next, there is a *chance node* with two possible outcomes: "condition worsens" and "condition resolves," which occur in the Treat branch with probability P_T and $1 - P_T$, respectively. In the Do Not Treat branch, the chance node outcomes are the same, but the probabilities of the outcomes are P_{NT} and $1 - P_{\text{NT}}$, respectively. Finally, the payoff nodes at the end of the process for the Do Not Treat Branch have a reward of R for the condition resolving on its own and a negative reward of $-C_W$, denoting a cost penalty if the condition worsens. The Treat Branch has the same payoffs, except for an additional negative reward, $-C_T$, representing the cost of treatment, which can be a monetary cost or a penalty for harm from treatment side effects. The following assumptions are relevant to understanding the model.

1. The treatment outcomes are uncertain, depending on a random variable with two outcomes; therefore, $0 < P_T$, $P_{\text{NT}} < 1$.
2. The treatment branch has an additional, strictly positive cost of treatment; therefore, $C_T > 0$.
3. The probability the condition resolves with treatment is higher than without treatment; therefore, $1 - P_T \geq 1 - P_{\text{NT}}$.

To understand these assumptions, apply the basic rules of probability to compute the expected payoffs as follows:

$$\text{Treat decision expected payoff: } P_T \times (-C_W - C_T) + (1 - P_T) \times (R - C_T),$$

$$\text{Do not treat decision expected payoff: } P_{\text{NT}} \times (-C_W) + (1 - P_{\text{NT}}) \times (R).$$

The optimal decision depends on the rewards R, $-C_T$, and $-C_W$, as well as the probabilities P_T and P_{NT}, collectively referred to as the *model data*. Now, consider the three assumptions, one by one. First, relaxing assumption 1 implies the treatment outcomes are certain. In this case, the chance node and everything to the right of it are replaced with a payoff, and the decision-maker's choice is trivially to treat if the expected payoff is higher than the no-treatment branch. Assumption 2 implies that there is no "free lunch." In other words, there must be a cost to choosing treatment; otherwise, there would be no reason not to treat and, therefore, no decision to make. Finally, assumption 3 ensures that the treatment is expected to yield some benefits in resolving the condition; otherwise, there would be no reason to consider it in the first place.

The above example illustrates key concepts for understanding many of the approaches to sequential decision-making discussed in this book. The following Sections 2.2 and 2.3 describe key concepts, starting with basic probability and moving on to approaches for quantifying decisions, respectively. Section 2.4 discusses how to extend decision trees across multiple stages, with examples from cancer detection and new product launch planning. The chapter concludes with references and recommendations for additional reading.

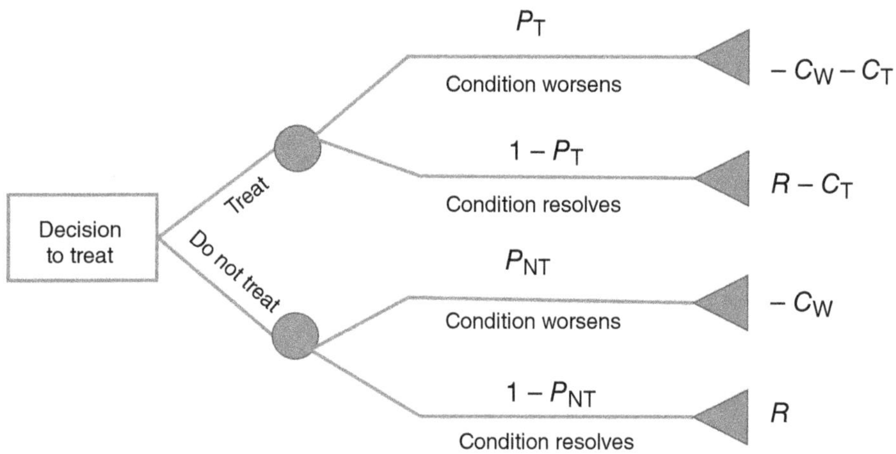

Figure 2.1 Example of a single-state decision tree for treatment under uncertainty. The rectangle indicates the decision to treat the medical condition; circles indicate uncertain outcomes of random variables that describe the uncertain response following the treatment decision; and triangles indicate final outcomes at the end of the decision process.

2.2 Basic Probability Concepts

Uncertainty will play an important role throughout this book. It is represented by random variables and probability distributions that can be incorporated into mathematical models based on historical observations, prospective data collection, or expert opinion. For simplicity, this book focuses on discrete, finite random variables, thereby avoiding the need for calculus and more abstract theoretical considerations that arise in a more general context. From a practical perspective, this assumption will have little impact because most problems—even those involving continuous random variables—can be well approximated by discrete, finite random variables, using methods such as Monte Carlo sampling covered in Chapter 6. Given this underlying assumption, the following summarizes the essential elements of probability theory necessary for this book.

First, consider a discrete and finite random variable, X, that has as possible outcomes, $X_1, X_2, \dots, X_R$ with corresponding probabilities $p_1, p_2, \dots, p_R$, respectively. Assume all outcomes have a nonzero probability of occurring, i.e., $0 < p_i \leq 1$, for all $i = 1, \dots, R$. The complete set of possible outcomes is referred to as the *universe* of possibilities, denoted by $\Omega = \{X_1, X_2, \dots, X_R\}$. Any subset of outcomes defines an *event*, A, with an associated probability of $P(A)$. This includes the extreme case of a null event, denoted by $\emptyset$, an empty set that represents no outcome.

Often, it is useful to summarize the characteristics of random variables, such as the *mean*, defined as the expectation of the random variable X as follows:

$$E[X] = \sum_{i=1}^{R} p_i X_i.$$

This is also referred to as the *first moment* of random variable X. Another commonly used measure is *variance*, which is related to the second moment, $E[X^2]$, as follows:

$$E[(X - E[X])^2) = \sum_{i=1}^{R} p_i (X_i - \sum_{i=1}^{R} p_i X_i)^2.$$

Consider the following example to solidify understanding of the mean and variance.

Example 2.1: *Expected value*

Consider the example of rolling a six-sided die again. Given that each side has an equal probability of landing up, the random variable representing the outcome of rolling the die has six possible outcomes, $\Omega = \{1, 2, 3, 4, 5, 6\}$, each with probability $1/6$. Thus, the expected value of the dice roll is

$$E[X] = \sum_{i=1}^{6}(1/6) \times i = 3.5.$$

Note that the mean value need not correspond to one of the possible outcomes, as in this case. Next, the variance of the mean can be computed as:

$$E[(X - E[X])^2) = \sum_{i=1}^{6}(1/6) \times (X_i - 3.5)^2 = 2.917.$$

Variance is a commonly used measure of the expected deviation of a random variable from its mean.

Often, decision-analytic problems involve multiple random variables, raising questions about whether they are related. Events may be related or independent. Two events, A and B, are referred to as independent if $P(A \cap B) = 0$, meaning that observing A and B simultaneously is impossible. With this notation in mind, the following are fundamental laws of probability used in this book:

$$P(\Omega) = 1, \tag{2.1}$$

$$P(\emptyset) = 0, \tag{2.2}$$

$$P(A \cup B) = P(A) + P(B) - P(A \cap B). \tag{2.3}$$

Equation 2.1 means the probability of observing some outcome from the universe of possibilities is 1. Equation 2.2 means the probability of observing no outcome is 0. Equation 2.3 denotes the probabilistic relationship between two events, A and B, which may have a nonempty intersection unless they are independent.

Example 2.2: *Union of events*

Consider an event A representing the outcome that an even number is observed when rolling a six-sided die (2, 4, or 6); thus, $P(A) = 1/2$. Consider a second event B, which represents 1, 2, or 3 turning up when the die is rolled; thus, $P(B) = 1/2$. These events are not independent because of the intersection $A \cap B = \{2\}$, thus $P(A \cap B) = 1/6$. The union of the events, on the other hand, $A \cup B = \{1,\ 2,\ 3,\ 4,\ 6\}$ can be obtained using Equation 2.3 as follows:

$$P(A \cup B) = 1/2 + 1/2 - 1/6 = 5/6.$$

Another important concept is *conditional probability*, which frequently arises in sequential decision-making with partial information. *Bayes' Rule*, a fundamental law of probability, describes conditional probabilities as follows:

$$P(A|B) = \frac{P(A \cap B)}{P(B)}.$$

$P(A|B)$ can be read as "the probability of event A occurring given that event B has already occurred." If events A and B are independent, then $P(A \cap B)$ is $P(A) \times P(B)$, in which case Bayes' rule naturally reduces to $P(A|B) = P(A)$. On the other hand, if there is an overlap in the outcomes associated with the events, then the occurrence of event B can inform the probability of event A occurring.

Example 2.3: *Conditional probability*

Building on the above example, one might ask: What is the probability that the outcome of a die roll is an even number, conditional on the outcome not being 2? This sort of question can be answered using Bayes' rule as follows:

$$\begin{aligned} P(\text{Even Roll} \mid \text{Roll Not 2}) &= \frac{P(\text{Even Roll} \cap \text{Roll Not 2})}{P(\text{Roll Not 2})} \\ &= \frac{1/3}{5/6} \\ &= \frac{2}{5}. \end{aligned}$$

Example 2.4: *Bayesian Updating*

Bayes' rule is frequently used in sequential decision-making when new information arises that affects the probability of an event and potentially future decisions. Consider a patient at risk of a hospital-acquired infection (I) during recovery from surgery. Each day they spend in the hospital, there is a probability $P(I \mid C) = 0.2$ and $P(I \mid \bar{C}) = 0.05$, where C and $\bar{C}$ denote compromised and not compromised immunity, respectively. Suppose patients have an initial probability $P(C) = 0.4$ of compromised immunity upon admission to the hospital, and a particular patient has no infection after one day of recovery. What is the new probability that their immunity is compromised? The conditional probability is

$$P(C \mid \bar{I}) = \frac{P(C \cap \bar{I})}{P(\bar{I})} = \frac{P(\bar{I} \mid C)P(C)}{P(\bar{I})} = \frac{0.8 \times 0.4}{P(\bar{I} \mid C)P(C) + P(\bar{I} \mid \bar{C})P(\bar{C})} = 0.36.$$

Each day that passes with the patient not acquiring an infection, the probability the patient was compromised at admission decreases.

2.3 Quantifying Decisions: Payoffs and Decision-maker Perspectives

Any quantitative model for choosing decisions must define the decision-maker's *criteria* as a mathematical (reward) function. In all the models discussed in this book, it is assumed that the objective can be expressed mathematically; however, the functional form depends on the decision-maker's beliefs and behavior. Moreover, units of measure may differ across models. For example, GPS-based navigation models aim to minimize the time required to move from point a to point b. Models of business problems may seek to minimize costs or maximize profits using monetary estimates, measured in dollars. Models in healthcare might consider measures of length or quality of life. At the same time, in other contexts, decision-makers may wish to minimize or maximize the probability of a bad or good event, respectively. Sometimes, a standard metric can be used, such as kilometers, years, dollars, or liters. In other cases, when multiple criteria exist, it may be necessary to define a conversion factor between them if they correspond to different metrics. This section summarizes several alternative approaches to defining a decision-maker's criterion.

2.3.1 Quality Adjusted Life Years

Often, decision problems arise in which the payoffs cannot be realistically evaluated in monetary terms. A common measure used in public health is the *Quality-Adjusted Life Year* (QALY). A QALY is a number in the range $[0,\ 1]$, with 1 corresponding to a year of perfect health and 0 corresponding to death. *Disutilities*, denoted as μ, define the impact of health conditions that reduce the quality of life for an individual. For example, patients who experience a non-fatal heart attack rate their quality of life after the heart attack at 0.93, according to a study by Clarke et al. [1], implying a disutility of 0.07 to account for the long-term lifestyle changes resulting from a heart attack. This illustrates how QALYs are used to quantify health conditions. In some cases, disutilities are used to quantify the burden of health interventions, such as blood tests, biopsies, or a surgery, where the magnitude of the disutility depends on the burden of the intervention on the patient. Moreover, QALYs may be time-dependent. For example, following surgery, a patient may experience high disutility during the initial recovery period, followed by lower disutility after recovery, reflecting long-term side effects.

Some studies seek to consider both QALYs and costs. A concept related to QALYs is *willingness to pay* (WTP), which specifies a conversion factor between QALYs and dollars that represents a societally acceptable monetary value associated with a year of life. This is particularly relevant in situations where QALYs and costs are both relevant to defining the decision-maker's objective. For example, if a decision-maker wishes to evaluate the cost vs. benefit of a new health intervention (e.g., a new medical treatment, device, or procedure), then the societal WTP may be used to convert expected QALYs gained from the intervention to dollars to be traded off against the cost of the intervention. This concept is routinely applied in cost-effectiveness analyses to help decision-makers (e.g., health insurers) decide whether to cover the costs of a new intervention. Drummond et al. [2] is a comprehensive reference on methods for evaluating healthcare programs, including WTP.

2.3.2 Probability of an Event

Sometimes, a decision-maker may be concerned with an event's likelihood. For instance, in a competitive game, the decision-maker (player) seeks to select a policy that maximizes their probability of winning. Alternatively, in a healthcare context, a decision-maker (the patient) may wish to select a policy that minimizes the probability of contracting an infectious disease. In some cases, the decision-maker's objective could be defined by probability. For example, consider *Value at Risk* (VaR), a widely used criterion in the banking industry, which represents the maximum value a financial portfolio could lose over a fixed time at a certain probability level. For instance, VaR quantifies the potential 24-hour financial loss with 5% probability, providing a measure of the likelihood of a major financial loss for a bank.

2.3.3 Utility

A decision-maker's satisfaction with their decisions can be defined as a function $u(x)$, where x denotes an outcome (e.g., a monetary reward) and $u(x)$ assigns it a numerical value (utility). The simplest example is a linear function $u(x) = a + bx$. However, this does not account for nonlinear variation in a decision-maker's satisfaction as x varies. For instance, at negligibly low values of x, the decision-maker might be indifferent to the size of the payoff, but as x grows, the decision-maker's satisfaction may increase at an increasing rate to reflect their ability to use the payoff to achieve specific personal goals (e.g., purchase a new car, a new house, etc.). In such cases, a quadratic utility function $u(x) = a + bx^2$, could be a more accurate representation of the decision-maker's utility. On the other hand, some decision-makers exhibit *diminishing returns*, meaning the relative utility of additional payoff is increasing at a decreasing rate, suggesting a functional form such as $u(x) = \sqrt{x}$, for example. Figure 2.2 illustrates the differences between these examples of utility functions.

Utility functions can reflect a decision-maker's risk attitude under uncertainty in order for the recommendations based on the decision tree to satisfy the decision-maker. Risk attitudes have been categorized as *risk-neutral*, *risk-taking*, and *risk-averse*. A risk-neutral decision-maker will agree to a bet with a positive expected return. For example, a gamble that returns a reward of W with probability p and a loss of L with probability $1 - p$ will appeal to the risk-neutral decision-maker if $pW - (1 - p)L \geq 0$. On the other hand, a risk-averse decision-maker will have a nonlinear utility function. They will agree to a bet if $pu(W) - (1 - p)u(L) \geq 0$. A risk-taking decision-maker will have a utility function increasing at an increasing rate, while a risk-averse decision-maker's utility will increase at a decreasing rate. Differences in risk attitude can translate into differences in optimal decisions that maximize expected utility.

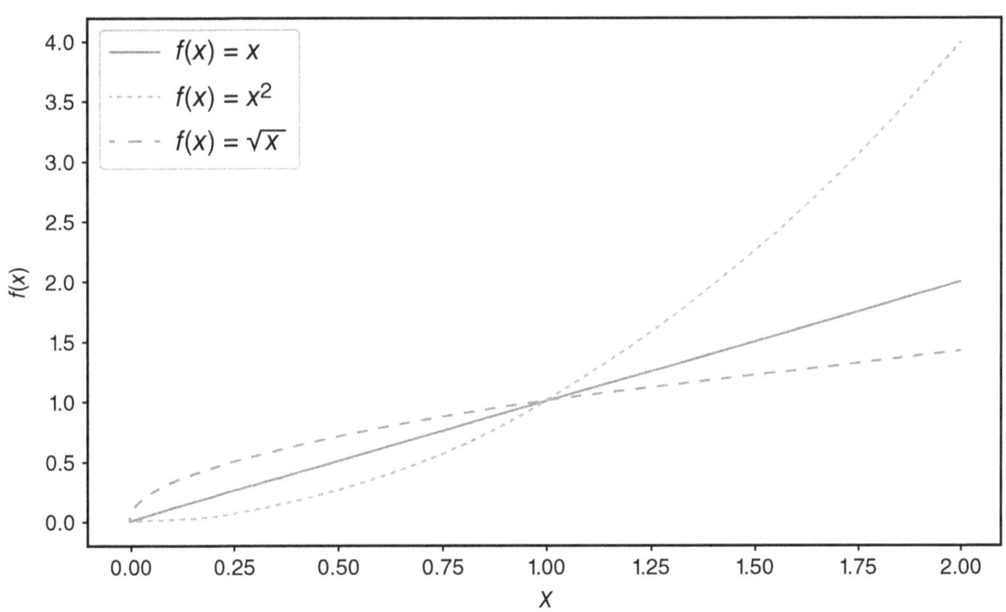

Figure 2.2 Illustration of alternative risk preferences, including risk-neutral (x), risk-taking (x^2), and risk-averse ($\sqrt{x}$).

Example 2.5: *Risk preferences*

Consider the case of a decision-maker who makes a bet that returns \$100 if they win, with probability p, and a loss of \$200 with probability $1 - p$. A risk-neutral decision-maker will be satisfied to take the bet if the expected value of the return is not less than 0, which occurs when $p \geq 2/3$, based on reorganizing the implied inequality $(100p - 200(1 - p) \geq 0)$. On the other hand, a risk-averse decision-maker will expect $p > 2/3$, and a risk-taking decision-maker will be satisfied with choices of $p < 2/3$. In each case, the precise threshold depends on the shape of the utility function. For instance, a risk-taking decision-maker, with $u(x) = x^2$, will accept the bet if $p > 4/5$.

There are many methods for estimating utility functions. One method, known as *revealed preferences*, infers utility from a decision-maker's previous choices as reflected in historical data. Another method, *stated preferences*, uses surveys with questions about various hypothetical comparisons designed to elicit preferences. In situations where uncertainty is present, the goal is to estimate expected utility, which can be done by presenting the decision-maker with "lotteries" that incorporate the effects of uncertainty on risk attitude. An example of a simple lottery is the decision-maker choosing between two options: (1) walking away with a guaranteed reward (e.g., \$100), and (2) a gamble that yields a greater reward (e.g., \$200) with probability p and no reward with probability $1 - p$. For example, if $p = 0.5$, a risk-averse decision-maker will take the guaranteed reward, a risk-neutral decision-maker will be indifferent between the guaranteed reward and the gamble, and a risk-taking decision-maker will choose to gamble. The probability p is varied until the decision-maker is indifferent between the two choices. At the point of indifference, the value of p quantifies the decision-maker's utility for \$100 relative to the lottery that yields \$200 with probability p and \$0 with probability $1 - p$. A utility function for the decision-maker can be fit by presenting a series of such lotteries. In this example, the rewards (\$100–200) and losses (\$100) are modest, with moderate probabilities. Alternatively, the rewards and losses may vary significantly, as seen in real lotteries, like PowerBall, where probabilities of winning are vanishingly small, the losses are modest, but the reward for winning is enormous.

2.3.4 Time Value of Rewards

All rewards, including monetary rewards, are subject to value adjustments when received at future dates. To illustrate this point, consider the following example with two options: (1) receive \$100 today; (2) receive \$100 in a year. A rational decision-maker will prefer to receive the \$100 today because they can invest it at a risk-free interest rate, λ, and receive $(1 + \lambda) \times \$100$ in one year. Thus, option 1 is financially superior to option 2. In the case of a recurring reward, common in sequential decision-making, the time value of the reward must be accounted for. This gives rise to the *net present value* of future rewards, which defines the decision-maker's current valuation (today) for the future rewards they will receive.

Example 2.6: *Valuation of a bond*

Suppose a bond provides a payment of \$500/month for a year. The total payments are \$6,000 over the year. Considering the time value of money, a decision-maker would expect to pay less than \$6,000 upfront for such an investment. Each payment must be discounted based on a preference for earlier versus later payments. Suppose λ is the *discount factor*. If payments are made at the end of each month, the value of the bond today is:

$$V_B = \lambda \times \$500 + \lambda^2 \times \$500 + \cdots + \lambda^{12} \times \$500 \tag{2.4}$$

If the monthly discount factor is $\lambda = 0.99$, then the value of the bond today is \$5,760.25. This value is called the *net present value* and is used in most contexts where rewards are paid over time, but a decision must be made here and now. This concept is most easily illustrated with monetary awards, where λ is often chosen to represent the annual return of a risk-free investment (often 3%, hence $\lambda \approx 0.97$). However, discounting future rewards is not restricted to monetary rewards. It arises in any situation in which a reward today is preferred over a reward at some future date.

2.3.5 Regret

Regret quantifies decisions using the difference between the decision and the optimal decision in retrospect, following the resolution of uncertainty. It is relevant in cases where the decision-maker learns the best possible decision after making their decision. Thus, a decision-maker may wish to consider their anticipated regret when making a decision. For example, consider an investor deciding whether to purchase shares in a startup. They may experience regret in two ways: (1) the decision-maker chooses not to purchase shares, and the shares increase in value over the next year, in which case the decision-maker regrets not purchasing more; (2) the decision-maker chooses to purchase shares, but the share price decreases over the next year, in which case the decision-maker regrets their purchase. Thus, the decision-maker's goal could be to minimize expected regret.

Cumulative regret is a related performance measure that arises commonly in the context of learning. Chapters 8 and 9 cover *model-free* decision-making where the mathematical model that governs system dynamics is not known to the decision-maker. The decision-maker is assumed to be able to observe states, make decisions, and receive rewards. An interactive learning process called *reinforcement learning*, similar to trial-and-error, is used to adjust and improve the policy over time. In such cases, there is an opportunity cost of learning, because suboptimal decisions are made in the search for the optimal policy. Thus, the goal of maximizing total rewards is replaced by minimizing cumulative regret to account for the losses incurred during learning.

2.4 Multi-stage Decision Trees

The number of decision stages depends on the context. Reading a decision tree from left to right, begin with the earliest decision, then observe the outcomes of a probabilistic event, and subsequently make new decisions, observe new outcomes, and so on. While it is natural to view a decision tree from left to right—in the direction of the flow of time—as will be seen throughout this book, it is equally important to view it from right to left because decisions to be selected in the future influence the ideal decisions to be made in the past. In short, anticipating how a decision-maker will respond to future decisions (sometimes called recourse decisions) is critical to making the best decision today.

Example 2.7: *Genetic biomarkers*

Consider a common medical decision relevant to cancer diagnosis. For patients suspected of having cancer, the standard procedure is to do a biopsy to collect a sample of tissue to be examined by a pathologist who renders a diagnosis; however, biopsies can sometimes be painful and invasive. Therefore, ideally, the patient would only be subjected to a biopsy if necessary. Imagine that a new biomarker, based on a simple blood test, has been discovered that provides probabilistic information about the likelihood that the patient has cancer. Thus, it can be used as part of the decision-making process to help determine whether a biopsy is warranted; however, the biomarker test is expensive, and its high cost must be weighed against its diagnostic value to patients. This example uses a decision tree to evaluate the incremental benefit of the biomarker.

Figure 2.3 illustrates the decision tree, with two branches emanating from the (leftmost) root decision node, corresponding to the decision about getting the biomarker test (BT). For each branch, there is a certain probability that the patient has an undiagnosed cancer, $P(C)$, or not, $1 - P(C)$. The following is a description of the notation in each branch.

- No Biomarker Branch: In the No Biomarker branch, the patient has a biopsy, which detects cancer (if the patient has cancer) with conditional probability $P(B+|C)$ and misses cancer with probability $P(B-|C) = 1 - P(B+|C)$; the payoffs for these branches are $R(\text{CT})$ and $R(\text{CNT})$, respectively, denoting an expected reward (e.g., patient survival in years) for a patient who receives cancer treatment (CT) and a patient who has cancer but is not treated (CNT), because of a missed diagnosis. Naturally, it is assumed that $R(\text{CT}) \geq R(\text{CNT})$; otherwise, there would be no point in considering a biopsy in the first place. The probability that the patient does not have cancer is $1 - P(C)$. Since this is not directly observable to the decision-maker, a biopsy is performed, and the patient receives a reward $R(\text{NC})$ for cancer-free survival. Finally, a disutility μ_B for the biopsy, in QALYS, is applied to all payoffs for branches in which the patient receives a biopsy.
- Biomarker Branch: There is a similar set of chance nodes and payoffs in the "Biomarker" branch of the decision tree. The most significant difference is that the biomarker test defines the chance nodes after the first chance node, which determines whether the patient has cancer. The biomarker is used to decide whether to perform a biopsy: $P(\text{BT}+|C)$ if the patient has cancer and the test is positive, and $P(\text{BT}-|C)$ if the test is negative, given that the patient has cancer. Finally, the remaining branches define the outcomes of a biopsy following a positive biomarker test. The payoffs at each terminal node in the tree are $R(\text{CT}) - \mu_B$ and $R(\text{CNT}) - \mu_b$ for diagnosed and treated patients and undiagnosed and untreated patients, respectively. Note that the disutility of biopsy, μ_B, is applied to each terminal node for which a biopsy is done to reflect the impact on QALYs from the procedure.

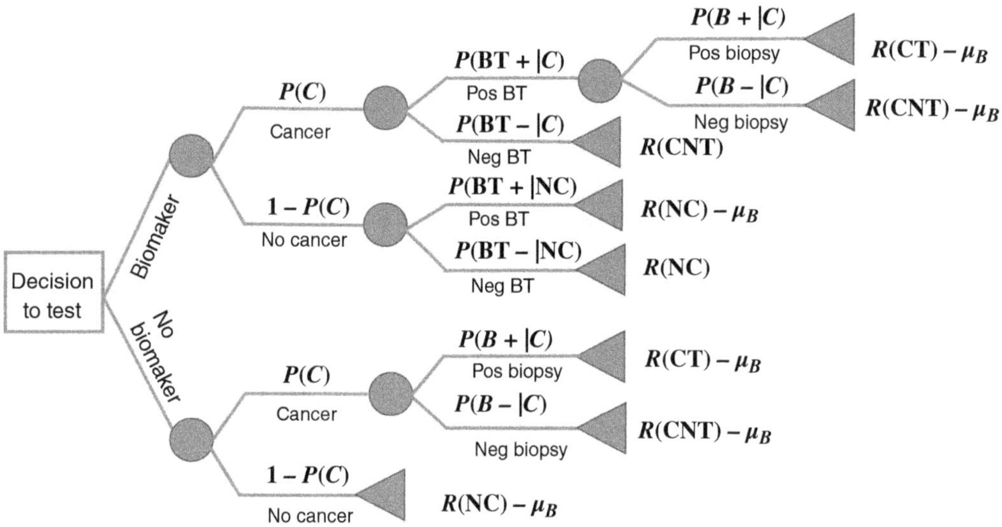

Figure 2.3 Example of a decision tree for evaluating whether or not to employ a biomarker test (BT) for a patient at risk of cancer. The probability that the patient has cancer is $P(C)$, also known as the disease's incidence. $P(\text{BT}+|C)$ and $P(\text{BT}-|\text{NC})$ are called the *sensitivity* and *specificity* of the biomarker test (BT), respectively. Similarly $P(B+|C)$ and $P(B-|\text{NC})$ are the *sensitivity* and *specificity* of the biopsy (*B*), respectively. Parameters for the payoffs at the leaf nodes are defined in Table 2.1.

Table 2.1 Decision tree model parameters for evaluating a new cancer biomarker test. Note that the following dependent parameters are omitted for brevity: $P(B+|\text{NC})=0$, $P(\text{NC})=1-P(C)$, $P(\text{BT}-|\text{NC})=1-P(\text{BT}+|\text{NC})$, and $P(B-|\text{NC})=1-P(B+|\text{NC})$.

R(NC)	*R*(CT)	*R*(CNT)	μ_b	$P(C)$	$P(\text{BT}+|C)$	$P(\text{BT}+|\text{NC})$	$P(B+|C)$
5.0	4.5	1.8	0.05	0.15	0.75	0.2	0.8

The decision tree can be used to determine which option has the higher expected payoff: the BT branch or the No BT branch. The analysis of decision trees proceeds from right to left, sometimes referred to as *collapsing the tree*, by computing the expected rewards at each chance node. If there are multiple chance nodes at the rightmost part of the tree, they can be evaluated in any order. Table 2.1 defines the model parameters, which are arbitrary for the purposes of this example but plausible and necessary to complete the calculations. Starting with the rightmost chance node, the expected payoff is computed as follows:

$$P(B+|C)\times(R(\text{CT})-\mu_B)+P(B-|C)\times(R(\text{CNT})-\mu_B)=3.91.$$

Next, move one step to the left and continue collapsing the decision tree for the next three rightmost chance nodes. Figure 2.4 shows the partially collapsed tree with the expected payoffs filled in. Ordering the three chance nodes from top to bottom and using the parameters in Table 2.1, the chance node evaluations are as follows:

$$P(\text{BT}+|C)\times(3.91)+P(\text{BT}-|C)\times R(\text{CNT})=3.38,$$

$$P(\text{BT}+|\text{NC})\times(R(\text{NC})-\mu_B)+P(\text{BT}-|\text{NC})\times R(\text{NC})=4.99,$$

$$P(B+|C)\times(R(\text{CT})-\mu_B)+P(B-|C)\times(R(\text{CNT})-\mu_B)=3.91.$$

Finally, the payoff node in the last remaining branch of the tree evaluates to:

$$R(\text{NC})-\mu_B=4.95.$$

Figure 2.5 illustrates the completely collapsed tree with the expected payoff for the BT and No BT branches, with expected QALYs of 4.75 and 4.96, respectively. Thus, the new biomarker test does more harm than good and should be rejected based on the provided parameters, regardless of cost. Naturally, different parameter choices may lead to different decisions, and thus identifying reliable estimates is very important. Often it is difficult, or even impossible, to get accurate objective

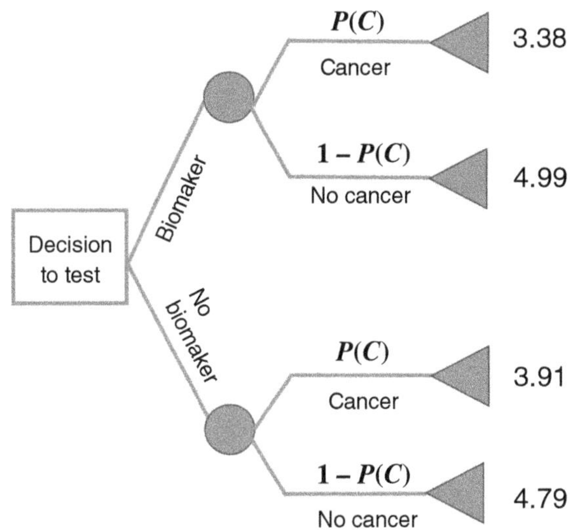

Figure 2.4 Partially collapsed decision tree for deciding whether to use a biomarker test (BT).

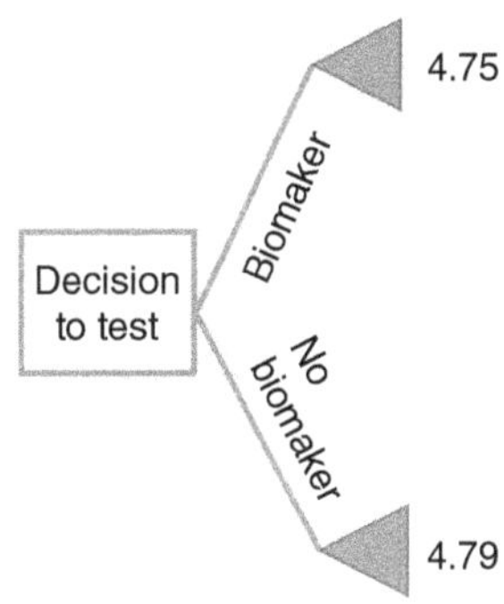

Figure 2.5 Fully collapsed decision tree with the expected payoffs for using and not using the biomarker test.

estimates of model parameters. Thus, experimentation with varying parameters is an important part of the analysis for real problems. The question of how to analyze the sensitivity of decisions to parameter choices is taken up in Chapter 6.

2.5 Expected Value of Perfect Information

Uncertainty in sequential decision-making comes at a cost. If the decision-maker could foresee which uncertain outcomes in the tree would be realized, the decision could be optimized by selecting the optimal choice based on the anticipated outcome. An often-asked question is: What is the *expected value of perfect information* (EVPI)? The EVPI quantifies the difference between decisions made under uncertainty versus those made when the uncertain outcomes can be predicted in advance. To illustrate this concept, consider the following example.

Example 2.8: *New product launch*

Consider a medium-sized company that manufactures inflatable stand-up paddleboards that pack for air travel, serving customers who want to take one on vacation. The company is competing for market share and has recently discovered a new material that is 60% lighter than their current product. Production costs are expected to remain the same; however, the new material has not been field-tested, and there is a risk of reliability issues that could reduce customer sales. Furthermore, there is uncertainty about the reliability of the new material and the resulting quality control issues that could harm customer satisfaction. Finally, there is a risk that the company will lose market share if it decides not to launch the new product and a competitor launches a new and similar product.

Consider the decision tree in Figure 2.6. Note from the tree that there are three sources of uncertainty: (1) uncertainty in sales if the new product is launched, depending on its reliability; (2) uncertainty in sales if the new product is not launched because of the potential for a competitor to steal market share by launching a superior product; (3) the degree of reliability problems (moderate or severe) if reliability issues are encountered with the new product.

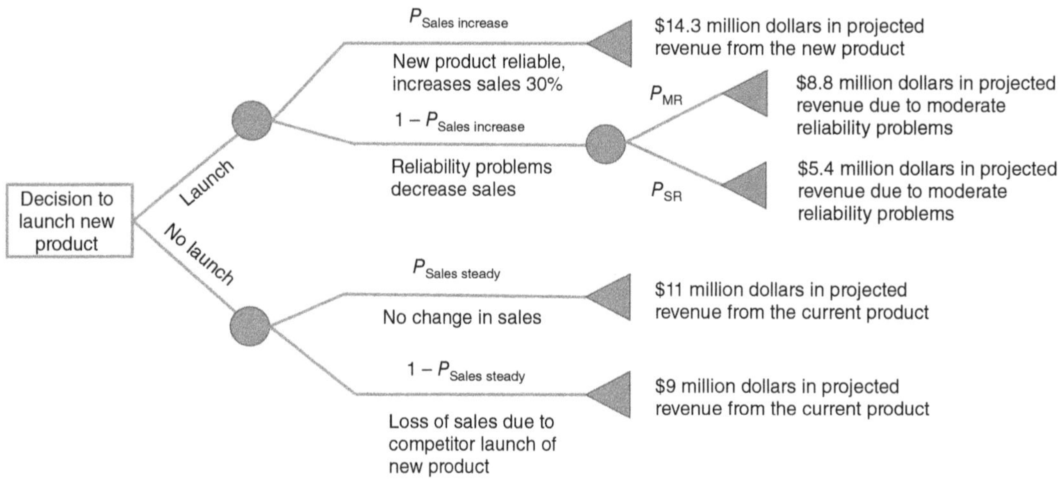

Figure 2.6 Decision tree for the decision to launch a new stand-up paddleboard using new but untested materials. The reliability problems branch has two possible outcomes: moderate reliability (MR) problems and severe reliability (SR) problems.

The data in the tree represent probabilities associated with the chance nodes, based on expert opinion and projected revenues forecasted by company analysts, who use historical data from previous product launches to inform their projections. These projections are linked to the payoff nodes. The company could conduct accelerated field testing to alleviate uncertainty about reliability. But the tight deadline and necessary resources come at a cost. What should the company be willing to pay? Consider the case where $P_{\text{Sales increase}} = 0.4$ in the Launch branch and coincidentally $P_{\text{Sales steady}} = 0.4$ in the No Launch branch as well. Assuming $P_{MR} = P_{SR} = 0.5$, collapsing the decision tree yields the following:

$$\text{Launch payoff: } P_{\text{Sales increase}} \times \$14.3\text{M} + (1 - P_{\text{Sales increase}}) \times \$7.1\text{M} = \$9.98\text{M},$$

$$\text{No launch payoff: } P_{\text{Sales steady}} \times \$11\text{M} + (1 - P_{\text{Sales steady}}) \times \$9\text{M} = \$10.2\text{M}.$$

Thus, the best choice is not to launch the new product with an expected payoff of \$10.2M. To evaluate EVPI, split the original decision tree into two decision trees. One in which moderate reliability is achieved with probability 1 and a second in which severe reliability problems occur with probability 1. In other words, the two new decision trees define the situation in which the resulting reliability is known in advance, i.e., the perfect information case. This prompts the following computation and comparison.

Decision Tree 1 (Moderate Reliability Problems Scenario):

$$\text{Launch payoff: } P_{\text{Sales increase}} \times \$14.3\text{M} + (1 - P_{\text{Sales increase}}) \times \$8.8\text{M} = \$11\text{M},$$

$$\text{No launch payoff: } P_{\text{Sales steady}} \times \$11\text{M} + (1 - P_{\text{Sales steady}}) \times \$9\text{M} = \$10.2\text{M}.$$

Decision Tree 2 (Severe Reliability Problems Scenario):

$$\text{Launch payoff: } P_{\text{Sales increase}} \times \$14.3\text{M} + (1 - P_{\text{Sales increase}}) \times \$5.4\text{M} = \$8.96\text{M},$$

$$\text{No launch payoff: } P_{\text{Sales steady}} \times \$11\text{M} + (1 - P_{\text{Sales steady}}) \times \$9\text{M} = \$10.2\text{M}.$$

The decision to launch is optimal under the moderate reliability (MR) problems scenario but not the severe reliability (SR) problems scenario. Thus, decision tree 1 has an expected payoff, $\text{EP}_1 = \$11\text{M}$, and decision tree 2 has an expected payoff, $\text{EP}_2 = \$10.2\text{M}$. Now, to calculate EVPI, take the expected payoff (EP) under the (perfect information) assumption that the Launch decision is made knowing the outcome of reliability by paying for the accelerated field testing:

$$\text{EVPI} = P_{\text{MR}} \times \text{EP}_1 + (1 - P_{\text{SR}}) \times \text{EP}_2 - \text{EP}.$$

Given the moderate and severe reliability problems are equally likely, then EVPI = \$10.6M − \$10.2M = \$0.4M. In other words, the decision-maker should be willing to pay up to \$0.4M for accelerated field testing, assuming they are confident in the data used in the decision trees.

2.6 Concluding Remarks

This chapter focused on decision trees, a widely used tool in *Decision Analysis*, which encompasses far more than just decision trees and the related topics covered in this chapter. Many decisions require substantial effort to frame the problem by working with decision-makers to understand their goals, the relevant factors, and how those factors influence one another. Much attention has been given to utility theory, risk preferences, and other topics related to how decision-makers value outcomes. There are excellent resources for learning more about decision analysis, including several listed in the references for this chapter: Raiffa [3], von Neumann and Morgenstern [4], Clemen [5], Abas and Raiffa [6], and Keeney [7]. Several examples were used to illustrate the application of decision trees, including a biomarker-based cancer-diagnosis example, motivated by a real-world decision analysis published by Merdan et al. [8]. The book by Drummond et al. [2] covers methods for the economic evaluation of health interventions, making it a useful resource for readers interested in this topic.

Decision trees are a stepping stone to more complex sequential decision-making models over extended time frames. Future chapters will cover problems too complex for decision trees and present alternative models and algorithms for making optimal decisions in such cases. Other key concepts in this chapter included a review of basic probability, alternative ways to model a decision-maker's objective, and the time value of rewards.

2.7 Practice Exercises

1. A consultant is asked to recommend a reservation policy for a restaurant that accounts for the fact that some customers are no-shows. Most reservations are for parties of two, with no-shows according to the following probability distribution on any given evening:

$$p^{\mathrm{NS}} = \begin{cases} 0.5 & 0\% \text{ of reservations no-show} \\ 0.3 & 15\% \text{ of reservations no-show} \\ 0.2 & 25\% \text{ of reservations no-show.} \end{cases}$$

   The restaurant has 20 two-person tables. How many bookings should they be willing to make for their first seating at 6 pm if the owners rate the regret from an unutilized table as half the regret of not being able to seat a party of two?

2. For the Bayesian updating example of Section 2.2, compute the probability the patient had compromised immunity conditional on not having an infection after 2, 3, 4, and 5 days. Describe the results and reasons for any trends observed.

3. For the genetic biomarker example, what choice of disutility for biopsy, μ_B, results in the expected payoffs being equal across branches? For what values should the biomarker test be used?

4. Consider the following situation. A student has received an offer of a full-time position upon graduation. They also have a second offer for a three-month consulting contract. The consulting project offers 50% more compensation but does not guarantee future work. There is a 60% chance of a follow-on consulting project for an additional three months. If that second project succeeds, there is an 80% chance of being hired full-time by the consultancy. If the follow-on project does not materialize, or if the second project is unsuccessful, there is a 70% chance of finding another full-time job at the same pay as the original full-time offer, but there is a 30% chance the student needs to settle for a job that pays 25% less. The student is planning three years out because they intend to return to college for an MBA after that time. Assuming all the jobs are equally interesting, what decisions will maximize their total earnings?

5. A decision-maker is contemplating selling their car and using public transportation and ride-sharing instead for the next three years of graduate school. They could sell their car now for \$6,500 or sell it in three years for \$3,000. They estimate that selling their car would also save \$800/year on insurance, \$250/year in maintenance, and \$900/year in gas. The cost of public transportation and ride-sharing is estimated to be \$2,400/year. Assuming a risk-neutral decision-maker, for what choice of annual discount factor would the decision-maker be indifferent to the two options? How might risk aversion affect decisions in this case?

6. Compute the EVPI for the genetic biomarkers example of Section 2.4. Conduct a sensitivity analysis on EVPI by varying each of the probabilities in the decision tree by ±20%. Which one has the greatest influence on the EVPI? Which has the least influence?

7. Consider the decision tree in Figure 2.7. Assume a risk-neutral decision-maker used the tree to decide what decision to make and chose A. What can be inferred about the probabilities of events X and Y?

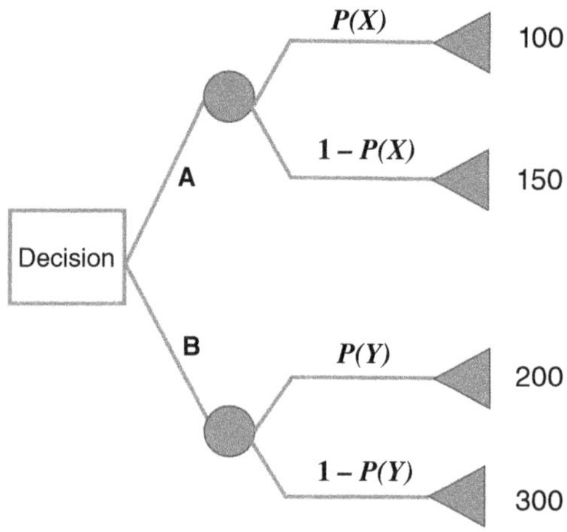

Figure 2.7 Decision tree example with missing data for probabilities.

8. Consider an alternative to the decision tree in Figure 2.7, in which the probabilities are known, but the payoff for each branch is unknown. The risk-neutral decision-maker chooses a decision B. The probabilities are $P(X) = 0.65$ and $P(Y) = 0.4$. What can be inferred about the payoffs?

9. Assume a decision-maker has two options. The first is to receive an immediate payoff of R_1. The second is to receive an unlimited number of payoffs, of size R_2, starting today and continuing indefinitely with payments at the end of each month. Assuming the discount factor $\lambda < 1$ if the decision-maker is risk-neutral and believes they will live forever, what conditions on R_1 and R_2 governs the optimal choice?

10. If a decision tree has T stages, N chance nodes at each stage, each with M possible random outcomes, how many unique branches does the decision tree have?

11. Consider a decision tree with 10 stages where each stage has two outcomes that occur with equal probability: (a) a reward of 10 is received, and the decision process terminates; (b) a reward of 2 is received, and the decision-maker has two choices; the first choice is to stop the decision process and receive a one-time reward of 5; the second is to continue the decision process, with the same outcomes and (stopping) decision opportunity repeated. This continues until the last decision epoch, when the decision-maker gets a guaranteed reward of 20. Draw the decision tree and compute the optimal policy as a function of the discount factor, $\lambda \in [0, 1]$.

References

1 Clarke, P., Gray, A. & Holman, R. (2002). Estimating utility values for health states of type 2 diabetic patients using the EQ-5D (UKPDS62). *Medical Decision Making, 22*, 340–349.
2 Drummond, M. F., Sculpher, M. J., Claxton, K., Stoddart, G. L. & Torrance, G. W. (2015). *Methods for the Economic Evaluation of Health Care Programmes*. Oxford University Press.
3 Raiffa, H. (1968). *Decision Analysis*. Random House.
4 von Neumann, J. & Morgenstern, O. (1944). *Theory of Games and Economic Behavior*. Princeton University Press.

5 Clemen, R. T. (1996). *Making Hard Decisions: An Introduction to Decision Analysis* 2nd. ed. Duxbury Press, Belmont, CA.

6 Abas, A. & Raiffa, H. (2015). *Foundations of Decision Analysis*. Pearson.

7 Keeney, R. L. (1982). Decision analysis: An overview. *Operations Research*, *30* (5), 803–838.

8 Merdan, S., Tomlins, S. A., Barnett, C. L., Morgan, T. M., Montie, J. E., Wei, J. T. & Denton, B. T. (2015). Assessment of long-term outcomes associated with urinary prostate cancer antigen 3 and TMPRSS2:ERG gene fusion at repeat biopsy. *Cancer*, *121* (22), 4071–4079.

3

Deterministic Dynamic Programs

3.1 Introduction

The decision trees of Chapter 2 illustrated some basic concepts of decision-making under uncertainty; however, their utility is limited because, as the number of random variables and decisions grows, the chance nodes and decision nodes grow exponentially. For instance, imagine a sequence of three stages, with three decisions and two random outcomes following each decision. This gives rise to a decision tree with $3 \times 2 \times 3 \times 2 \times 3 \times 2 \times 3 \times 2 = 1{,}296$ leaf nodes. As the number of stages, decisions, and outcomes increases, the size of the decision tree grows prohibitively large, ruling out storing it in computer memory, let alone finding optimal policies.

This chapter introduces *dynamic programming* (DP), a fundamental approach for avoiding exponential growth in solving certain types of sequential decision-making problems. It focuses on deterministic dynamic programs—temporarily eliminating uncertainty—to establish key concepts and intuition in sequential decision-making. Chapter 4 extends these models to stochastic dynamic programs that account for uncertainty, such as that illustrated by decision trees with chance nodes, albeit in a more efficient way using *Markov chains*.

DP can efficiently solve a wide range of sequential decision-making problems. The key principle underlying DP is the *principle of optimality* often credited to Richard Bellman's work beginning in the 1940s and, more specifically, an article from 1954 titled "The Theory of Dynamic Programming," [1] which describes efficiencies for solving certain types of problems by breaking up a large problem into a series of smaller, easier-to-solve problems. This falls into a category of problem-solving methods known as "divide and conquer."

The principle of optimality applies to many sequential decision-making problems. This chapter first introduces the deterministic context to familiarize readers with mathematical notation and model formulations and provides examples that develop intuition about the approach and its application to various problems, thereby preparing readers for future chapters. Second, the principle of optimality is discussed by introducing a fundamental DP problem, the *shortest path problem*, which seeks the minimum-distance path from a start to a finish node in a network. As will be seen, it can have many other interpretations beyond minimizing the distance traveled. The chapter continues with discussions of production planning, resource allocation, and pattern recognition before returning to the most general form of the shortest path problem and concluding with DP's shortcomings.

3.2 Mathematical Formulation of Dynamic Programs

A DP has several common attributes that are shared across model formulations. Decisions are made at discrete points in time, called *decision epochs*, indexed by $t = 1, \ldots, T$ where T denotes the end of the decision horizon. States and decisions are denoted as $s_t \in S$ and $a_t \in A$, respectively, with S and A denoting sets of possible states and decisions ($\in$ can be read as "is a member of"). In situations in which there is more than one state at a decision epoch, the decision will be expressed as $a_t(s_t)$ to reference the state in which the decision is to be applied. Moreover, for situations where the available decisions depend on the state of the system, the decision set, A, will be written as $A(s_t)$. The reward function, $r_t(s_t, a_t)$, defines the decision maker's reward at decision epoch, t, when the system is in state s_t and decision a_t is taken. A *state transfer function* $g(s_t, a_t)$, which, when given a state and decision, returns the next state the system will be in at epoch $t + 1$, i.e., $s_{t+1} = g(s_t, a_t)$. The final critical element is the *optimal value function* $v_t(s_t)$, which is the maximum value of the total discounted rewards for all future decision epochs, $t, t+1, \dots, T$, given the system is in state s_t and optimal decisions are taken at each epoch remaining in the decision horizon. It is also called the optimal *value-to-go* function.

Collectively, the above elements define the necessary ingredients to formulate a DP, which is expressed as a set of *optimality equations* (also known as *Bellman's Equations*), which can be solved to obtain an optimal policy. These equations can be expressed as follows:

$$v_t\left(s_t\right) = \max_{a_t \in A} \left\{r_t\left(s_t, a_t\right) + \lambda v_{t+1}\left(s_{t+1}\right)\right\}, \forall s_t \in S, t = 1, \dots, T-1, \tag{3.1}$$

$$v_T\left(s_T\right) = r_T\left(s_T\right), \forall s_T \in S, \tag{3.2}$$

where the symbol ∀ (read as "for all") indicates that Equation 3.1 is repeated "for all states, s_t, of set S."

The method for finding a solution to the optimality equations begins with the last decision epoch, T, for which the optimal value function is fixed according to the boundary condition in Equation 3.2. Subsequently, iterative backward substitution, also known as *backward induction*, is used to obtain $v_{T-1}(s_{T-1})$ for all s_{T-1}, followed by $v_{T-2}(s_{T-2})$ for all s_{T-2} and so on until the solution is completed with the evaluation of $v_t\left(s_1\right)$ for all s_1. This process is analogous to collapsing the decision trees of chapter 2. The next step is to use the solution, $v_t(s_t)$, $s_t \in S$, $t = 1, \dots, T$ to obtain an optimal decision at each state and decision epoch where

$$a_t^*(s_t) \in \arg\max_{a_t \in A} \left\{r_t\left(s_t, a_t\right) + \lambda v_{t+1}\left(s_{t+1}\right)\right\}, \forall s_t \in S, t = 1, \dots, T-1.$$

Note that multiple decisions may maximize the argument to the arg max operation, and thus, $a_t^*(s_t)$ is an arbitrarily selected element of the set of optimal decisions when more than one exists.

To recap, a DP comprises the following elements: states, decisions, reward function, discount factor, state transfer function, and the optimal value function, which together define the optimality equations that can subsequently be solved to obtain an optimal policy. Several examples are used to develop intuition about DPs for sequential decision-making, beginning with the best-known example, the *shortest path problem.*

3.3 Shortest Path Problems on Directed Acyclic Networks

DP can be used to find the shortest path that joins two points in a network. This is one of the most common applications because mapping programs, such as Google Maps, are widely used, enabled by the routine availability of GPS services on smartphones and other devices. In addition, the *shortest path problem* is an important DP model because many seemingly distinct applications can be formulated as shortest path problems, as shown later. This section introduces the shortest path problem for networks with directed arcs and no cycles. An algorithm for the more general case of *cyclic networks* is discussed in Section 3.7.

To begin, consider the network shown in Figure 3.1, composed of *nodes* and *arcs* between the nodes representing feasible movements by a decision maker. Each arc has a *weight* that denotes the cost of traversing the arc (e.g., time, distance traveled). In this example, assume the weights represent distances. While the optimal policy is unknown, the input data, e.g., rewards and state transfer functions, are known. For this example, they have been arbitrarily selected and could have been different, e.g., the reward (weight) between E and G could be 2, not 4, or an arc could exist between B and C. Needless to say, collecting the data and organizing the network is a necessary precursor to any optimal decision-making. (When the data are not known, and uncertainty arises, the approach is different, which is covered in the next chapter.) Reviewing the elements of the DP formulation described above, the following are the key elements:

States: nodes of the network, $\{S, A, B, C, D, E, F, G\}$

Decisions: which arc to select to move to the next node

Rewards: distance associated with each arc

State Transfer Function: binary function indicating an arc between nodes (1) or not (0)

Optimal Value Function: minimum distance from a given node to node G

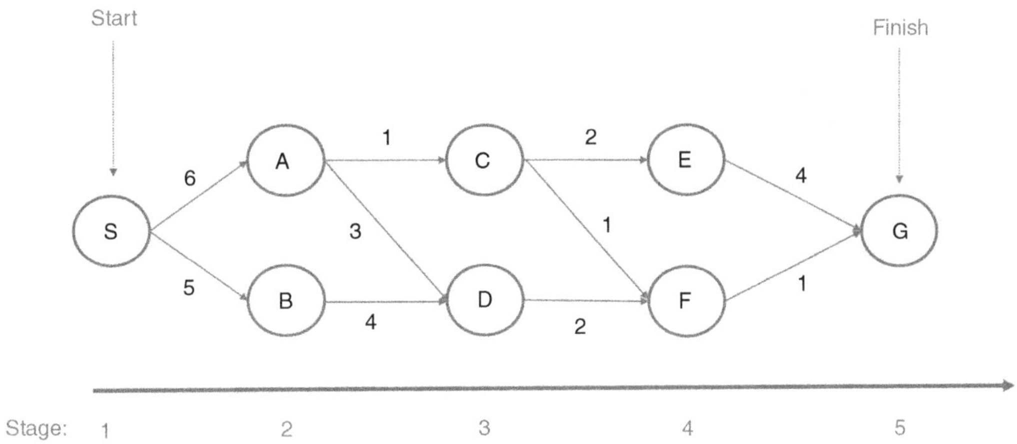

Figure 3.1 Example of a network defining possible pathways to illustrate the *shortest path problem* on a directed network with no cycles.

Overall, the goal is to identify the shortest path from node S to node G, i.e., the path minimizing the total distance traveled to reach node G. Given the objective of minimizing distance traveled, the arc weights are referred to as "costs" rather than rewards.

Finding the minimum length path requires determining the policy that minimizes the total travel distance from S to G. To find the best decision at the initial node S, compare the following two possible decisions:

1. Distance from S to A + Shortest Path from A to G
2. Distance from S to B + Shortest Path from B to G

To compute these distances, it is necessary to know the shortest path length from A to G and from B to G. For such a small example, it is easy to enumerate all paths and find the shortest one. For example, starting at node A, there are three options to reach node G: $A \to C \to E \to G$, $A \to C \to F \to G$, $A \to D \to F \to G$, which have the following path lengths, respectively: 7, 3, 6. Thus, a distance of 3 is the shortest path length from A to G.

Enumerating the paths suffices in this case because the network is small, but one can imagine the complexity of enumerating all paths for a much larger network, such as a city, which could have thousands of nodes and many potential paths between them. The optimality equations can be solved to find the optimal solution. The key to solving the optimality equations is to work backward through the network, associating each node with a decision epoch, t, as shown in Figure 3.1. This can be done as follows.

Epoch 5: Start by noting the final stage ($T = 5$) optimal value function $v_5(G) = 0$; next, move backward through the network, from stage 5 to 4 to 3, and so on, evaluating the optimal value function, $v_t(\cdot)$, for each node recursively as follows.

Epoch 4: Compute the shortest path from E and F to G:

$$v_4(E) = 4 + v_5(G) = 4 + 0, \; v_4(F) = 1 + v_5(G) = 1 + 0,$$

Epoch 3: Compute the shortest path from C and D to G:

$$v_3(C) = \min\{2 + v_4(E), 1 + v_4(F)\} = 2, \; v_3(D) = 2 + v_4(F) = 3,$$

Epoch 2: Compute the shortest path from A and B:

$$v_2(A) = \min\{1 + v_3(C), 3 + v_3(D)\} = 3, \; v_2(B) = 4 + v_3(D) = 7,$$

Epoch 1: Compute the shortest path from S:

$$v_1(S) = \min\{6 + v_2(A), 5 + v_2(B)\} = 9.$$

Having reached the start of the network and the first decision epoch, the solution to the optimality equations, i.e., the value of $v_t(s_t)$ for all epochs and states, is complete, and the minimal path length is 9. This provides the data for determining the optimal policy (the decision at each node), which in turn generates the shortest path from A to G. The decision selected at each state (node) is recorded in Table 3.1.

The above representation of the optimal policy illustrates why the term "lookup table" is often used to describe a policy, because it can be expressed as a table with columns (epoch, state, and decision). Working from node S and moving forward, using the optimal policy, the shortest path is $S \to A \to C \to F \to G$, with a minimal path length of 9, answering the original question about the shortest path. However, note that the optimal policy contains much more information. In particular, it provides the optimal decision that yields the shortest path from any node where the decision maker may be located, whether or not that node is on the shortest path. So, for example, if the decision maker accidentally deviates from the shortest path, finding themselves at node B, say, then the optimal policy guides the decision maker to node D, then F, and finally G. (Mapping software such as Google Maps frequently uses such lookup tables when it initiates rerouting, as often happens when drivers deviate from the initially suggested route.)

Table 3.1 The optimal policy for the shortest path problem in Figure 3.1 expressed as a lookup table.

Epoch	State	Decision
1	S	Go to A
2	A	Go to C
2	B	Go to D
3	C	Go to F
3	D	Go to F
4	E	Go to G
4	F	Go to G
5	G	Stay at G

The thought process for solving the above shortest-path problem using DP is summarized by the following steps:

Step 0: Define the states, decisions, transfer function, rewards, and optimality equations.
Step 1: Solve the optimality equations using backward recursion, starting at the last stage and moving backward to the first stage, to compute the optimal value function for each state.
Step 2: Use the solution to the optimality equations to compile the optimal policy by making note of which decisions minimized (or maximized) the optimal value function.

In this example, the states, the possible decisions, and the transfer function are readily identified in Figure 3.1. However, identifying these elements can be challenging and requires patience and practice with multiple examples to become comfortable formulating DP models. The next example is a well-known problem in production planning, known as the *lot-sizing problem.*

3.4 Production Lot-sizing

The dynamic lot-size problem arises when a decision-maker must decide how much of a product to produce to meet forecasted demand. The decision-maker must consider the production costs, including a fixed cost of initiating production and a variable cost per unit produced. Given the fixed production cost, producing enough at epoch t to cover future epochs' demands by carrying inventory may be desirable to reduce the number of times fixed costs are incurred. However, carrying inventory incurs *holding costs* by tying up capital that could be used for other productive purposes.

As with all models, several assumptions are necessary, as described next. Demand d_t occurs at decision epochs $t = 1, 2, \dots, T$. Demand for epoch t must be met either using inventory held over to epoch t from epoch $t-1$ inventory or manufacturing a new product in the period between epochs t and $t+1$ (referred to simply as period t production) or a combination of both. The cost of producing a_t units in period t is denoted by $c\,(a_t)$ defined as follows: for $a_t = 0$, $c\,(0) = 0$, and for $a_t > 0$, $c\,(a_t) = c^S + ba_t$, where c^S is the fixed cost incurred to initiate production (e.g., due to the setup of machines and other resources needed to produce the product) and b is a variable (per-unit) cost incurred for the materials and other resources needed to produce each unit. At epoch t, the inventory level s_t is reviewed by the decision-maker. An inventory holding cost per unit of inventory, c^H, is incurred for any inventory left at the end of period t that is carried over to period $t+1$.

To formulate the lot-sizing problem as a DP model, the same three-step approach described above for the shortest-path problem is appropriate: gather the data, formulate the model, and solve the optimality equations to find the optimal policy. The details of the lot-sizing problem are presented below, with extended elaboration for the second and third steps. First, identify the decision epochs, states, decisions, and rewards. According to the problem description, the firm's inventory level is reviewed at the start of each period. The production decision is then made in light of future demand. The state in a particular epoch is the inventory level, and the decision is the number of units to produce. The rewards are the cost of production and inventory holding, described in the problem description, which can be expressed as follows:

$$r_t\,(s_t, a_t) = h \times (s_t + a_t - d_t) + c\,(a_t),$$

where the first term is the cost of holding net inventory left over after meeting the demand for period t, and the second is the total production cost. The transfer equation that governs transitions from one state to the next is the following recursive equation, sometimes referred to as the *inventory balance equation* in production planning problems:

$$s_{t+1} = s_t + a_t - d_t.$$

With all the key elements of a DP model in place, the optimality equations can be written as follows:

$$v_t\,(s_t) = \min_{a_t \in A} \{h_t \times (s_t + a_t - d_t) + c\,(a_t) + v_{t+1}\,(s_t + a_t - d_t)\}\,, \forall s_t \in S, t = 1, \ldots, T, \tag{3.3}$$

and the boundary condition is: $v_{T+1}\,(s_{T+1}) = 0, \forall s_{T+1} \in S$.

Notice that the boundary condition is defined at a "dummy" epoch, $T + 1$, denoting the end of the horizon. The state, s_t, at each decision epoch, can be any integer value; thus, the number of optimality equations could make things very difficult because there is one equation for every combination of state and epoch, (s_t, t). In practice, the number of states is limited by product demand, as it is never optimal to produce and hold inventory beyond what is needed to meet customer demand. Additional observations by Wagner and Whitin [5] on the lot-sizing problem show that it can be simplified considerably under certain conditions. The first key insight is the following:

Theorem 3.1 [2]: *If it is optimal to produce during period, t, i.e., if $a_t > 0$ then $s_t = 0$.*

Proof: Suppose an optimal production plan suggests $s_t > 0$ and $a_t > 0$. Then, inventory holding cost can be reduced by delaying the production that resulted in s_t until period t. □

Here, it is assumed that the initial inventory state is $s_1 = 0$. This can be assumed without loss of generality, because any initial inventory could be used to cover demand and the model adjusted, so the first epoch is the start of the first period with nonzero demand. Theorem 1 states that it is only optimal to produce in a period if the available inventory is zero. The second important insight by Wagner and Whitin is the following:

Theorem 3.2 [2]: *Suppose it is optimal to produce in stage t. Then, for some $j = 0, 1, \ldots, T - t$, it is optimal to produce an amount that exactly suffices to meet the demands for stages $t, t + 1, \ldots, t + j$.*

Proof: Since all demands must be met, by assumption, any other decision implies there exists a period t' such that $s_{t'} \times a_{t'} > 0$, which contradicts Theorem 1. □

Given Theorems 3.1 and 3.2, production will occur only during periods in which the beginning inventory is zero; moreover, during each stage in which the beginning inventory is zero (and $d_t \neq 0$), production must occur. The following example illustrates the utility of Wagner and Whitin's insights.

Example 3.1: *Production lot-sizing*

Consider the following case of a company that assembles CT Scanners for hospitals, with the following demand over the next four months (Table 3.2).

At the beginning of each month, the company decides how many CT scanners to assemble. During a month in which any units are produced, there is a setup cost of $c^S = \$3,000$ for activities needed to prepare for assembly. In addition, there is a variable cost of $c^V = \$1,000$ per unit produced to cover external contractors for part of the assembly process for each unit. At the end of each month, there is an inventory holding cost of $c^H = \$500$ per unit per period for each unit carried over in inventory. The optimality equations can be written compactly as follows:

$$v_i = \min_{j=i,\ldots,T} \{c_{ij} + v_{j+1}\}, \quad v_{T+1} = 0. \tag{3.4}$$

Table 3.2 Monthly demand for CT Scanners for four months.

Month	Demand
1	1
2	3
3	2
4	4

Thanks to Theorems 3.1 and 3.2, it is known that production is only optimal when inventory is zero. Therefore, each epoch has only a single state (zero inventory), and the decision becomes how many periods of demand to cover. The optimality equations simplify considerably by aggregating costs of production and inventory holding for all combinations of epochs i to j where $i \leq j$ denoted c_{ij}. Since production only occurs when $s_t = 0$, the argument s_t can be dropped from the optimal value function to simplify the Equation 3.4. The optimality equations can be solved, yielding the following results:

$$v_5 = \$0 \text{ (boundary condition)}$$

$$v_4 = c_{44} = \$7,000 \text{ (cost to produce for period 4 demand)}$$

$$v_3 = \min\{c_{33} + v_4, c\} = \$11,000 \text{ (minimum of cost to produce for periods 3 and 4, or just period 3)}$$

$$v_2 = \min\{c_{22} + v_3, c_{23} + v_4, c_{24}\} = \$16,000 \text{ (minimum of cost to produce for period 2, period 2, 3, or period 2, 3, 4)}$$

$$v_1 = \min\{c_{11} + v_2, c_{12} + v_3, c_{13} + v_4, c_{14}\} = \$19,500 \text{ (minimum cost over all production options).}$$

According to the solution to the optimality equations, the minimum total cost is \$19,500, and the optimal policies are as follows: In the first period, produce for either 2 or 3 periods of demand (both options have the same cost, \$19,500, i.e., there are two optimal decisions at this epoch, and therefore at least two possible policies for the problem). If production is carried out over two periods, then in period 3, produce to meet the demand for periods 3 and 4. If, on the other hand, production is for three periods, then in period 4, produce to cover the demand in period 4.

The above example illustrates two important points. The first point is that analysis of DP models, such as Wagner and Whitin's theorems, can sometimes greatly simplify the solution process. Thus, carefully considering the model is often the best course of action before solving the optimality equations. The second point is that multiple optimal policies can exist for the same problem. Note that according to the model, either solution is equally good from a cost perspective, but in practice, other considerations not reflected in the model may make the decision-maker choose one option over another.

Python Example 3.2: Production lot-sizing

The following Python code implements the Wagner–Whitin algorithm to solve the lot-sizing problem solved above. The steps include collecting the input data, formulating the optimality equations, solving for the optimal policy, and finally printing the results in a form a decision-maker can understand.

```
import numpy as np

# Input data
demand = [1, 3, 2, 4] # Monthly demand
c_s = 3 # Setup cost for each production run
c_v = 1 # Variable cost per unit produced
c_h = 0.5 # Holding cost per unit inventory per month

# Initialize variables
T = len(demand)
cost = np.zeros((T, T))
decision = np.zeros((T))
v = np.zeros(T + 1) # Optimal value function

# Precompute cost to meet demand from i to j periods ahead
for i in range(0, T):
  for j in range(i, T):
    cost[i, j] = c_s + c_v * sum(demand[i:j + 1])
    inventory_cost = 0
    for s in range(i + 1, j + 1):
      inventory_cost = inventory_cost + c_h * (s - i) * demand[s]
    cost[i, j] = cost[i, j] + inventory_cost
```

```
# Solve the optimality equations to optimal decision each period
for i in range(T-1, -1, -1):
  min_value = float('inf')
  for j in range(i, T):
    v[i] = cost[i, j] + v[j + 1]
    if v[i] < min_value:
      min_value = v[i]
      optimal_decision = j


  v[i] = min_value
  decision[i] = optimal_decision

print("\nOptimal Production Schedule:")
print(f"\nTotal Cost: {min_value}")
period = 0
while period < T:
  periods_ahead = int(decision[period] - period)
  produce = sum(demand[period:period + periods_ahead + 1])
  if (periods_ahead == 0):
    print(f"Month {period}: Produce {produce} unit(s) for Month {period} demand")
  else:
    print(f"Month {period}: Produce {produce} unit(s) for Months {period} \
          to {period + periods_ahead} demand")
  if period < T:
      period = period + periods_ahead + 1
```

The output yields one of the two optimal policies obtained in the worked example above, confirming the optimal solution.

```
Optimal Production Schedule:

Total Cost:19.5
Month 0: Produce 4 unit(s) for Months 0           to 1 demand
Month 2: Produce 6 unit(s) for Months 2           to 3 demand
```

where the cost is in units of $1, 000 and production of 4 units occurs in the first month and 6 units two months later. Inventory is carried over from the first to the second month to cover demand (avoiding a setup cost in the second month) and from the third to the fourth month, again avoiding a setup cost in month 3.

3.5 Resource Allocation

Building on the previous examples, this section shows how DP models can be used to solve *resource allocation problems*. Such problems arise commonly when the total demands of multiple activities exceed the limited resources. For instance, an organization might have multiple types of projects to invest in, each with an estimated payoff upon completion, but the budget limits the selection.

The general problem of deciding how to allocate the budget in a way that maximizes total rewards subject to the limited resource availability is as follows:

$$\max \sum_{t=1}^{T} r_t(a_t). \tag{3.5}$$

Subject to the constraint: $\sum_{t=1}^{T} f_t(a_t) \le w$,

where the decision, $a_t \in \{0, 1, 2, ...\}$, is the amount of activity t, a non-negative integer, and $r_t(a_t)$ is the reward for choosing activity t at level a_t. Finally, $f_t(a_t)$ is the utilization of activity t given level a_t, and w is a limit on the available capacity for all activities (e.g., financial budget, time, and machine capacity). Unlike previous examples, t indexes the sequence in which activities are selected rather than a particular time epoch. For this reason, in this context, the term *stage* is used in place of it epoch. Any sequence of stages for solving the optimality equations yields the same optimal solution(s).

To formulate this problem as a DP model, define $v_t(s_t)$ to be the maximum reward that can be obtained if s_t units of the resource are allocated optimally to the remaining activities $t, t+1, \ldots, T$. The optimality equations are as follows:

$$v_t(s_t) = \max_{a_t \in A_t(s_t)} \{r_t(a_t) + v_{t+1}(s_t - f_t(a_t))\}, \forall s_t \in S, t = 1, ..., T,$$

$$v_{T+1}(s_{T+1}) = 0, \forall s_{T+1} \in S,$$

where $a_t \in A_t(s_t)$ and $A_t(s_t)$ is the set of non-negative integers satisfying the following constraint:

$$f_t(a_t) \le s_t.$$

Example 3.3: *Resource allocation*

Suppose an organization has a budget of \$10,000 that can be used to invest in an integer number of three types of projects. Each investment may comprise an integer number of projects of a given type, from zero to infinity (budget allowing). The project types under consideration are described in Table 3.3.

Therefore, the rewards and budget utilization are as follows:

$$r_1(a_1) = 11a_1, r_2(a_2) = 7a_2, r_3(a_3) = 12a_3, \tag{3.6}$$

$$f_1(a_1) = 4a_1, f_2(a_2) = 3a_2, f_3(a_3) = 5a_3. \tag{3.7}$$

Now that the model DP elements are defined, the equations can be solved in three stages, with each stage corresponding to a project. In this example, the selected sequence of stages corresponds to the ordering of project types: 1, 2, 3. Thus, the first step in solving the optimality equations (starting from the last stage) is to compute $v_t(s_t)$, which involves determining how many projects of type 3 can be selected given s_t elements of capacity. Thus, the third stage optimal value functions are determined as follows: $v_3(s_3) = \max\{12a_3\}$ where $5a_3 \le s_3$. It follows that in stage 3: $v_3(10) = 24$, $v_3(5) = v_3(6) = v_3(7) = v_3(8) = v_3(9) = 12$, $v_3(0) = v_3(1) = v_3(2) = v_3(3) = v_3(4) = 0$. Proceeding according to the selected sequence, optimal decisions at stage 2 are governed by:

$$v_2(s_2) = \max_{a_2 \in A_2(s_2)} \{7a_2 + v_3(s_2 - 3a_2)\},$$

where a_2 is an integer satisfying $3a_2 \le s_2$ and $A_2(s_2)$ is the set of decisions that satisfy this requirement. Thus,

$$v_2(10) = \max_{a_2 \in \{0,1,2,3)} \{7(0) + v_3(10), 7(1) + v_3(7), 7(2) + v_3(4), 7(3) + v_3(1)\}$$

$$= 24.$$

Table 3.3 Project costs and rewards.

Project type	Cost	Reward
Project 1	\$4,000	\$11,000
Project 2	\$3,000	\$7,000
Project 3	\$5,000	\$12,000

Because the first term in the max{ } operator achieves the maximum it follows that, $a_2^* = 0$. Next, for the state with 9 units of available resource,

$$v_2(9) = \max_{a_2 \in \{0,1,2,3)} \{7(0) + v_3(9), 7(1) + v_3(6), 7(2) + v_3(3), 7(3) + v_3(0)\}$$
$$= 21.$$

Since the third term in the *max* operator achieves the maximum, it follows that $a_2^* = 3$. Continuing in the same way, the remaining optimal value functions for $s_2 = 8,\ 7,\ 6, \ldots, 0$ are as follows:

$v_2(8) = 19, a_2^* = 1$
$v_2(7) = 14, a_2^* = 2$
$v_2(6) = 14, a_2^* = 2$
$v_2(5) = 12, a_2^* = 1$
$v_2(4) = 7, a_2^* = 1$
$v_2(3) = 7, a_2^* = 1$
$v_2(2) = 0, a_2^* = 0$
$v_2(1) = 0, a_2^* = 0$
$v_2(0) = 0, a_2^* = 0.$

With stage 2 complete, solve for the optimal decisions in stage 1, which are determined by:

$$v_1(s_1) = \max\{11a_1 + v_2(s_1 - 4a_1)\},$$

where a_1 is an integer satisfying $4a_1 \leq s_1$. Stage 1 begins with all 10 units of resources available. Therefore, at stage 1, only one state needs to be evaluated, the state $s_1 = 10$, as follows:

$$v_1(10) = \max_{a_1 \in \{0,1,2\}} \{11 \times 0 + v_2(10), 11(1) + v_2(6), 11(2) + v_2(2) = 22\} = 25,$$

where $a_1^* = 1$ is the decision that maximizes the optimal value function. Inspecting the results above and *backtracking* through the optimality equations from the first to the last stage yields the following optimal policy: $a_1^* = 1,\ a_2^* = 2,\ a_3^* = 0$, and the total reward for this policy is $v_1(1) = 25$. Note that, in this case, all available resources are consumed, although this need not be the case for optimal policies in other examples.

Recall that any sequence of project types would have worked. Question 3 from the practice examples is intended to verify this claim by solving the example above using an alternative sequence of project types to confirm that the same optimal solution is obtained.

Python Example 3.4: Resource allocation

The following Python code implements a DP model for the resource allocation problem, assuming projects can be selected an integer number of times, subject to the capacity constraint. It also assumes that the project weights and the resource capacity are integer-valued. The latter assumption is unnecessary but simplifies the implementation of the DP optimality equations, as integer state values can serve as indices for the optimal value function.

```
import numpy as np

# Input data
costs = [4,3,5] # per unit costs of the items
rewards = [11,7,12] # per unit reward for projects
capacity = 10 # Total capacity for projects (budget)
T = len(costs) # Number of items

# Optimal value function initialization
v = np.zeros(capacity + 1)
```

```
# Compute the maximum benefit for each capacity
for w in range(capacity + 1):
  for i in range(T):
    if costs[i] <= w:
      v[w] = max(v[w], v[w - costs[i]] + rewards[i])

# Reconstruct the optimal solution
solution = np.zeros(T)
w = capacity
while w > 0:
  for i in range(T):
    if w >= costs[i] and v[w] == v[w - costs[i]] + rewards[i]:
      solution[i] += 1 # Include this item in the solution
      w -= costs[i] # Reduce the remaining capacity
      break # Break to restart the item selection

# Print the maximum value and the optimal solution
print(f"Maximum rewards: {v[capacity]}")
print(f"Optimal solution: {solution}")
```

Running the code yields the following optimal solution, which matches the worked example above.

```
Maximum rewards: 25.0
Optimal solution: [1. 2. 0.]
```

In Chapter 5, constrained linear optimization models, called *linear programs*, are discussed, providing an alternative approach to resource allocation problems. However, linear programs require the following three assumptions to hold:

- *Assumption 1*: The amount of a resource assigned to an activity may be any nonnegative number
- *Assumption 2*: The benefit obtained from each activity is proportional to the resource assigned to the activity
- *Assumption 3*: The total utilization for each activity is proportional to the resources assigned to the activity

Even if assumptions 1, 2, and 3 do not hold, DP can solve resource-allocation problems. Thus, DP can have an advantage over other approaches, such as linear programming, in certain cases.

3.6 Pattern Recognition

Another application of DP is to identify patterns by computing a measure of similarity between strings of characters. This problem arises in a wide range of contexts. For example, in genetics, the goal is to compare DNA sequences to quantify their similarity. It may be necessary to do this many times with a new target string of interest evaluated against entries in a database of previously obtained DNA samples. Because DNA is subject to minor variations caused by the loss or addition of elements of the string, often no perfect match is found. Thus, a similarity measure is defined as the maximum number of character matches obtained by optimally inserting spaces to represent lost characters. The DP model can be formulated as follows:

States: pairwise combinations of positions in String 1, i, and String 2, j.

Decisions: whether to modify the strings by adding spaces between characters:

$$a_t \in \{\text{No Insert, Insert Space in String 1, Insert Space in String 2}\}.$$

Rewards: one point for each character match achieved and 0 otherwise.

$$r(i, j) = \begin{cases} 1 & \text{if } c_2(i) = c_1(j), \\ 0 & \text{if } c_2(i) \neq c_1(j). \end{cases}$$

Optimality equations: given the above definitions of the states, decisions, and rewards, the optimality equations are as follows:

$$v(i, j) = \max \begin{cases} r(i, j) + v(i-1, j-1) & (\text{decision: no space inserted}), \\ v(i-1, j) & (\text{decision: space inserted in string 1}), \\ v(i, j-1) & (\text{decision: space inserted in string 2}). \end{cases}$$

The addition of a space in a string is denoted by " - " and the boundary condition is $v(i, j) = 0$ for positions on the boundary, assuming a leading "dummy" character. Thinking of the problem in matrix form makes it easy to organize the optimality equations.

Example 3.5: *Pattern recognition in sequences*

This example illustrates the application of DP with the following example from Abbas and Holmes, 2005, based on the following two strings of characters:

1) C O U N T I N G

2) N T I G

Hypothesizing that strings 1 and 2 were, at some point, the same, but some characters were lost over time due to signal loss, the problem becomes one of finding the minimal number of character losses necessary to match the strings. So, if spaces can be added to either string, the question is: what is the optimal alignment that maximizes the number of matches? To illustrate, the following are possible alignments of the two sequences obtained by adding spaces to string 2:

Solution 1 : C O U N T I N G

– – – N T I G-

Solution 2 : C O U N T I N G

N---T I G–

Solution (1) has three matches, while solution (2) has only two matches. Thus, solution (1) is considered better than solution (2) according to the objective criterion used. Clearly, for this small problem, four matches are possible by inspection. This small example illustrates how a DP model can be formulated to solve such pattern recognition problems, including extremely large-scale problems that arise in practice.

The optimality equations can be solved using a matrix representation of the problem. Letting string 1 define the columns and string 2 the rows of a matrix, each element, (row i, column j), corresponds to the optimal value function, $v(i, j)$ for state (i, j), based on Equations 3.8. To initiate solving the DP, begin by filling in the table with the known elements, i.e., those that define the boundary condition, as illustrated in Figure 3.2 where the first row and the first column are all zero. Continue filling in the matrix starting with element (2, 2), because the necessary information for computing $v(2, 2)$ is available, and

Figure 3.2 Illustration of the first step of the solution process for the pattern recognition problem. The optimality equations are expressed as a matrix, with each element corresponding to a position in string 1 and string 2. The boundary condition is based on a dummy character, Ø, that initiates each string.

	Ø	C	O	U	N	T	I	N	G
Ø	0	0	0	0	0	0	0	0	0
N	0								
T	0								
I	0								
G	0								

	Ø	C	O	U	N	T	I	N	G
Ø	0	0	0	0	0	0	0	0	0
N	0	0	0	0	1	1	1	1	1
T	0	0	0	0	1	2	2	2	2
I	0	0	0	0	1	2	3	3	3
G	0	0	0	0	1	2	3	3	4

Figure 3.3 The results for the optimal value function for each state obtained by solving the optimality equations by moving down one row and right one column at a time to fill in uncompleted elements of the matrix.

	Ø	C	O	U	N	T	I	N	G
Ø	0	0	0	0	0	0	0	0	0
N	0	0	0	0	1	1	1	1	1
T	0	0	0	0	1	2	2	2	2
I	0	0	0	0	1	2	3	3	3
G	0	0	0	0	1	2	3	3	4

Figure 3.4 Illustration of backtracking to obtain a solution defining where to insert spaces to achieve maximum character matches.

then continue to complete row 2 ((2, 3), (2, 4), ...) or column 2 ((3, 2), (4, 2), ...). Continuing to move row-wise or column-wise generates the complete matrix as illustrated in Figure 3.3.

Next, backtracking through the optimality equations, starting from the lower right element, $v(5,9) = 4$ to determine the optimal decision(s) that led to the solution of Equations 3.8. The path of decisions that defines the optimal solution is illustrated in Figure 3.4 and the optimal placement of spaces that achieves a maximum alignment score of 4 is

Optimal Solution : C O U N T I N G

– – – N T I - G

This approach can be applied to string comparison problems in many contexts, such as spell-checking, speech recognition, web-based search tools, and song matching in music databases. In such cases, a binary representation of strings is used, but the approach is the same.

3.7 Generalization of Shortest Path Problems to Include Cycles

This section generalizes the shortest path problem first introduced in Section 3.3. The previous shortest path example was in the context of a directed acyclic network, illustrated in Figure 3.1, which had an obvious start and end node. Each node was associated with a stage of the decision-making process, which mapped directly to the sequence of stages required to formulate the DP model and the corresponding optimality equations. In this section, networks that may not possess these special properties are considered; they may be directed or undirected and may contain cycles. The algorithm described next is suitable for these general shortest path problems. Despite the generalization in the previous example in Section 3.3, the notion of finding a shortest path from one node to another remains well-defined; however, applying DP requires modifications.

The DP model formulation for the shortest path problem in the cases of an *acyclic directed network* is considered first. Such networks, by definition, have no cycles. Therefore, there is a clear sequence of nodes, meaning the nodes can be ordered according to the arcs (e.g., A comes before C and D, and C comes before E and F); thus, there is a well-defined notion of a path from one node to another in the network provided the network is connected, i.e., there exists at least one path between any two nodes in the network. To formulate the DP model for this situation, let v_i denote the length of the shortest path from node i to some predetermined destination node N, and d_{ij} denotes the distance from node i to j. The optimality equations for the shortest path problem are then as follows:

$$v_i = \min_{j>i}\{v_j + d_{ij}\} \text{ for } i = 1, 2, \dots, N-1, \quad (3.8)$$

$$v_N = 0 \text{ (boundary condition)}. \quad (3.9)$$

As noted previously, in Section 3.3, the boundary condition is defined at the destination (end) node. The convention is to assume $d_{ij} = \infty$ if there is no arc between nodes i and j to imply that a sub-path from i to j cannot be part of the shortest path.

The shortest path problem of any type (e.g., directed or undirected, with or without cycles) can be solved via an alternative definition of the optimal value function and the optimality equations, known as *forward dynamic programming*. This alternative assumes v_i denotes the length of the shortest path from the start node (node 1) to node i. This is the reverse of the definition in Section 3.3, where the optimal value function was the shortest path from the current node to the destination node. This modification is equivalent to the backward DP approach and, as will become clear shortly, is also ideally suited to handling cyclic networks.

Optimality Equations (Forward DP):

$$v_i = \min_{j<i}\{v_j + d_{ji}\} \text{ for } i = 2, 3, \dots, N,$$

$$v_1 = 0 \text{ (boundary condition)},$$

$$d_{ij} = \infty \text{ (if no arc between nodes } i \text{ and } j).$$

In this case, the sequence of steps in Equation 3.7 is reversed. The boundary condition is defined at the first node, and the optimality equations iteratively evaluate successor nodes.

Next, return to the case of general networks that may contain *cycles*. For example, a map of a city naturally contains many cycles (e.g., a two-way street between two locations) or a cyclic path of nodes (e.g., A to B to C to D to A, such as four right turns in a road network). If these situations are not treated carefully, they will cause the backward and forward DP formulations to fail because, in such cases, there is no longer a consistent ordering of the nodes to define the recursion for the optimality equations. In other words, it is no longer obvious what constitutes the start and end of the network, and it is possible that the previously defined algorithms may enter an infinite loop. The following modification of the forward DP approach addresses this issue.

Let N_i denote the set of nodes consisting of the i closest (in terms of the shortest path) immediately connected nodes to node 1. Let $v_i(j)$ denote the length of the shortest path from node 1 to j when restricted to using nodes in N_i. With these definitions, the algorithm for solving any shortest path problem is presented in Algorithm 3.1.

Algorithm 3.1 General Algorithm for Solving Shortest Path Problem with Cycles

Step 0: Based on the boundary condition at Stage 1 set $N_1 = \{1\}$, $v_1(j) = d_{1j}$. ▷ $d_{1j} = \infty$ if there is no arc from node 1 to j
Set $i = 2$.
while there are remaining nodes **do**
    Step 1: Find $k_i = \arg\min_{j \notin N_{i-1}} v_{i-1}(j)$.
    Step 2: Set $N_i = N_{i-1} \cup \{k_i\}$.
    Step 3: Compute $v_i(j)$ as:

$$v_i(j) = \begin{cases} v_{i-1}(j), & \text{if } j \in N_i, \\ \min\{v_{i-1}(j),\ v_{i-1}(k_i) + d_{k_i,j}\}, & \text{if } j \notin N_i. \end{cases}$$

    Step 4: Set $i = i + 1$.
end while

Step 0 initializes the algorithm. Step 1 identifies a node, k_i, the next closest node to node 1 that has yet to be considered in previous algorithm iterations. Thus, for example, in the first iteration, it is the node closest to node 1. Step 2 adds the new node from Step 1 to the set of nodes, N_i. Step 3 finds the shortest path from node 1 to j for all unvisited nodes, i.e., nodes not yet added to N_i. Step 4 checks a stopping condition and either stops the algorithm or continues it (via Step 1) until all nodes have been considered.

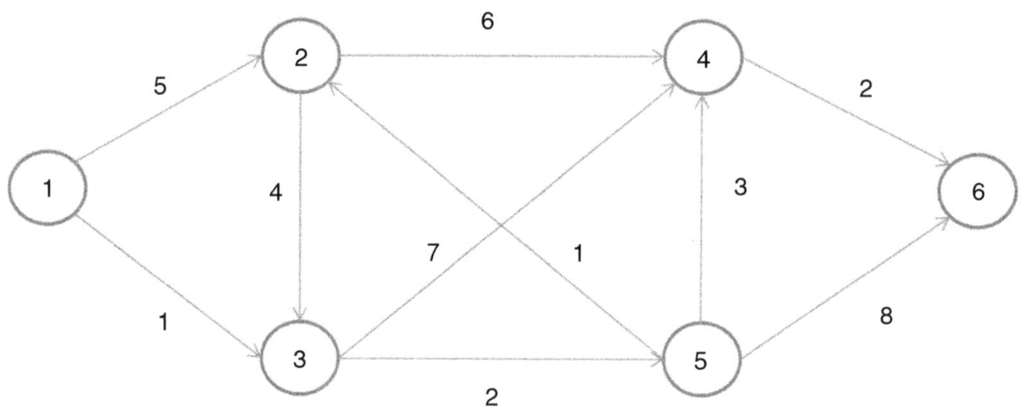

Figure 3.5 An example of a cyclic network.

The algorithm for solving general shortest path problems described above is effectively the well-known *Dijkstra's algorithm*, named after its inventor, who published it [3]. It can solve any shortest path problem discussed so far.

Example 3.6: *Shortest path on a cyclic network*

Consider the *cyclic network* in Figure 3.5. Assume the goal is to find the shortest path from node 1 to node 6. The numeric node labels, $\{1,\ 2,\ 3,\ 4,\ 5,\ 6\}$, are arbitrarily selected. For Stage 1, use the boundary condition to initiate the set of i closest nodes to node 1 to be $N_1 = \{1\}$ and the optimal value function to $v_1(j) = d_{1j}$. Note that $d_{1j} = \infty$ if there is no arc from node 1 to j. Thus, for Stage 1, it follows that:

Stage 1:

Step 0: $N_1 = \{1\}$, $k_1 = 1$, $v_1(j) = d_{1j}$

$v_1(1) = 0$, $v_1(2) = 5$, $v_1(3) = 1$, $v_1(4) = v_1(5) = v_1(6) = \infty$.

Stage 2:

Step 1: $k_2 = \arg\min_{j \notin N_1}\{v_1(j)\} = 3$

Step 2: $N_2 = \{1, 3\}$

Step 3: $v_2(1) = 0$, $v_2(3) = 1$,

$v_2(2) = \min\{v_1(2), v_1(3) + d_{32}\} = \min\{5, \infty\} = 5$,

$v_2(4) = \min\{v_1(4), v_1(3) + d_{34}\} = \min\{\infty, 8\} = 8$,

$v_2(5) = \min\{v_1(5), v_1(3) + d_{35}\} = \min\{\infty, 3\} = 3$,

$v_2(6) = \min\{\infty, \infty\} = \infty$.

Step 4: $i = 3$; return to Step 1.

Stage 3:

Step 1: $k_3 = \arg\min_{j \notin N_2}\{v_2(j)\} = 5$

Step 2: $N_3 = \{1, 3, 5\}$

Step 3: $v_3(1) = 0$, $v_3(3) = 1$, $v_3(5) = 3$,

$v_3(2) = \min\{v_2(2), v_2(5) + d_{52}\} = \min\{5, 4\} = 4$,

$v_3(4) = \min\{v_2(4), v_2(5) + d_{54}\} = \min\{8, 6\} = 6$,

$v_3(6) = \min\{v_2(6), v_2(5) + d_{56}\} = \min\{\infty, 11\} = 11$.

Step 4: $i = 4$; return to Step 1.

Stage 4:

Step 1: $k_4 = \arg\min_{j \notin N_3}\{v_3(j)\} = 2$

Step 2: $N_4 = \{1, 2, 3, 5\}$

Step 3: $v_4(1) = 0$, $v_4(2) = 4$, $v_4(3) = 1$, $v_4(5) = 3$,

$v_4(4) = \min\{v_3(4), v_3(2) + d_{24}\} = \min\{6, 10\} = 6$,

$v_4(6) = \min\{v_3(6), v_3(2) + d_{26}\} = \min\{11, \infty\} = 11$.

Step 4: $i = 5$; return to Step 1.

Stage 5:

Step 1: $k_5 = \arg\min_{j \notin N_4}\{v_4(j)\} = 4$

Step 2: $N_5 = \{1, 2, 3, 4, 5\}$

Step 3: $v_5(1) = 0$, $v_5(2) = 4$, $v_5(3) = 1$, $v_5(4) = 6$, $v_5(5) = 3$,

$v_5(6) = \min\{v_4(6), v_4(4) + d_{46}\} = \min\{11, 8\} = 8$.

Step 4: $i = 6$; All nodes explored; therefore stop with the optimal solution.

The shortest path length is 8. Backtracking through the solution to find the optimal decision at each stage gives the optimal path: $1 \rightarrow 3 \rightarrow 5 \rightarrow 4 \rightarrow 6$.

Python Example 3.7: Shortest path problem with cycles

All of the examples in this chapter were based on small problems that can be worked out by hand to develop intuition for DP models and to understand how and why they work in certain circumstances. In practice, DP applies to much more complex problems. The following Python code solves the shortest-path problem in the general case of a cyclic network for a small example, which can be extended to solve large-scale shortest-path problems.

```
from math import inf

# Input data
network = {  # directed, weighted adjacency list
```

```
    '1': {'2': 5, '3': 1},
    '2': {'3': 4, '4': 6, '5': 7},
    '3': {'4': 1, '5': 2},
    '4': {'6': 2},
    '5': {'4': 3, '6': 8},
    '6': {}
}
source = '1' #start node

def shortestpath_dp(adj, source, target=None):
    v = {u: inf for u in adj}      #initialize value function
    parent = {u: None for u in adj}
    v[source] = 0.0
    N = set() #initialize to empty set
    stage = 0
    while len(N) < len(adj):
        unvisited = [u for u in adj if u not in N]
        k = min(unvisited, key=lambda u: v[u]) #Find nearest node not in N
        if v[k] == inf: break  # break if remaining nodes unreachable
        N.add(k) #add nearest node to set of visited nodes
        stage += 1
        if target is not None and k == target: break
        for j, w_kj in adj[k].items():  #Update path length from k to neighbors

            if j in N:
                continue
            if v[k] + w_kj < v[j]:


                v[j] = v[k] + w_kj
                parent[j] = k
    return v, parent

def reconstruct_path(parent, target):
    path = []
    cur = target
    while cur is not None:
        path.append(cur)
        cur = parent[cur]
    return list(reversed(path))

#Find shortest path from start node to all nodes in network
v, parent = shortestpath_dp(network, source=source)

print(f"\nShortest paths from {source}:")
for node in sorted(network):
    if v[node] < inf:
        path = reconstruct_path(parent, node)
        length = int(v[node]) if float(v[node]).is_integer() else v[node]
        print(f"{source} → {node}: {'-->'.join(path)} (length={length})")
    else: print(f"  {source} → {node}: unreachable")
```

The above Python code generates the solution to the shortest path problem in Figure 3.5, with the following output:

```
Stage 1: choose k=1, N=['1']
  v: {'1': 0.0, '2': None, '3': None, '4': None, '5': None, '6': None}
Stage 2: choose k=3, N=['1', '3']
  v: {'1': 0.0, '2': 5.0, '3': 1.0, '4': None, '5': None, '6': None}
Stage 3: choose k=4, N=['1', '3', '4']
  v: {'1': 0.0, '2': 5.0, '3': 1.0, '4': 2.0, '5': 3.0, '6': None}
Stage 4: choose k=5, N=['1', '3', '4', '5']
  v: {'1': 0.0, '2': 5.0, '3': 1.0, '4': 2.0, '5': 3.0, '6': 4.0}
Stage 5: choose k=6, N=['1', '3', '4', '5', '6']
  v: {'1': 0.0, '2': 5.0, '3': 1.0, '4': 2.0, '5': 3.0, '6': 4.0}
Stage 6: choose k=2, N=['1', '2', '3', '4', '5', '6']
  v: {'1': 0.0, '2': 5.0, '3': 1.0, '4': 2.0, '5': 3.0, '6': 4.0}

Shortest paths from 1:
1 → 1: 1 (length=0)
1 → 2: 1-->2 (length=5)
1 → 3: 1-->3 (length=1)
1 → 4: 1-->3-->4 (length=2)
1 → 5: 1-->3-->5 (length=3)
1 → 6: 1-->3-->4-->6 (length=4)
```

Thus, the intermediate results are provided as well as the shortest path from the specified start node, 1, to all other nodes, and the shortest path and path length from node 1 to the terminal node 6. The code can easily be modified to solve other shortest path problems by providing alternative networks as defined by a user-supplied network in the form of an adjacency list.

3.8 Counter Example: When DP Does Not Work

Just as no single tool is appropriate for every task, it is essential to acknowledge that DP models, too, have limitations. Consider the network illustrated in Figure 3.6 and assume that instead of the sum of the arc lengths, the length of a path is defined to be equal to the product of the arc lengths on the path. Blindly applying DP based on the optimality equations in Equations 3.1 and 3.2 leads to the following product form of the optimality equation:

$$v_t\left(s_t\right) = \min_{a_t \in A}\left\{r_t\left(s_t, a_t\right) \times \lambda v_{t+1}\left(s_{t+1}\right)\right\}, \forall s_t \in S, t = 1, \ldots, T-1, \tag{3.10}$$

$$v_T\left(s_T\right) = 1, \forall s_T \in S. \tag{3.11}$$

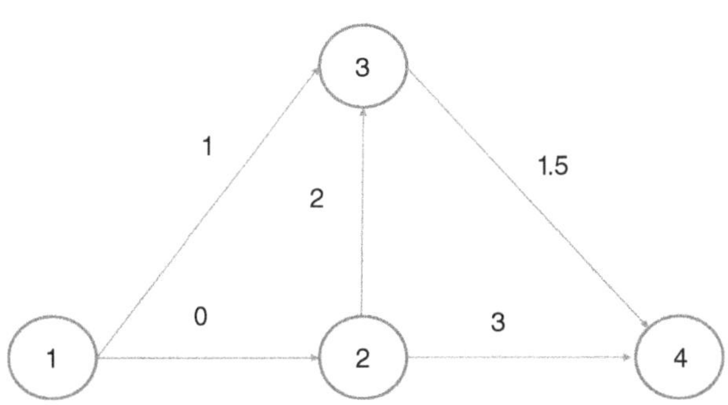

Figure 3.6 Example of a shortest path problem that violates the principle of optimality when the rewards are the product of arc weights.

However, this formulation fails to solve this problem, as shown in Figure 3.6, where any path through nodes 1 and 2 has an optimal value function of zero. The problem is that the principle of optimality is invalid: the optimal value function at a given state (node) for a particular subpath is not separable, because the rewards are defined as a product rather than a sum. For example, at node 2, it is optimal to go directly to node 4, and avoid the path through node 3. But, starting at node 1, the subpath from 2 to 3 to 4 has the same optimal value function as the path from 2 to 4, which violates the principle of optimality. This illustrates the importance of keeping the fundamental assumptions of DP in mind before jumping to what seems like a logical model.

3.9 Concluding Remarks

This chapter introduced the basic idea of DP in a deterministic setting, including the principle of optimality, model formulations, optimality equations, and algorithms, followed by examples that draw on applications to situations in which a series of sequential decisions must be made. Several sources informed the examples, including [4] for shortest-path and resource-allocation problems, Wagner and Whitin [2] for the production lot-sizing problem, and [5] for the pattern recognition problem.

While it is natural to think of sequential decision-making problems from the first to the last decision, according to the passage of time, DP reveals that, for many problems, working backward in time enables finding optimal solutions efficiently. In other words, the answer to the question, "What shall I do next?" depends not only on the current decision at hand but also on all the future decisions to be made.

The deterministic DP models in this chapter provide a foundation for handling non-deterministic settings. The next chapter covers sequential decision-making problems arising from uncertainty about the future, which apply across all areas where a system's future state is influenced by the decision maker's choices and subject to randomness.

3.10 Practice Exercises

1. Assume a University of Michigan engineering student is traveling from her house to her IOE 512 Lecture in the Electrical Engineering and Computer Science (EECS) building on North Campus. Use the data from Table 3.4 to find the shortest path she could take to go from home to class. What is the shortest path if she wants to include a stop for coffee at Starbucks?

Table 3.4 Distances measured in time between locations in Ann Arbor.

From	To	Time	From	To	Time
Home	Big House	4	Home	IM Building	2
Home	Friend's House	3	Home	Diag	7
Big House	Ashley's	3	Big House	Bivouac	4
IM Building	Ashley's	2	IM Building	Bivouac	4
IM Building	Law School	1	Friend's House	Bivouac	1
Friend's House	Law School	5	Friend's House	UMMA	3
Diag	Law School	1	Diag	UMMA	2
Diag	Hist. Museum	4	Ashley's	Starbucks	5
Ashley's	CCTC	3	Bivouac	Starbucks	4
Bivouac	CCTC	4	Law School	Starbucks	5
Law School	CCTC	6	UMMA	CCTC	3
Hist. Museum	Starbucks	2	Hist. Museum	CCTC	1
Starbucks	Medical School	3	Starbucks	The Arb	2
Starbucks	Hospital	4	Starbucks	Farmer's Market	2

From	To	Time	From	To	Time
Starbucks	Bao Space	1	CCTC	The Arb	2
CCTC	Hospital	1	CCTC	Farmer's Market	4
CCTC	Bao Space	3	CCTC	7/11	2
Medical School	Mitchell Field	2	Medical School	Broadway Cafe	3
The Arb	Mitchell Field	1	The Arb	Broadway Cafe	3
Hospital	Mitchell Field	3	Hospital	Lowertown	2
Hospital	No Thai!	3	Farmer's Market	Broadway Cafe	2
Farmer's Market	Lowertown	1	Farmer's Market	No Thai!	2
Bao Space	Broadway Cafe	4	Bao Space	Lowertown	3
Bao Space	No Thai!	2	7/11	Lowertown	4
7/11	No Thai!	2	Mitchell Field	NCRB	2
Mitchell Field	EWRB	3	Mitchell Field	IOE	2
Mitchell Field	Mujo	4	Broadway Cafe	NCRB	1
Broadway Cafe	EWRB	3	Broadway Cafe	IOE	2
Broadway Cafe	Mujo	4	Broadway Cafe	Sand Pit	3
Lowertown	EWRB	4	Lowertown	IOE	5
Lowertown	Mujo	3	Lowertown	Sand Pit	2
Lowertown	NCRC	4	No Thai!	IOE	3
No Thai!	Mujo	2	No Thai!	Sand Pit	3
No Thai!	NCRC	1	NCRB	EECS Building	3
EWRB	EECS Building	1	IOE	EECS Building	2
Mujo	EECS Building	3	Sand Pit	EECS Building	1
NCRC	EECS Building	3			

2. Show that you can ignore the variable per unit production cost in the optimality equations when solving the production lot-sizing problem if the cost is constant over the decision.

3. Test your understanding of the approach for the resource allocation problem by solving this problem with a different ordering of items (e.g., 3, 1, 2) and confirm you get the same solution.

4. Solve the production lot-sizing problem in Example 3.2 using the same cost parameters, but set the decision-making frequency to 1 day and the horizon to 1 year, i.e., $T = 365$ days. Assume demand in the first period is 100 units, increases by 5% for the first 25 weeks, and then drops by 4% each week for the remainder of the time horizon (round demand to the closest integer value each day). Experiment with varying the setup cost and inventory holding cost. What do you observe about how the model parameters influence the optimal solution?

5. A vaccine manufacturer is contracted to deliver the following number of doses over the next six quarters: 2000, 3000, 1000, 2500, 3000, 2000. Each dose has a variable production cost of \$10 per dose. The inventory holding cost is \$0.05 per quarter per dose. The cost of setting up production in a given quarter is \$200. Any doses produced in a given quarter may be used to meet the demand for that quarter or any future quarter. Use the Wagner-Whitin algorithm to find all optimal production policies.

6. Consider the following three strings:

1. ZOMBIEATTACK
2. ZOOBEATCKAT
3. OMBBIATATCAK

Use DP to determine which of strings (2) and (3) is most similar to string (1).

7. Propose a transformation of the optimality equations for problems with a product form of rewards, like in Figure 3.6. Apply it to the following design problem. An electronic device consists of three main components that can fail. The components are in series, so the failure of any one component causes the device to fail. The device's reliability (probability of no failure) can be improved by installing one or two backup units for each component. Table 3.5 provides the reliability, r, and the cost, c. The total budget for the construction of the device is \$10,000. Formulate and solve the optimality equations you proposed to maximize the device's reliability. Explain why this approach will fail for examples with one or components with perfect reliability.

Table 3.5 Reliability data for a device design problem.

No. of parallel units	Component 1		Component 2		Component 3	
	r_1	c_1(\$)	r_2	c_2(\$)	r_3	c_3(\$)
1	0.6	1000	0.7	3000	0.5	2000
2	0.8	2000	0.8	5000	0.7	4000
3	0.9	3000	0.9	6000	0.9	5000

8. Consider the *social network* in Figure 3.7. Each person is represented as a node. An arc between two nodes indicates friendship. The "distance" between any two people is the length (# of edges) in the shortest path between them. Write code in Python that computes, for each node, the average minimum distance to all other nodes. Which nodes correspond to the most important influencer(s)?

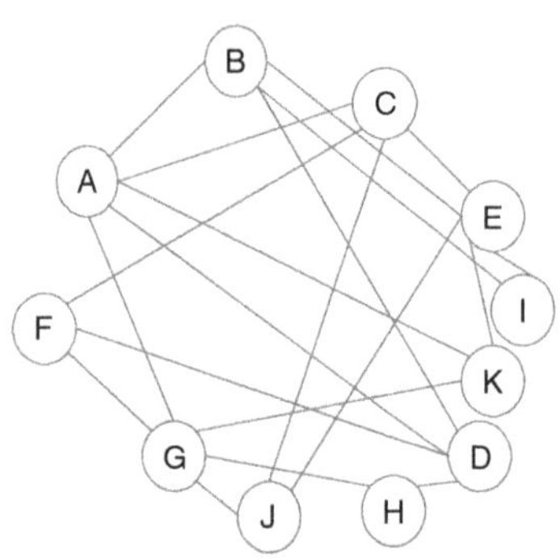

Figure 3.7 Social network with nodes representing people and arcs representing friendships.

9. When using DP to find the shortest path on an acyclic network, you begin with a node that has only outward-pointing arcs. Prove that every acyclic network contains at least one node with only outward-pointing arcs. (Hint: assume the contrary and show that this implies the existence of a cycle.)

10. A semiconductor manufacturer can purchase a new photolithography tool for \$1M. The annual operating costs and resale values are provided in Table 3.6. Purchases are paid at the beginning of a period, and operating costs and resales occur at the end of a period. Determine a purchase and replacement policy that minimizes total cost of owning and operating the tool over the next 7 years given the manufacturer must have a tool in all years.

Table 3.6 Resale value of a used photolithography tool and annual operating costs.

Tool age at end of year	Resale value ($Millions)	Operating cost (Millions)
1	0.50	0.10
2	0.40	0.15
3	0.30	0.25
4	0.25	0.40
5	0.20	0.55
6	0.15	0.65
7	0.10	0.70

11. Consider a game where there are N coins on a table, and you and your opponent must pick up 1, 2, 3, ... , or M coins at each stage. The person who is forced to pickup the last coin loses. Define the states, decisions, rewards, and optimality equations. Assume your opponent's policy is to pick up the same number of coins you do each time. Under what conditions are you guaranteed to win, independent of your opponent's policy? What is the optimal policy?

12. Prove that any $v_t(s_t), \forall s_t \in S$ that solves the optimality equations:

$$v_t(s_t) = \min_{a \in A}\{r_t(s_t, a) + \lambda v_{t+1}(g_t(s_t, a))\}, \ \forall s_t \in S, t = 1, 2, \ldots , T - 1,$$

with boundary condition $v_{T+1}(s_T) = r_T(s_T), \forall s_T \in S$ yields an optimal policy. (Hint: consider a proof by induction).

13. Consider the shortest path network in Figure 3.8. Assume that the cost of a path is the largest arc cost on the path. Formulate and solve a dynamic program for this problem and provide all optimal paths. Do the optimality equations generate all optimal paths for the problem? If not, explain why.

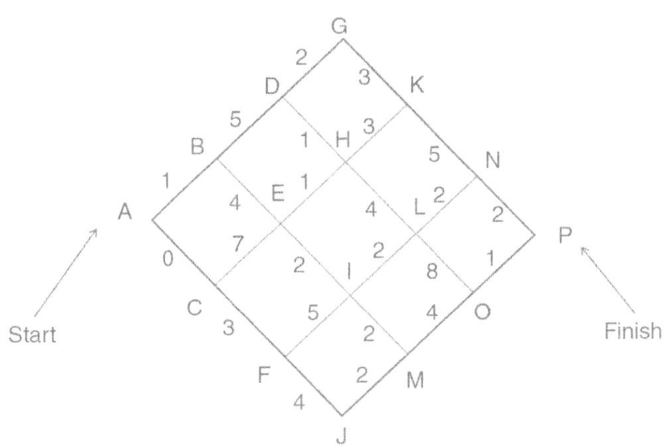

Figure 3.8 Network for a shortest path problem on a grid. Nodes are labeled $A, B, C, \ldots, P$.

14. Consider a *greedy* policy that is defined as follows:

$$a_t^{\text{Greedy}} = \arg\min\{r_t(s_t, a_t)\}$$

Derive sufficient and necessary conditions for a greedy policy to be optimal for a DP.

References

1 Bellman, R. (1954). The theory of dynamic programming. *Bulletin of the American Mathematical Society*, *60* (6), 503–515.
2 Wagner, H. M. & Whitin, T. M. (1958). Dynamic version of the economic lot size model. *Management Science*, *5* (1), 89–96.
3 Dijkstra, E. W. (1959). A note on two problems in connexion with graphs. *Numerische Mathematik*, *1* (1), 269–271.
4 Dreyfus, S. E. & Law, A. M. (1977). *The Art and Theory of Dynamic Programming*. Academic Press.
5 Abbas, A. E. & Holmes, S. P. (2004). Bioinformatics and management science: Some common tools and techniques. *Operations Research*, *52* (2), 165–190.

4

Markov Decision Processes

4.1 Introduction

This chapter revisits the case of uncertainty in decision-making, first introduced in Chapter 2, using decision trees with chance nodes to model uncertainty. As noted in the introduction to Chapter 3, extending decision trees over long periods, with many decision epochs and uncertainty, can lead to computationally intractable decision trees that are too large to evaluate. For example, a decision tree with 52 decision epochs (e.g., weeks), 10 possible decisions at each epoch, and 10 possible random outcomes at each epoch would have $10^{2\times 52}$ nodes, i.e., a 1 followed by 104 zeros.

As discussed in this chapter, a model of uncertainty based on a *Markov chain*—to be defined in Section 4.2—leads to models that can be solved efficiently by using the principles of DP. Markov chains are a special class of stochastic processes that describe state transitions over time, assuming that only the current state and the decision determine the probability of reaching the next state. As demonstrated below, even problems of the size described above can be solved computationally when Markov chains govern the uncertain state transitions.

Unlike decision trees, Markov chains do not grow exponentially in size with respect to time, decisions, or outcomes of random variables (states). For this reason, they have proven useful in many practical applications. The sequential decision-making models discussed in this chapter use Markov chains to formulate stochastic versions of the DP models of Chapter 3. Such models are known as *Markov decision processes* (MDPs). They share similarities with deterministic DP models, which are a special case of an MDP in which the state at each decision epoch is known with certainty. As will become clear, MDPs are the foundation for many variants of sequential decision-making models covered in future chapters.

This chapter begins with an introduction to Markov chains, the assumptions they imply, and examples illustrating their application. Next, it is shown how to evaluate predefined policies to compute expected rewards for a *Markov reward chain*, which combines a Markov chain with a reward function. This is followed by an introduction to algorithms for solving MDPs to find optimal policies. MDPs are discussed in multiple contexts, including finite-horizon models, with a defined endpoint and boundary conditions, similar to the DP models in Chapter 2 and infinite-horizon models, which have no defined endpoint. Examples illustrate the application of these models and methods for solving them.

4.2 Markov Chains

Suppose some characteristics of a system are observed at discrete epochs. Let X_t be the value associated with the characteristic at epoch t. A discrete-time stochastic process describes the relation between the random variables $X_0, X_1, X_2, \dots, X_T$. Such a process is a Markov chain if the following is true:

$$P\,(X_{t+1} = i_{t+1}|X_t = i_t, X_{t-1} = i_{t-1}, \dots, X_1 = i_1, X_0 = i_0) = P\,(X_{t+1} = i_{t+1}|X_t = i_t).$$

Thus, the probability distribution for the state at decision epoch $t+1$ depends only on the state at epoch t; the history of states visited before t is not informative about the future distribution. This is known as the Markov assumption, or the memoryless property of Markov chains. At the first decision epoch, the system could be in one of the available states with certainty, or a vector $\vec{q} = (q_1,\ q_2, \dots, q_N)$ may be specified to define the initial probability distribution for the Markov chain at epoch 0, where $P(X_0 = i) = q_i$. The conditional probability, $P\,(X_{t+1} = j|\, X_t = i) = p_{ij}^t$, is referred to as the *transition probability* at epoch t from state i to j. If the transition probabilities do not change over time, the Markov chain is referred to as a *stationary* Markov chain; otherwise, it is a non-stationary Markov chain. In the former case, the index t will be suppressed to simplify the mathematical notation when possible.

The transition probabilities can be expressed as a matrix, called a *transition probability matrix*, with each row representing a "from" state and each column representing a "to" state. For a given row i and column j of a stationary transition probability, the matrix element $p(j|i)$ can be expressed concisely as p_{ij}, which is the probability that when the system is in state i it transitions to state j. The probabilities in each row must sum to 1 because they enumerate all possible outcomes of a transition.

The transition probability matrix for an N-state stationary Markov chain can be written as follows:

$$P = \begin{bmatrix} p_{11} & p_{12} & \cdots & p_{1N} \\ p_{21} & p_{22} & \cdots & p_{2N} \\ \vdots & \vdots & & \vdots \\ p_{N1} & p_{N2} & \cdots & p_{NN} \end{bmatrix}.$$

The transition probability matrix P records all the one-step transition probabilities. In some cases, it may be desirable to know the probability of multi-step transitions. If a stationary Markov chain is in a state i at epoch m, the probability that n epochs later, the Markov chain is in state j is:

$$P\left(X_{m+n} = j \mid X_m = i\right) = P\left(X_n = j | X_0 = i\right) = p_{ij}(n),$$

where $p_{ij}(n)$ is referred to as the *n-step* transition probability, which is the ijth element of matrix P^n, which is obtained by successively multiplying the transition probability matrix by itself n times. For example, for states, i and j, the two-step transition probability from state i to j is:

$$p_{ij}(2) = \sum_{k=1}^{N} p_{ik} p_{kj}. \tag{4.1}$$

Example 4.1: *Two-step transition probabilities*

For a two-state stationary Markov chain with the following transition probability matrix (see Figure 4.1 for an illustration), what is the probability of going from state 2 to state 1 in two steps?

$$P = \begin{bmatrix} 0.90 & 0.10 \\ 0.20 & 0.80 \end{bmatrix}. \tag{4.2}$$

To compute this, $p(X_2 = 1|X_0 = 2) = p_{21}(2) =$ element (2, 1) of matrix P^2:

$$P^2 = \begin{bmatrix} 0.90 & 0.10 \\ 0.20 & 0.80 \end{bmatrix} \begin{bmatrix} 0.90 & 0.10 \\ 0.20 & 0.80 \end{bmatrix} = \begin{bmatrix} 0.83 & 0.17 \\ 0.34 & 0.66 \end{bmatrix}. \tag{4.3}$$

Thus, the probability of going from state 2 to state 1 in two steps is 0.34.

Markov chains are often conveniently illustrated as networks, such as in Figure 4.1, composed of nodes (states) and arcs (allowable state transitions). Such networks are characterized in detail by classifying their components, e.g., states, based on certain attributes of the transition probability matrix. To begin, given two states, i and j, a *path* from i to j is a sequence of transitions that begins at state i and ends at state j, such that each transition has a positive probability. A state j is *reachable* from state i if a path leads from i to j and two states, i and j, *communicate* if j is reachable from i, and i is reachable from j. A subset of states S in a Markov chain is called a *closed set* if no state outside of S is reachable from any state in S. A state i is an *absorbing state* if $p_{ii} = 1$ and therefore, once the state is entered, the system remains in that state for all future epochs.

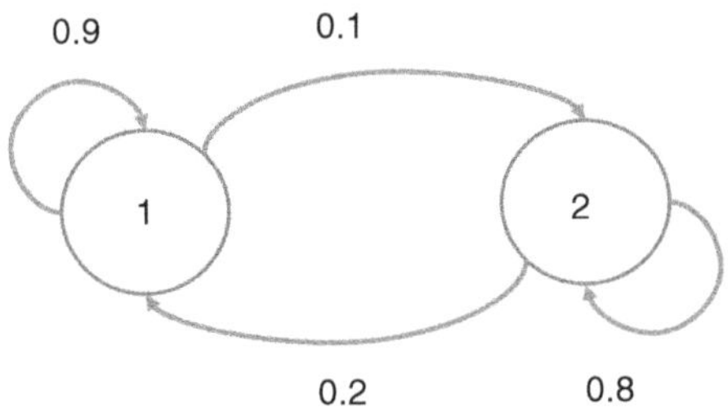

Figure 4.1 Illustration of a simple two-state stationary Markov chain. This illustration leverages the stationary nature of the Markov chain to express it compactly, without explicitly incorporating time steps.

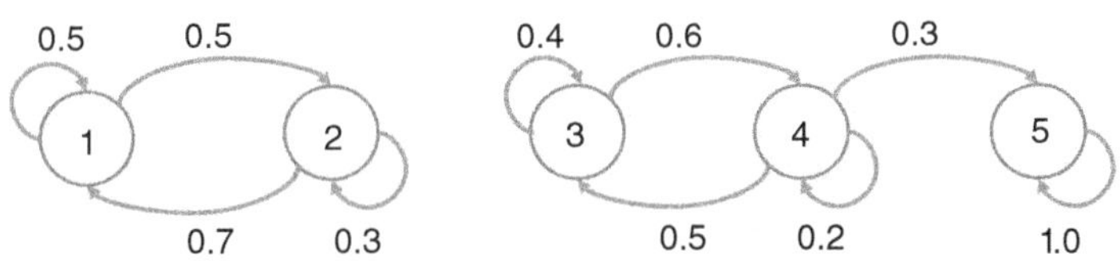

Figure 4.2 Illustration of a Markov chain with two classes, two recurrent states, four transient states, and an absorbing state.

Alternatively, a state i is a *transient state* if a state j exists that is reachable from i, but the state i is not reachable from state j; all other states are referred to as *recurrent* states.

Example 4.2: *Visualizing a Markov chain*

For the following transition probability matrix, draw a graphical representation of the Markov chain (Figure 4.2) and identify the "type" for each state.

$$P = \begin{bmatrix} 0.5 & 0.5 & 0 & 0 & 0 \\ 0.7 & 0.3 & 0 & 0 & 0 \\ 0 & 0 & 0.4 & 0.6 & 0 \\ 0 & 0 & 0.5 & 0.2 & 0.3 \\ 0 & 0 & 0 & 0 & 1 \end{bmatrix}. \tag{4.4}$$

From Figure 4.2, the different types of states are classified as follows. First, states 1 and 2 are reachable from one another, and therefore they communicate, and they are recurrent. States 3 and 4 are reachable from one another and therefore communicate; however, state 5 is absorbing, and therefore states 3 and 4 are transient because the system will eventually enter state 5 without returning to states 3 or 4. Finally, the sets $\{1, 2\}$ and $\{3, 4, 5\}$ are closed sets, often referred to as *classes* because no communication occurs between states in different sets.

4.3 Markov Decision Processes

This section defines the basic elements of MDPs and their mathematical notation. The elements are the same as those in the deterministic DP models of Chapter 3, except for the initial state probability vector, $\vec{q}$, which defines the probability distribution of states at epoch 0, and the transition probabilities that govern state transitions. Numerous small examples are provided to help build intuition, with plausible model parameters selected to illustrate the methods covered. In Section 4.6, the question of estimating model parameters is covered from the perspective of real applications.

The formal definition of an MDP model starts with defining the complete set of elements as follows:

Decision epochs: $t = 0, 1, 2, \ldots, T$,

State at epoch $t : s_t \in S = \{s_1, s_2, \ldots, s_N\}$,

Decisions at epoch $t : a_t \in A = \{a_1, a_2, \ldots, a_M\}$,

Rewards for decision epoch t given decision a_t is taken in state $s_t : r_t\ (s_t, a_t)$,

Discount factor: λ,

Initial state probability distribution: $\vec{q} = (q_{S_1}, \ldots, q_{S_N})$,

Transition probabilities from state s_t to s_{t+1} given decision $a_t : p_t(s_{t+1}|s_t, a_t)$.

Collectively, these seven elements define an MDP. In this stochastic context, decisions influence the transition probabilities. The policy $\pi = (\vec{d}_1, \vec{d}_2, \ldots, \vec{d}_T)$ is a collection of decision rules, one for each decision epoch, t, where $\vec{d}_t = (d_t(s_1), d_t(s_2), \ldots, d_t(s_N))$ prescribes the decision to make in each state. A policy, π, is said to *induce* a particular Markov chain because the decisions defined by the policy determine the transition probabilities.

The set of all possible sequences of states and decisions, called *sample paths*, is:

$$\Omega = \{S \times A \times S \times A \times \cdots \times A \times S\} = \{S \times A\}^T \times S.$$

In this case, the symbol $\times$ is the *Cartesian product* applied to sets. For example, given two sets S_1 and S_2, the Cartesian product, $S_1 = \{A, B\} \times S_2 = \{C, D\}$, denotes all ordered pairs: $(A, C),\ (A, D),\ (B, C),\ (B, D)$. Thus, the set Ω is the set of all possible sample paths. The individual sample paths, denoted by $\omega \in \Omega$, define the sequence of states and decisions for a particular realization of the stochastic process induced by the decision-maker's policy:

$$\omega = (s_0, a_0, s_1, a_1, s_2, a_2, \ldots, a_{T-1}, s_T). \tag{4.5}$$

Each sample path occurs with *path probability* $p(\omega)$ which is defined as follows:

$$p(\omega) = q(s_0) \prod_{t=0}^{T-1} p(s_{t+1}|s_t, a_t).$$

With these definitions of notation, the expected rewards can be defined by first defining the reward for a single sample path ω as follows:

$$W(\omega) = \sum_{t=0}^{T-1} \lambda^t r_t\ (s_t, a_t) + \lambda^T r_T(s_T). \tag{4.6}$$

Next, the expected rewards can be expressed as the expected value of $W(\omega)$ as follows:

$$E_\omega^\pi\ [W\ (\omega)] = \sum_{\forall \omega \in \Omega} p\ (\omega)\ W\ (\omega), \tag{4.7}$$

where the expectation operator, $E_\omega^\pi[\cdot]$ implies that the expectation of sample paths is taken with respect to the Markov chain induced by policy π.

Equation 4.7 can be viewed as *forward policy evaluation* because it maps all possible sample paths, ω, for the induced Markov chain, from the first decision epoch through the end of the horizon. An alternative approach to evaluate policies is to use the *backward policy evaluation algorithm* in Algorithm 4.1. A comparison reveals that both forward and backward policy iterations are correct; however, in the case of Markov chains, backward policy evaluation can be done far more efficiently.

The reader may have noticed a subtle inconsistency here when comparing Algorithm 4.1 to the backward evaluation of decision trees in Chapter 2. In backward evaluation of decision trees, each leaf node at the end of the tree has a unique history of states and decisions that led to it. In Algorithm 4.1, however, the evaluation at any given decision epoch, t, considers only the state, s_t, and ignores the history that led to that state, which can be denoted as $h_t(s_t) = (s_0, a_0, s_1, a_1, s_2, a_2, \ldots, a_{t-1}, s_t) \in H_t(s_t)$, where $H_t(s_t)$ denotes the set of all possible histories leading up to state s_t and $h_t(s_t)$

Algorithm 4.1 Policy Evaluation for Finite Horizon MDP

Input: MDP data elements: decision epochs, states, decisions, transition probability matrix, initial state distribution, rewards, discount factor, and a policy, $\pi \in \Pi$
Boundary Condition at T: $v_T(s_T) = r_T(s_T), \forall s_T \in S$
Backward Induction:
for $t = T - 1$ to 0 **do**
  for all $s_t \in S$ **do**

$$v_t^\pi(s_t) = r_t(s_t, d_t(s_t)) + \lambda \sum_{s \in S} p_t(s_{t+1}|s_t, d_t(s_t)) v_{t+1}^\pi(s_{t+1})\}$$

  end for
end for
Return: Expected discounted rewards, $v_0^\pi(s_0), \forall s_0 \in S$

indexes a specific history leading up to state s_t. This important difference holds for Markov chains because their memoryless property allows the value function to be evaluated at each state independently of its history. For readers interested in a rigorous explanation, the verification that Algorithm 4.1 generates the expected discounted future rewards for a given policy $\pi \in \Pi$ can be proven as follows.

Theorem 4.1 (*Policy Evaluation*): *For any policy $\pi \in \Pi$ the policy evaluation algorithm (Algorithm 4.1) generates $v_t^\pi(h_t(s_t))$ for all $h_t(s_t)$ and $v_0^\pi(s_0) = E_{\omega_{s_0}}^\pi \left[W\left(\omega_{s_0}\right)\right]$ for all $s \in S$ where $\omega_{s_0} \in \Omega_{s_0}$ and Ω_{s_0} is the subset of all sample paths starting with state s_0.*

Proof: The proof is by induction. The base case is established for $t = T$, $v_T(h_T) = r_T(s_T)$ for all $s_T \in S$ by definition. Now, for the induction hypothesis, assume that the policy evaluation algorithm generates $v_n^\pi(h_n(s_n))$ for all $h_n(s_n)$ at decision epochs $n = t+1, t+2, \ldots, T$, then it follows that

$$\begin{aligned}
v_t^\pi(h_t(s_t)) &= r_t(s_t, d_t(h_t(s_t))) + E_{h_t(s_t)}^\pi \left[E_{h_{t+1}}^\pi \left[\sum_{n=t+1}^{T-1} r_n(X_n, Y_n) + r_N(X_T) \right]\right] \quad \text{(by the induction hypothesis)} \\
&= r_t(s_t, d_t(h_t(s_t))) + E_{h_t(s_t)}^\pi \left[\sum_{n=t+1}^{T-1} r_n(X_n, Y_n) + r_N(X_T) \right] \quad (\text{since } H_t \subset H_{t+1}) \\
&= E_{h_t(s_t)}^\pi \left[\sum_{n=t}^{T-1} r_n(X_n, Y_n) + r_N(X_T) \right] \quad (\text{since } s_t \text{ and } d_t \text{ are known at } t),
\end{aligned}$$

where X_n and Y_n are random variables for the state and decision induced by policy π at epoch n. Now, observe that the expectation on the right-hand side of the last equality depends only on s_t since X_n, Y_n are history independent. Therefore $v_t^\pi(h_t(s_t)) = v_t^\pi(s_t), \forall s_t \in S, t = 0, 1, \ldots, T-1$, completing the proof. □

The history-independent nature of Algorithm 4.1 for computing expected rewards in Markov chains significantly reduces the computational burden relative to the decision trees in Chapter 2. This is one of the reasons Markov chains are so popular and MDP models are widely used in sequential decision-making.

4.3.1 Estimating Computational Complexity of Policy Evaluation

In light of Theorem 4.1, how much computational effort is needed to evaluate a policy for a Markov reward chain? Evaluating the number of operations (e.g., +, −, ∗, ÷) and equating each to a time for execution gives an estimate in terms of computation time. Still, times will vary based on the computing platform, the efficiency of code implementation, and other factors. For this reason, when evaluating computational effort, the goal is to understand how it scales with model size. To do this, first recognize that the size of a Markov reward chain is determined by the number of states, $|S|$, and the number of decision epochs, T. Based on the steps in Algorithm 4.1, the computation time will be proportional to T, as the optimal value function must be evaluated at each decision epoch. Additionally, at each decision epoch, the value function for each state must be evaluated; thus, the time complexity is proportional to $|S|$. Therefore, the overall computational effort required to evaluate a policy is proportional to $T \times |S|$. This can be expressed as $\mathcal{O}(T|S|)$, where the $\mathcal{O}(\cdot)$ implies

$$\mathcal{O}(T|S|) \leq cT|S|,$$

for some finite choice of constant c. In other words, the $\mathcal{O}(\cdot)$ notation denotes the proportional growth with respect to changes in the argument.

Computationally, policy evaluation is a problem that is solvable in *polynomial time*, meaning the effort grows proportional to a polynomial function of the model size, i.e., the product of T and $|S|$. Often, problems of this type, which exhibit polynomial growth, are referred to as *easy*, in contrast to *hard* problems that exhibit exponential growth, such as encountered for decision trees in Chapter 2. This classification, albeit frequently used to categorize problems, is coarse. Hard, or computationally challenging, policy evaluation problems certainly arise when T and $|S|$ grow large. Although the problem class is referred to as "easy," this classification is relative to problems that require effort that increases exponentially with respect to the number of epochs and states. Conversely, sometimes "hard" problems are small enough to require little computational effort. Nevertheless, this classification of problems (easy vs. hard) characterizes how computational effort grows with problem size.

4.3.2 Finding Optimal Policies Efficiently

Evaluation of policies is often important, but more importantly, policy evaluation is a stepping stone to policy optimization. Indeed, alternative policies can be compared via forward policy evaluation using the expectation operator, $E^{\pi}_{\omega_0}[\cdot]$, to determine which is the best. Thus, an optimal policy is defined as follows:

Definition: An optimal policy π^* satisfies

$$E^{\pi^*}_{\omega}[W(\omega)] \geq E^{\pi}_{\omega}[W(\omega)], \text{ for all polices } \pi \in \Pi. \tag{4.8}$$

The definition of an optimal policy, π^*, in Equation 4.8 is correct, but it is not very practical because it requires evaluating all possible policies, which for all but very small problems is impractical.

The following set of (Bellman's) optimality equations follows from the application of the dynamic programming principle of optimality discussed in Chapter 3:

$$v_t(s_t) = \max_{a_t \in A} \left\{ r_t(s_t, a_t) + \lambda \sum_{s_{t+1} \in S} p(s_{t+1}|s_t, a_t) v_{t+1}(s_{t+1}) \right\}, \forall s_t \in S, t = 0, \dots, T-1, \tag{4.9}$$

$$v_T(s_T) = r(s_T), \forall s_T \in S. \tag{4.10}$$

Equations 4.9 and 4.10 can be solved by backward induction to determine the optimal expected value function for each epoch and each state. Note the similarity to the optimal value function in Chapter 3, which is modified here to optimize *expected rewards* as defined by the underlying Markov chain induced by the policy selected by way of the maximization step in Equation 4.9. Similar to the objective in Chapter 3, the goal here is to find an optimal policy. Importantly, the optimality equations in Equations 4.9 and 4.10 assume optimal policies depend only on the current state s_t at each epoch, t. An unrestricted version of the problem could be expressed based on the complete *history* up until decision epoch t, and this would be necessary for problems where the optimal decision depends on the history leading up to state s_t. Fortunately, however, Theorem 4.1 established history independence for policy evaluation when the underlying stochastic model is a Markov chain, and this simplification carries over to MDPs, for which it can be proven that there always exists a Markovian policy (one that ignores prior history and considers only the current state) that is optimal. This follows from the memoryless property of Markov chains and the additive nature of the rewards that satisfy the principle of optimality. Algorithm 4.2 describes the implementation of backward induction to solve Equations 4.9–4.10. It parallels backward induction for deterministic DP models, but it must also compute expectations over future states to obtain the optimal value function for each state at each decision epoch.

Algorithm 4.2 Backward Induction for a Finite Horizon MDP

Input: MDP data elements: decision epochs, states, decisions, transition probability matrix, rewards, discount factor
Boundary Condition: $v_T(s_T) = r_T(s_T), \forall s_T \in S$
Backward Induction:
for $t = T - 1$ to 0 **do**
  for all $s_t \in S$ **do**

$$v_t(s_t) = \max_{a_t \in A_t}\{r_t(s_t, a_t) + \lambda \sum_{s \in S} p_t(s_{t+1}|s_t, a_t)v_{t+1}(s_{t+1})\}$$

$$d_t^*(s_t) \in \arg\max_{a_t \in A_t}\{r_t(s_t, a_t) + \lambda \sum_{s \in S} p_t(s_{t+1}|s_t, a_t)v_{t+1}(s_{t+1})\}$$

  end for
end for
Return: Optimal Policy, π^*

4.3.3 Analysis of the Backward Induction Algorithm

Readers who are satisfied with letting their intuition guide the legitimacy of using Algorithm 4.2 to solve MDPs can skip this section. Those wishing a more rigorous mathematical explanation will benefit from the formal details presented herein.

The following theorem, attributed to Bellman, establishes a method for solving more general stochastic dynamic programs, beyond MDPs, for which the memoryless assumption of Markov chains is not a valid modeling assumption. Once the method is shown to be valid, it is used to demonstrate that MDPs are significantly easier to solve than general stochastic dynamic programs with non-Markovian stochastic processes, i.e., those that exhibit history dependence, and can be solved by Equations 4.9 and 4.10 above. The expectation of rewards for the optimal policy in the following, Theorem 4.2, $E_{\omega_0}^{\pi^*}[W(\omega)], \forall s_0 \in S$, is defined over the subset of sample paths $\Omega_{s_0} \subseteq \Omega$ that begin with state s_0. Thus, Theorem 4.2 establishes that the optimality Equations 4.9 and 4.10 yield an optimal solution.

Theorem 4.2 (*General History-dependent Optimality Equations*): *Suppose $v_t(h_t)$, for all decision epochs t and histories h_t, is a solution to the following equations, then $v_t(h_t) = v_t^*(h_t), \forall h_t \in H_t, t = 0, 1, \dots, T$ and $v_0(s_0) = E_{\omega_0}^{\pi^*}[W(\omega)], \forall s_0 \in S$, and $\omega_{s_0} \in \Omega_{s_0}$.*

$$v_t(h_t) = \max_{a_t \in A}\left\{r_t(s_t, a_t) + \lambda \sum_{j \in S} p_t(j \mid s_t, a_t)v_{t+1}(h_{t+1})\right\}, \quad t = 0, \dots, T-1, \quad \forall h_t \in H_t,$$

$$v_T(h_T) = r_T(s_T), \quad \forall h_T \in H_T.$$

Proof: Given that $v_t(h_t), t = 1, \dots, T, h_t \in H_t$ is a solution to the above equations, it suffices to prove it is an optimal solution. The following is a proof by induction that $v_t(h_t) = v_t^*(h_t)$, for all $h_t \in H_t, t = 1, 2, \dots, T$. First, since $v_T(h_T) = r_T(s_T)$, it follows trivially that $v_T(h_T) = v_T^*(h_T)$, for all h_T in the last decision epoch. Next, assume (from the induction hypothesis) that $v_t(h_t) = v_t^*(h_t)$, for $t = n+1, n+2, \dots, T$. Using the optimality equations, the following can be established:

$$v_n(h_n) = \max_{a_n \in A}\left\{r_n(s_n, a_n) + \lambda \sum_{j \in J} p_n(j \mid s_n, a_n)v_{n+1}^*(h_{n+1})\right\}, \tag{4.11}$$

$$\geq \max_{a_n \in A}\left\{r_n(s_n, a_n) + \lambda \sum_{j \in S} p_n(j \mid s_n, a_n)v_{n+1}^\pi(h_{n+1})\right\}, \quad \text{for any } \pi, \tag{4.12}$$

$$= v_n^\pi(h_n), \tag{4.13}$$

where π is any arbitrary feasible policy. The first equality follows from the induction hypothesis that $v_{n+1}(h_{n+1}) = v_{n+1}^*(h_{n+1})$ and $p_n(j \mid s_n, a_n) \geq 0$, for all $j \in S$ because it is a probability and must be nonnegative. The second inequality follows from the fact that $v_{n+1}^*(h_{n+1}) \geq v_{n+1}^\pi(h_{n+1}), \forall \pi \in \Pi$, thus completing the proof. □

The next theorem demonstrates that evaluating all HD policies for the more specialized case of MDPs introduces considerable redundancy that can be avoided for computational advantage. This is achieved by proving that, for every MDP, an optimal *Markov Deterministic* (MD) policy exists, where an MD policy requires only the current state to select an optimal decision. This proves why Algorithm 4.2 generates an optimal solution for an MDP.

Theorem 4.3 (*Optimality of Markov Deterministic Policies for MDPs*): *Let $v_t^*(h_t)$, for all $t = 1, \dots, T$ and $h_t \in H_t$, be a solution of the following optimality equations. If $v_t^*(h_t)$ depends on h_t only through s_t, then an MD optimal policy can be obtained by solving the following equations:*

$$v_t(s_t) = \max_{a_t \in A} \left\{ r_t(s_t, a_t) + \lambda \sum_{j \in S} p_t(j \mid s_t, a_t) v_{t+1}(s_{t+1}) \right\}, \quad t = 0, \dots, T-1, \quad \forall s_t \in S, \tag{4.14}$$

$$v_T(s_T) = r_T(s_T), \quad \forall s_T \in S. \tag{4.15}$$

Proof: First, it must be shown that $v_t^*(h_t)$ only depends on s_t. This can be proven by induction as follows. Since $v_T^*(h_T) = v_T^*(h_{T-1}, a_{T-1}, s_T) = r_T(s_T)$, for all h_{T-1}, a_{T-1}, it follows that $v_T^*(h_T) = v_T^*(s_T)$. Next, assume $v_n^*(h_n) = v_n^*(h_{n-1}, a_n, s_n)$, for all $h_n \in H_n$, $n = t+1, \dots, T$. Then it follows:

$$v_t^*(h_t) = \max_{a_t \in A} \left\{ r_t(s_t, a_t) + \lambda \sum_{j \in S} p_t(j \mid s_t, a_t) v_{t+1}(h_t, a_t, j) \right\}$$

$$= \max_{a_t \in A} \left\{ r_t(s_t, a_t) + \lambda \sum_{j \in S} p_t(j \mid s_t, a_t) v_{t+1}(j) \right\}.$$

The last equality follows from the induction hypothesis. Therefore, it follows that:

$$d_t^*(h_t) = \arg\max_{a_t \in A} \left\{ r_t(s_t, a_t) + \lambda \sum_{j \in S} p_t(j \mid s_t, a_t) v_{t+1}(j) \right\},$$

and $d_t^*(h_t) = d_t^*(s_t)$, for all $h_t \in H_t$, $t = 0, \dots, T$. □

Theorem 4.3 establishes that Algorithm 4.2 determines the optimal value function and the optimal policy for an MDP, with improved efficiency due to the memoryless property of Markov chains. In fact, it establishes that finding the optimal policy to an MDP is easy, i.e., it can be obtained in polynomial time given problem instance size parameters, T, $|S|$, $|A|$, rather than the exponential effort required if all histories needed to be considered.

Note the following important point, which will be used frequently. The second equation in Step 5 of Algorithm 4.2 extracts an optimal decision, although, as mentioned earlier, it is not uncommon to have more than one optimal decision. Algorithm 4.2 can be modified to collect all optimal decisions at Step 5, thereby constructing all alternative optimal policies by retaining the set of optimal decisions for each state and epoch. However, the number of optimal policies may grow large because each tie results in a bifurcation of the optimal policy, doubling the total number each time. Unless necessary for a particular problem, it is assumed that ties are broken arbitrarily to obtain a single optimal policy, possibly from a larger set of alternative optimal policies.

4.3.4 Shortest Path Problem Revisited

Stochastic variations of the deterministic shortest path problems discussed in Chapter 3 can be modeled as an MDP. The following example illustrates the application of the backward induction algorithm established in Algorithm 4.2 and verified by Theorem 4.3.

Example 4.3: *Stochastic shortest path problem*

Consider the stochastic shortest path problem illustrated in Figure 4.3 in which a football player, currently at node S, needs to make a sequence of decisions to get to the goal line F while avoiding tacklers at nodes A, B, C, D, E. Each split point corresponds to a node at which a decision must be made to proceed left or right (while continuing to make forward progress

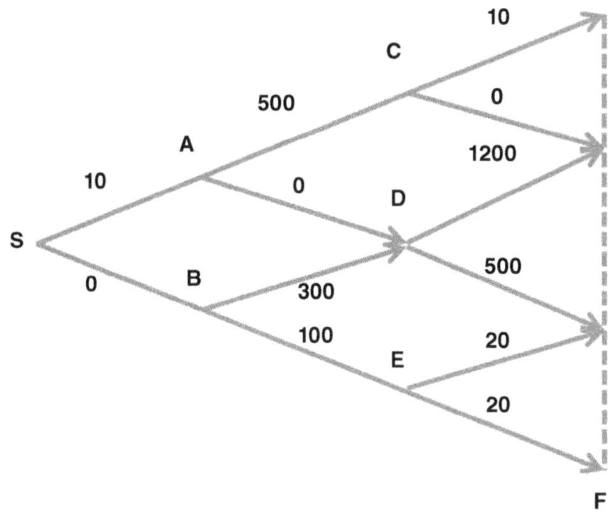

Figure 4.3 A directed network with paths moving from left to right, starting at node S. The dashed line represents the goal line.

toward the goal line). The objective is to cross the finish line (dashed line) with minimal cost penalties, which could serve as a measure of the effort expended by the player. Assume the state transition outcomes depend probabilistically on the choice of decision:

$$\Pr(u|a_t = u) = p^u \text{ and } \Pr(d|a_t = d) = p^d,$$

where u denotes the decision to move up to the right, and d is the decision to move down to the right. Assume outcomes of the decisions are uncertain and have the following probabilities: $\Pr(u|a = u) = 0.8$ and $\Pr(d|a = d) = 0.7$. Given the start is at node S, and the goal is to reach the "finish line" F, what is the expected distance traveled under policy π?

Assume $\lambda = 1$, as it was not specified in the problem statement, and the nature of the problem suggests a short time frame (traversing a small network) for which discounting is not typically necessary. Given the symmetric structure of the network, assign all four nodes along the dotted line (the "finish line") to epoch 4, nodes (C, D, E) to epoch 3, nodes (A, B) to epoch 2, and node S to epoch 1. The reward can be written as follows:

$$r_t(s_t, a_t) = \begin{cases} Pr(u \mid a_t = u)L^u(s_t) + Pr(d \mid a_t = u)L^d(s_t) & \text{if } a_t = u \\ Pr(d \mid a_t = d)L^d(s_t) + Pr(u \mid a_t = d)L^u(s_t) & \text{if } a_t = d, \end{cases}$$

where $L^u(s_t)$ and $L^d(s_t)$ are the arc weights for the up and down branches, respectively, at state s_t. Note that the nature of the problem requires computing expected rewards to account for the uncertainty in the outcome of the selected decisions. Applying Algorithm 4.2 yields the following.

Epoch 4 (boundary condition):

$$v_4(F) = 0, \text{ for all 4 nodes at the finish line, } F.$$

Epoch 3:

$$\begin{aligned} v_3(C) &= min\{0.8(10) + (0.2)(0), 0.7(0) + (0.3)(10)\} = 3, \\ v_3(D) &= min\{0.8(1200) + (0.2)(500), 0.7(500) + (0.3)(1200)\} = 710, \\ v_3(E) &= min\{0.8(20) + (0.2)(20), 0.7(20) + (0.3)20\} = 20. \end{aligned}$$

Table 4.1 Two optimal policies expressed as a look-up table for the shortest path problem in Figure 4.3.

State	Decision
A	u
B	d
C	d
D	d
E	u or d

Epoch 2:

$$v_2(A) \quad = \min\{0.2(v_3(D)) + 0.8(500 + v_3(C)), 0.7(v_3(D)) + 0.3(500 + V_3(C))\} = 544.4,$$

$$v_2(B) \quad = \min\{0.8(300 + v_3(D)) + 0.2(100 + v_3(E)), 0.3(300 + v_3(D)) + 0.7(100 + v_3(E))\} = 387.$$

Epoch 1:

$$v_1(S) = \min\{0.8(10 + v_2(A)) + 0.2(v_2(B)), 0.7(v_2(B)) + 0.3(10 + v_2(A))\} = 437.22.$$

Thus, the expected distance traveled under the optimal policy is 437.22. Backtracking through the solution to identify the selected decision for each state and decision epoch yields the optimal policies in Table 4.1.

Note that the policy is deterministic, but the path followed is stochastic because the outcomes of decisions to move up or down are random. As a result, when following the optimal policy through the network multiple times, the actual path will vary randomly from one problem instance to another. The optimal policy is guaranteed only to minimize the expected distance traveled across many attempts to traverse the network. It does not necessarily yield the optimal path for any given problem instance. Another way to express this is that the expected travel distance converges to the minimum expected travel distance as the number of times the network is traversed using the optimal policy increases.

4.4 Optimal Stopping Time Problems

Many sequential decision-making problems arise in the context of a decision-maker who can observe the outcomes of a stochastic process and choose to stop it at their discretion. Consider an investor who holds a stock option that allows her to purchase a stock at a fixed *strike price* over a specific time frame, after which the option expires (e.g., 30 days). If the stock's market price exceeds the strike price, the investor can exercise the option, buy the stock at a price below the market value, and immediately sell it at the market price for a profit. For example, if the strike price is \$100 and the stock is currently trading for \$112, then the investor can purchase the stock for \$100 and immediately sell it at the market price for a profit of \$12.

The investor wants to exercise the option when the market value exceeds the strike price, but how high is high enough? If the investor sells today, she may regret it if the stock price continues to increase; however, if she holds the option, the price could decrease, perhaps dipping below the strike price and remaining there until the option expires. Thus, the investor seeks to determine the optimal policy for exercising the option to maximize expected profit.

The stopping time problem can be modeled as an MDP with discrete decision epochs, at which the investor can choose to exercise the option by quitting the process (Q) or continuing the process (C) for at least one more epoch until the option expires at the end of the decision horizon, at epoch T. The following is the MDP formulation for this problem.

Decision epochs: $t = 0, \dots, T$,

States: $S = S' \cup \{\Theta\}$, where S' is the set of states from which decision Q or C can be made and Θ is an absorbing state,

$$\text{Decisions: } A = \begin{cases} \{C, Q\}, & s \in S', \\ \{\emptyset\}, & s = \Theta, \end{cases}$$

$$\text{Rewards: } r_t(s_t, a_t) = \begin{cases} f_t(s_t) & \text{for } s_t \in S',\ a_t = C, \\ q_t(s_t) & \text{for } s_t \in S',\ a_t = Q, \\ 0 & \text{for } s_t = \Theta, \\ q_T(s_T) & \text{for } s_t \in S', \end{cases}$$

Transition probabilities:

$$p_t(j|s_t, a_t) = \begin{cases} p_t(j|s_t) & \text{for } s_t \in S',\ j \in S',\ a_t = C, \\ 1 & \text{for } s_t \in S',\ j = \Theta,\ a_t = Q \text{ or } s_t = j = \Theta, \\ 0 & \text{otherwise.} \end{cases}$$

The above MDP elements yield the following optimality equations:

$$v_t(s_t) = \max_{a_t \in A} \left\{ q_t(s_t), f_t(s_t, a_t) + \lambda \sum_{j \in S} p(j|s_t, C) v_{t+1}(j) \right\}, \forall s_t \in S, t = 0, \dots, T-1,$$

$$v_T(s_T) = q_T(s_T), \forall s_T \in S.$$

From the optimality equations, the decision-maker's choices are apparent as either waiting for a given expected reward $q_t(s)$ or taking an immediate reward, $f_t(s, a)$, plus possible future expected rewards generated by the continuation of the process.

Stopping time problems arise in many contexts. The following example illustrates its use for optimizing the time of medical treatment decisions.

Python Example 4.4: Optimal time to treat

Some medical decisions can be viewed as "stopping time" problems. For example, a decision to start preventive treatment involves a decision-maker (some combination of the patient and physician) choosing to initiate treatment and receiving an expected reward or deferring the decision to start treatment for one additional decision epoch. Consider the case of a new drug that reduces the risk of heart disease. Starting treatment lowers the probability of a future heart attack or stroke; however, the drug has side effects, causing a moderate degree of burden to the patient. Moreover, once started, the treatment is intended to continue for the rest of their lives. The decision process can be described as follows. Patients receive annual risk assessments and choose between two options: (1) initiate treatment now or (2) defer the decision for one more decision epoch. Assuming that at age T, patients who have not initiated treatment start treatment, the optimality equations can be written as follows:

$$v_t(s_t) = \max_{a_t \in \{Q, C\}} \left\{ q_t(s_t), r(s_t, C) + \sum_{j \in S} p(j|s_t, C) v_{t+1}(j) \right\}, \forall s_t \in S, t = 0, \dots, T-1,$$

$$v_T(s_T) = q_T(s_T), \forall s_T \in S.$$

Figure 4.4 illustrates the MDP's states, transitions, and decisions. Health states are recurrent states that the patient transitions through over time until one of the following events: (1) the decision is made to start treatment; (2) the occurrence of heart attack, stroke, or death from any other cause, which are referred to collectively as *adverse events*; (3) the decision maker reaches the final epoch T at the end of the decision horizon. Health states are defined by four risk categories: low (L), medium (M), high (H), and very high (V). Two absorbing states represent treatment initiation and the occurrence of an adverse event. The decision at each epoch is either C, the decision to defer treatment, or Q, the decision to start treatment. The latter results in the patient entering the treatment-absorbing state with probability 1 and receiving a reward

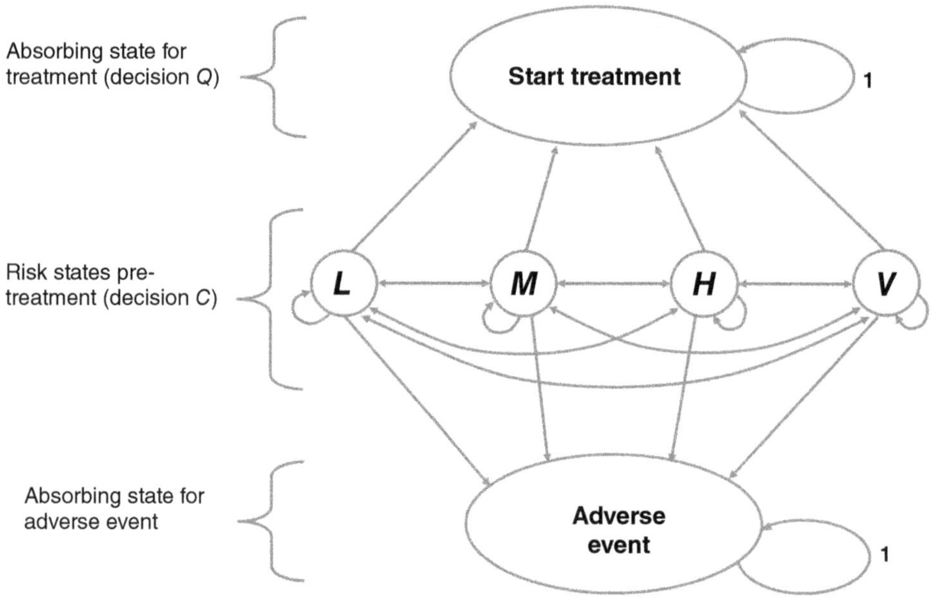

Figure 4.4 Illustration of an optimal stopping time problem for drug treatment initiation to prevent adverse health outcomes. Choosing to start treatment at epoch t in any of the risk states, $s_t \in \{L, M, H, V\}$, automatically moves the patient to the treatment state, resulting in a reward $q_t(s_t)$.

Table 4.2 Transition probabilities among states: low (*L*), medium (*M*), high (*H*), very high (*V*), and adverse event (*E*). The transition probability matrix is nonstationary (time-dependent) to reflect the increasing risk of an adverse event with age.

State	*L*	*M*	*H*	*V*	*E*
L	$0.74\times(1-p_t(E\|L))$	$0.20\times(1-p_t(E\|L))$	$0.05\times(1-p_t(E\|L))$	$0.01\times(1-p_t(E\|L))$	$p_t(E\|L)$
M	$0.10\times(1-p_t(E\|M))$	$0.70\times(1-p_t(E\|M))$	$0.17\times(1-p_t(E\|M))$	$0.03\times(1-p_t(E\|M))$	$p_t(E\|M)$
H	$0.01\times(1-p_t(E\|H))$	$0.24\times(1-p_t(E\|H))$	$0.62\times(1-p_t(E\|H))$	$0.13\times(1-p_t(E\|H))$	$p_t(E\|H)$
V	$0.01\times(1-p_t(E\|V))$	$0.06\times(1-p_t(E\|V))$	$0.30\times(1-p_t(E\|V))$	$0.63\times(1-p_t(E\|V))$	$p_t(E\|V)$
E	0	0	0	0	1.0

$q_t(s_t)$, denoting the expected quality-adjusted survival for a patient starting treatment at epoch t. The absorbing state Θ represents the occurrence of an adverse event with no expected reward. Assume decision epochs $t = 0, \dots, 25$ correspond to ages 50 ($t = 0$) through $T = 75$ ($t = 25$), defined by the approved clinical treatment window for the drug.

Assume that the annual rewards before treatment are $r_t(s_t, C) = 1$ for all s_t and $t = 0, \dots, 24$, corresponding to a perfect year of health, and the expected quality-adjusted life span for patients starting treatment at epoch t can be estimated using the following equation medical researchers have approximated from a previous study:

$$q_t(s_t) = -0.721 \times (t + 50) + 64.36, \forall s_t \in S, t = 1, \dots, T.$$

Transition probabilities between health states for decision C are computed from electronic medical record data on a cohort of patients observed over time, based on the proportion of transitions between each pair of states. Table 4.2 presents maximum-likelihood estimates computed by estimating the proportion of patients moving between the discrete states $\{L, M, H, V\}$ at each epoch.

Assume the probability of an adverse event can be estimated from national health data as follows:

$$p_t(E|s_t) = C^R(s_t)0.0003 \times e^{8.91\times(t+50)} \tag{4.16}$$

where $C^R(s_t)$ is a parameter determined by the health state of the patient that reflects the risk differences among the states as follows:

$$C^R(s_t) = \begin{cases} 0.1 & \text{for } s_t = L, \\ 0.4 & \text{for } s_t = M, \\ 2.0 & \text{for } s_t = H, \\ 10.0 & \text{for } s_t = V. \end{cases}$$

The following Python code initializes and solves the optimality equations for the optimal stopping time MDP.

```
import numpy as np

# Input data
T = 25 # Time horizon (age 50 through 75 years, T=25 boundary condition
nstates = 5
ndecisions = 2
states = ['L', 'M', 'H', 'V', 'E'] # Health states and adverse event (E)
decisions = ['C', 'Q']            # Decisions: Continue ('C'), Quit/Treat ('Q')
c_R = [0.1, 0.4, 2, 10, 0]        # Risk multipliers by state L, M, H, V, E
d_T = 0.05                      # Annual disutility of treatment

# Transition probabilities for decision 'C', excluding E state for now
P_C = np.array([
    [0.74, 0.20, 0.05, 0.01],   # From L
    [0.10, 0.70, 0.17, 0.03],   # From M
    [0.01, 0.24, 0.62, 0.13],   # From H
    [0.01, 0.06, 0.30, 0.63]    # From V
])

# Transition probability matrix: TPM[t, s, d, s'] = P(s' | s, d) at time t
P = np.zeros((T + 1, nstates, ndecisions, nstates))
for t in range(T + 1):
    for s in range(nstates - 1):  # Exclude E
        p_e = c_R[s] * 0.0003 * np.exp(0.0891 * (t + 50))
        P[t, s, 0, :-1] = P_C[s] * (1 - p_e)
        P[t, s, 0, -1] = p_e
    P[t, -1, :, -1] = 1.0  # E is absorbing
    P[t, :, 1, :] = 0.0

# Reward function: rewards[t, s, d]
rewards = np.zeros((T + 1, nstates, ndecisions))
for t in range(T + 1):
    rewards[t, :, 0] = 1  # Decision 'C'
    rewards[t, :, 1] = -0.721 * (t + 50) + 64.36 - (25 - t) * d_T #Decision 'Q'
    rewards[t, -1, :] = 0  # E state reward is 0

# Initialize
value = np.zeros((T + 1, nstates, ndecisions))  # Value-to-go function
v = np.zeros((T + 1, nstates))      # Optimal value function
optimal_policy = np.zeros((T + 1, nstates), dtype=int)  # Policy
```

```
# Boundary condition
v[T] = rewards[T, :, 1]
optimal_policy[T, :-1] = 1  # decision 'Q'
optimal_policy[T, -1] = -1  # E: no decision

# Backward induction
for t in range(T - 1, -1, -1):
    for s in range(nstates):
        if s == nstates - 1:  # E state
            v[t, s] = 0
            optimal_policy[t, s] = -1
            continue
        for d in range(ndecisions):
            value[t, s, d] = rewards[t, s, d] + np.dot(P[t, s, d], v[t + 1])
        optimal_policy[t, s] = np.argmax(value[t, s, :])
        v[t, s] = value[t, s, optimal_policy[t, s]]

# Print results
print("Optimal Policy:")
for t in range(T + 1):
    print(f"Optimal policy at age {t + 50}:")
    for s in range(nstates):
        if optimal_policy[t, s] == -1:
            decision = '-' # For E state


        else:
            decision = decisions[optimal_policy[t, s]]
        print(f"  State {states[s]}: {decision}")
```

Running the code yields the following optimal policy, which dictates when a decision maker who has not yet initiated treatment should start treatment:

Ages 50–56: Initiate treatment in states V, H.

Ages 57–74: Initiate treatment in states V, H, M.

Age 75: Initiate treatment in all states (by assumption).

Therefore, the optimal policy prescribes that a decision maker who does not initiate treatment in a given decision epoch (and has not initiated treatment in a prior epoch) should start treatment if they are in states H or V; otherwise, they should defer treatment. Furthermore, starting at age 59, a decision maker who has not yet started treatment should initiate treatment for states V, H, M. All decision-makers who have not begun treatment by age 75, including those in the low-risk state L, are recommended to initiate treatment.

4.5 Production Planning with Uncertain Demand

Production planning is an early application of MDPs because of the dynamic nature of decisions and the challenges of balancing production costs and benefits to meet customer demand. The production lot-sizing problem of Chapter 3, solved using the Wagner-Whitin algorithm, can be extended to the more realistic case of uncertain demand. To do this, consider the following assumptions:

1. A Markov chain is used to represent demand uncertainty.
2. The decision maker chooses how much to produce at each decision epoch.
3. If inventory exceeds demand, a holding cost is incurred for inventory held until the next decision epoch.
4. Any demand that is not filled immediately in the epoch it arises is lost.

Note the last assumption, known as "lost sales," allows for the possibility that not all demand is covered during the planning horizon. Without this adjustment to the deterministic case considered in Chapter 3, where it is assumed all demand must be covered, the decision maker would be forced to produce enough for the maximum possible level of demand, regardless of the probability that it materializes. Given the above assumptions, the MDP model can be formulated as follows:

Decision epochs: $t = 0, 1, \dots, T$.

States: the inventory level $s_t \in \{0, 1, 2, \dots, M\}$ defines the number of units in inventory up to some maximum M at epoch t, defined by the following state transfer equation:

$$s_t = \max\{s_{t-1} + a_t - D_t, 0\}.$$

The states are uncertain because demand, D_t, is a random variable that induces randomness in the inventory levels. The inventory at the first epoch, s_0, is defined in the problem statement and serves as a boundary condition for the recursion.

Decisions: the production order $a_t = \{0, 1, 2, \dots, M - s_t\}$ at epoch t, where M is the maximum available space for inventory.

Transition probabilities:

$$p_t(j|s_t, a_t) = \begin{cases} 0, & j > s_t + a_t, \\ p_{s_t + a_t - j}, & \text{if } s_t + a_t \geq j > 0, \\ q_{s_t + a_t}, & \text{if } j = 0, \end{cases}$$

where $q_{s_t+a_t}$ is the probability demand exceeds supply.

Rewards: At each epoch, rewards encompass sales revenue, production cost, and inventory holding cost. The sales revenue is the following:

Expected revenue: $E_{D_t}[r \times \min(s_t + a_t, D_t)]$,

where r denotes the per unit revenue for satisfying demand. The production cost is the sum of the fixed (setup) cost and the variable per unit cost:

Production cost (fixed + variable): $c^S \times I(a_t) + c^V \times (a_t)$

where $I(a_t)$ is an indicator function defined as

$$I(a_t) = \begin{cases} 0 \text{ if } a_t = 0, \\ 1 \text{ if } a_t > 0. \end{cases}$$

Finally, the inventory holding cost at epoch t can be written as $c^H \times (s_{t-1} + a_t - D_t)$.

With all the elements of the rewards (revenues and costs) established, the reward function can be written as:

$$r_t(s_t, a_t) = E_{D_t}[r \times \min(s_t + a_t, D_t)] - c^S \times I(a_t) - c^V(a_t) - c^H \times (s_t + a_t), \forall s_t \in S, a_t \in A, \quad (4.17)$$

$$r_T(s_T) = R(s_T), \forall s_T \in S.$$

The last equation, corresponding to the boundary condition at the end of the horizon, could reflect a salvage value that depends on the number of units remaining in inventory (e.g., if they can be sold on a secondary market) or simply impose a zero reward to discourage excess inventory at the end of the horizon. Alternatively, $R_T(s_T)$ could encourage achieving some target inventory level if the finite decision horizon approximates a longer-term planning problem by penalizing inventory falling below or above the target.

The following example, from Puterman [1], shows the steps involved in constructing the elements of the MDP from data and its solution.

Example 4.5: *Production planning under uncertain demand*

Consider the following specific example of the production planning MDP (from Puterman, 1994) with four decision epochs and an inventory space for at most three units. What is the expected profit if zero inventory is available at the first stage? What is the optimal production policy?

Assume $r = 8$, $c^S = 4$, $c^V(a_t) = 2a_t$, $c^H = 1$, $R(s_T) = 0, \forall s_T$, $M = 3$, $T = 3$, $\lambda = 1$. The parameters mirror those of the deterministic problem, except that a new parameter, M, is introduced to define the maximum available space for inventory. The probability distribution for demand, d, in each period is:

$$p_d = \begin{cases} 0.25, & \text{for } d = 0, \\ 0.50, & \text{for } d = 1, \\ 0.25, & \text{for } d = 2. \end{cases}$$

States: $s_t = 0,\ 1,\ 2,\ 3$ inventory level at the start of epoch t.

Decisions: $a_t = 0, \ldots, 3 - s_t$ denoting the order quantity during epoch t.

Rewards: $r_t(s_t, a_t)$ can be expressed as an $|S| \times |A|$ matrix using 4.17 to compute the matrix elements as follows:

$$R = \begin{bmatrix} 0 & -1 & -2 & -5 \\ 5 & 0 & -3 & \text{NP} \\ 6 & -1 & \text{NP} & \text{NP} \\ 5 & \text{NP} & \text{NP} & \text{NP} \end{bmatrix},$$

where the subscript t is omitted because the rewards are stationary. Rewards for state-decision pairs that are not possible are denoted by "NP" in the reward matrix. The transition probability matrices for each of the possible decisions, $a_t = 0,\ 1,\ 2,\ 3$, are computed using Equation 4.5 and the maximum inventory level $M = 3$:

$$P(0) = \begin{bmatrix} 1 & 0 & 0 & 0 \\ 0.75 & 0.25 & 0 & 0 \\ 0.25 & 0.50 & 0.25 & 0 \\ 0 & 0.25 & 0.5 & 0.25 \end{bmatrix}, \quad P(1) = \begin{bmatrix} 0.75 & 0.25 & 0 & 0 \\ 0.25 & 0.50 & 0.25 & 0 \\ 0 & 0.25 & 0.50 & 0.25 \\ \text{NP} & \text{NP} & \text{NP} & \text{NP} \end{bmatrix},$$

$$P(2) = \begin{bmatrix} 0.25 & 0.5 & 0.25 & 0 \\ 0 & 0.25 & 0.5 & 0.25 \\ \text{NP} & \text{NP} & \text{NP} & \text{NP} \\ \text{NP} & \text{NP} & \text{NP} & \text{NP} \end{bmatrix}, \quad P(3) = \begin{bmatrix} 0 & 0.25 & 0.5 & 0.25 \\ \text{NP} & \text{NP} & \text{NP} & \text{NP} \\ \text{NP} & \text{NP} & \text{NP} & \text{NP} \\ \text{NP} & \text{NP} & \text{NP} & \text{NP} \end{bmatrix},$$

where the subscript t is dropped because the transition probabilities are stationary. Similar to the reward matrix, entries for infeasible transitions—due to the maximum allowable inventory level of 3, which limits production—are denoted by "X." Now, with all of the model data specified, the backward induction algorithm can be applied to solve the problem.

Epoch 3 (Boundary Condition): Set $t = 3$, $v_3(s) = 0$ for $s = 0, 1, 2, 3$.

Epoch 2: Since $t \neq 1$ continue. Set $t = 2$.

$$v_2(s) = \max_{a \in A}\{r(s, a) + \sum_{j \in S} p(j|s, a)\, v_4(j)\},\ s = 0, 1, 2, 3 \tag{4.18}$$

$$= \max_{a \in A}\{r(s, a)\}. \tag{4.19}$$

It follows that $v_2(0) = 0,\ v_2(1) = 5,\ v_2(2) = 6,\ v_3(3) = 5$ and $a_2(s) = 0,$ for $s = 0, 1, 2, 3$.

Epoch 1: Since $t \neq 1$ continue. Set $t = 1$,

$$v_1(s) = \max_{a \in A}\{r(s, a) + \sum_{j \in S} p(j|s, a)\, v_2(j)\},\ s = 0, 1, 2, 3$$

$$= \max_{a \in A}\{v_1(s, a)\},$$

where $v_1(s, a)$ can be written as follows, with X denoting decisions that are not feasible due to the limit on inventory, $M = 3$:

$$v_1(s, a) = \begin{bmatrix} 0 & 0.25 & 2 & 0.5 \\ 6.25 & 4 & 2.5 & X \\ 10 & 4.5 & \text{NP} & \text{NP} \\ 10.5 & \text{NP} & \text{NP} & \text{NP} \end{bmatrix}.$$

Thus, it follows that $v_1(0) = 2,\ v_1(1) = 6.25,\ v_1(2) = 10,\ v_1(3) = 10.5$ and $a_1(0) = 2,\ a_1(1) = 0,\ a_1(2) = 0,\ a_1(3) = 0$.

Epoch 0: Having reached the first decision epoch, the remaining value functions can be computed:

$$v_0(s) = \max_{a \in A}\{v_0(s, a)\}.$$

The choices of decisions for each state can be expressed as a matrix with rows corresponding to states and decisions corresponding to columns, as follows:

$$v_0(s, a) = \begin{bmatrix} 2 & 2.06 & 4.12 & 4.19 \\ 8.06 & 6.12 & 6.19 & \text{NP} \\ 12.12 & 8.19 & \text{NP} & \text{NP} \\ 14.19 & \text{NP} & \text{NP} & \text{NP} \end{bmatrix}.$$

It follows by taking the maximum for each row that $v_0(0) = 4.19,\ v_0(1) = 8.06,\ v_0(2) = 12.12,\ v_3(3) = 14.19$ and $a_0(0) = 3,\ a_0(1) = 0,\ a_0(2) = 0,\ a_0(3) = 0$. Thus, given there is no initial inventory at epoch 0, it is optimal to order 3 units in period 1 and nothing after that.

4.6 MDP Parameter Estimation

This section briefly discusses how to estimate data for MDPs. A complete treatment of this important topic is outside the scope of this book because the range of questions and challenges that can arise from one application to another is wide. Nevertheless, the following discussion provides background on what to expect when estimating parameters in real-world MDP applications and provides references to statistical resources for further learning.

The two main categories of data for MDPs are rewards and transition probabilities. The choice of reward parameters depends on the decision-maker's criterion, which can vary significantly depending on their perspective. Common examples described in Chapter 2 include monetary rewards, QALYs, probability of a favorable or unfavorable event, and regret. Reward estimates are sometimes derived from cost estimates, revenue data, and product prices, or, in the case of QALYs, elicited through patient surveys or published medical studies. Alternatively, in some cases, the reward choice may need to be based on expert judgment due to insufficient data. In such cases, an MDP model may not be reliable for determining an exact policy; rather, it is useful for understanding how factors influencing the input data affect the optimal policy. Sensitivity analysis for MDPs is discussed in Chapter 6.

Transition probabilities are typically estimated from longitudinal data that capture the system state at discrete time intervals. The estimates depend on the definitions of the states in the Markov chain, which in turn are influenced by the available data. For example, for a production lot-sizing problem, if customer demand forecasts are generated weekly, then the choice of decision epochs may also be weekly. The ideal choice may be clear in situations with ample data and a discrete outcome (e.g., inventory level and network location). In other problems, conversely, the natural choice may have too many states to enumerate, or the state could be continuous. In these cases, practicality demands that states be aggregated into a smaller (or discrete) set of states. For example, in the optimal time-to-treat example in this chapter, risk—a continuous measure—was discretized into four states: low, medium, high, and very high. These modeling decisions are important because they ultimately influence the system dynamics and the optimal policy generated by an MDP model.

If the number of states is small, longitudinal data may contain many observed transitions, making accurate estimation of the transition probability matrix straightforward; however, a small number of states may also mean that state classification is coarse, and therefore, the resulting MDP model's accuracy may be poor. Alternatively, using many states may be useful to boost model accuracy and, hence, the accuracy of the resulting optimal policy, but it comes at the expense of a larger number of transition probabilities to be estimated, which increases the statistical error in the model parameters and the computational effort required to solve the model. In general, when making decisions about MDP model formulations, the choice of state must carefully weigh expert knowledge and constraints caused by the limited availability of sample data for model estimation.

For any two consecutive states, s and s', the total number of transitions from state s to state s' is denoted as $n_{s,s'}$. The maximum likelihood estimates of the transition probabilities are estimated as follows:

$$p(s' \mid s) = \frac{n_{s,s'}}{\sum_{i \in S} n_{i,s'}}, \qquad \forall\, s, s' \in S. \tag{4.20}$$

where S is the set of states selected by the model builder. The following small example illustrates the approach.

Example 4.6: *Treatment for Type 2 diabetes*

Consider the case of a transition probability matrix in the context of type 2 diabetes, where the states are defined by HbA1c, a commonly used continuous biomarker for estimating long-term blood sugar exposure. HbA1c is measured by a periodic blood test (e.g., every 3–6 months). To estimate the 3-month HbA1c transition probabilities, anonymized longitudinal patient data for HbA1c tests over a multi-year time horizon can be used. HbA1C is continuous, but it can be approximated by discretizing values into 10 discrete states using deciles of the empirical distribution to define the interval for each discrete state. This was the approach used by Zhang et al. [2] in a study of diabetes treatment options based on a very large longitudinal dataset. Consider this small example based on the following longitudinal data for a patient's HbA1C lab results over a five-year period. A model builder must determine an appropriate set of states:

(5.2, 6.5, 4.1, 5.3, 5.5, 5.8, 5.5, 6.5, 6.2, 6.6, 5.4, 6.7, 7.4, 7.8, 7.2, 6.3, 4.5, 6.5, 6.2, 5.4).

If the goal is to optimize treatment decisions over time, one possibility is to define a two-state Markov chain with "low" HbA1c (≤ 6) and "high" HbA1c (> 6). With the state defined, the raw data can be transformed to the selected state space, $S = \{L, H\}$, as follows:

$$(L, H, L, L, L, L, L, H, H, H, L, H, H, H, H, H, L, H, H, L).$$

Next, transition probabilities are estimated as the proportion of transitions based on the number of counts of each transition:

$$n_{LL} = 4, \quad n_{LH} = 4, \quad n_{HL} = 4, \quad n_{HH} = 7.$$

and, finally, using Equation 4.20, the statistical estimates of the transition probabilities are:

$$\tilde{p}(L|L) = \frac{1}{2}, \quad \tilde{p}(H|L) = \frac{1}{2}, \quad \tilde{p}(L|H = \frac{4}{11}, \quad \tilde{p}(H|H) = \frac{7}{11}.$$

and the resulting transition probability matrix estimate is:

$$\tilde{P} = \begin{pmatrix} \tilde{p}(L|L) & \tilde{p}(L|H) \\ \tilde{p}(H|L) & \tilde{p}(H|H) \end{pmatrix} = \begin{pmatrix} 0.5 & 0.5 \\ 0.36 & 0.64 \end{pmatrix}.$$

The elements of matrix $\tilde{P}$ are statistical estimates with a limited sample size, so a decision maker would be well advised to question the confidence intervals. The most obvious approach for this two-state example is to use the binomial distribution; however, because it yields a single-point estimate, it does not account for the fact that the transition probabilities must sum to 1. More advanced statistical methods exist for this. They are beyond the scope of this book but are implemented in many software packages and discussed in detail in introductory statistics texts, such as Wasserman [3], which provides a thorough introductory treatment of more advanced topics like this.

Some experimentation is often necessary to determine the best way to balance the trade-off between accurately differentiating alternative states of the system and the accuracy of the resulting statistical estimates of the transition probabilities (i.e., sufficient usable data for each "bin"). Regardless of the problem for which a model is being considered, it is crucial that input data be carefully scrutinized for potential biases or sources of error. Moreover, any attempts to use models such as MDPs to solve real-world problems should include a comprehensive list of potential sources of error and be accompanied by a sensitivity analysis that provides numerical evidence of how variation in model parameters influences "optimal" solutions. The book by Hernan and Robins [4] provides coverage of many important topics, including various sources of bias in observational data.

4.7 Infinite-horizon MDPs

Until now, the problems introduced as examples entail decisions made over a discrete and finite time horizon, with decision epochs indexed as $t = 0, \ldots, T$ beyond which the decision-making process terminates. While this is often a reasonable assumption, in some special cases, imposing an end to the decision horizon is unnecessary; moreover, in certain cases, this fact can be leveraged to make the resulting MDP easier to formulate and solve.

Infinite-horizon MDPs are models with unbounded decision epochs. The expected rewards for infinite-horizon models can be expressed as follows:

$$\lim_{T\to\infty} E^{\pi}[\sum_{t=0}^{T} \lambda^t r_t(s_t, a_t)].$$

An initial concern with this expression is that it may be unbounded because it sums infinite rewards. However, as long as the discount factor λ is strictly less than one, and the rewards themselves have an upper bound, the maximum expected reward is upper bounded as follows:

$$\lim_{T\to\infty} E^{\pi}\left[\sum_{t=0}^{T} \lambda^{t-1} r_t(s_t, a_t)\right] \leq M \sum_{t=0}^{\infty} \lambda^t = \frac{M}{1-\lambda},$$

where $M = \max_{s\in S, a\in A}\{r(s, a)\} < \infty$, and the bound follows from the sum being a geometric series that is finite provided $\lambda < 1$. Therefore, the limit is finite, and the MDP is well-defined.

Example 4.7: *Two-state infinite-horizon MDP*

Consider the two-state infinite-horizon MDP of Figure 4.5, which corresponds to the smallest meaningful instance of the problem. Assume the system starts in state s_1 and the decision-maker has two decisions to choose from $a_1(s_1)$, $a_2(s_1)$. If the decision-maker opts for $a_1(s_1)$ a reward $r(s_1, a_1(s_1))$ is received, and with probability $p(s_1|s_1, a_1(s_1))$ the system remains in state s_1 at the next decision epoch, and with probability $p(s_2|s_1, a_1(s_1))$ the system transitions to state s_2; alternatively, if the decision maker chooses decision $a_2(s_1)$ a reward $r(s_1, a_2(s_1))$ is received, and the system transitions to state s_2 in the next epoch with probability 1. Once the system is in state s_2 (an absorbing state), a reward $r(s_2, a_2(s_2))$ is received continuously at every future epoch. All rewards are discounted by $\lambda \in [0, 1)$ at each epoch, thus guaranteeing finite expected rewards over the infinite horizon.

The simplicity of this problem makes its analysis straightforward. Figure 4.5 illustrates this two-state problem for transitions across decision epochs, where R denotes the total future expected reward after entering s_2. Two alternative stationary policies are available, distinguished by the decision in state s_1, $a_1(s_1)$ or $a_2(s_1)$, and are referred to as policy π_1 and policy π_2, respectively. If the decision maker selects π_1, the expected rewards are

$$
\begin{aligned}
v^{\pi_1}(s_1) &= \sum_{t=0}^{\infty} \lambda^t \, p(s_1 \mid s_1, a_1(s_1))^t \, r(s_1, a_1(s_1)) \\
&\quad + \sum_{t=0}^{\infty} \lambda^{t+1} \, p(s_1 \mid s_1, a_1(s_1))^t \, p(s_2 \mid s_1, a_1(s_1)) \, R \\
&= \frac{r(s_1, a_1(s_1))}{1 - \lambda \, p(s_1 \mid s_1, a_1(s_1))} + \frac{\lambda \, R \, p(s_2 \mid s_1, a_1(s_1))}{1 - \lambda \, p(s_1 \mid s_1, a_1(s_1))}.
\end{aligned}
$$

where the first term is the expected reward received in state s_1, and the second is the expected reward received if the system transitions to s_2 summed over different decision epochs at which the transition occurs. R can be expressed as follows:

$$
R = \sum_{j=1}^{\infty} \lambda^j \, r(s_2, a_2(s_2)) = \frac{\lambda \, r(s_2, a_2(s_2))}{1 - \lambda}.
$$

Thus, for any choice of rewards, transition probabilities, and discount factor, the expected rewards for policy π_1 can be evaluated. A similar analysis for policy π_2 yields this outcome:

$$
v^{\pi_2}(s_1) = r(s_1, a_2(s_1)) + \frac{\lambda \, r(s_2, a_2(s_2))}{1 - \lambda}. \tag{4.21}
$$

Thus, a complete description of the expected rewards for each policy can be expressed in closed form for this simple example. Generally, this type of analysis is not possible for more complex MDPs, which motivates the following computational methods.

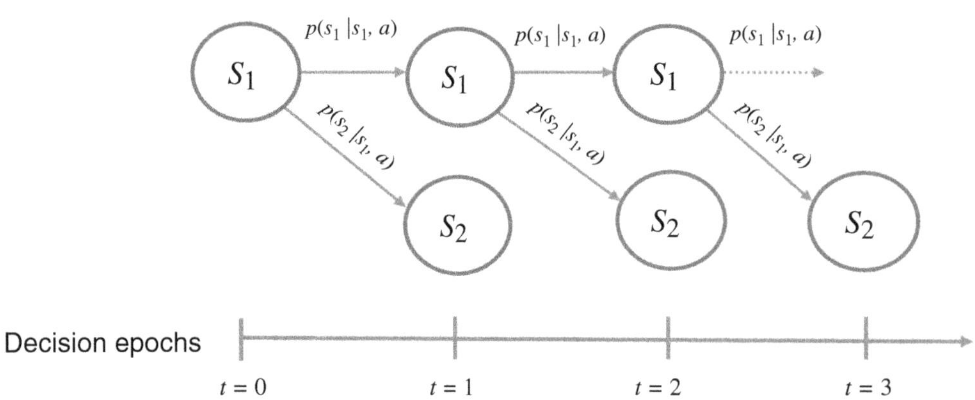

Figure 4.5 Illustration of a two-state Markov chain, with an absorbing state, in the infinite horizon context.

4.7.1 Policy Evaluation Over an Infinite Horizon

Now that the conditions for which finite expected rewards for any given policy are established, the next step is policy evaluation. For the case of stationary infinite-horizon MDPs, for which the model data (rewards, transition probabilities) do not vary over time, the parameters can be written without the subscript t, simplifying things considerably. Assuming a stationary policy, π, whereby the decision rule (the decisions for each state) at each epoch is the same, i.e., $\pi = (d, d, d, ...)$, the expected rewards over the infinite horizon can be written using the following vector notation:

$$\vec{v}^{\pi} = \sum_{t=1}^{\infty} \lambda^{t-1} P_{\pi}^{t-1} \vec{r}_{\pi} \tag{4.22}$$

$$= \vec{r}_{\pi} + \lambda P_{\pi} \vec{r}_{\pi} + \lambda^2 P_{\pi}^2 \vec{r}_{\pi} + \cdots), \tag{4.23}$$

$$= \vec{r}_{\pi} + \lambda P_{\pi} (\vec{r}_{\pi} + \lambda P_{\pi} \vec{r}_{\pi} + \lambda^2 P_{\pi}^2 \vec{r}_{\pi} \ldots, \tag{4.24}$$

$$= \vec{r}_{\pi} + \lambda P_{\pi} \vec{v}^{\pi}, \tag{4.25}$$

where the final equality follows from recognizing that the term in brackets is identical to the original sum defining $\vec{v}^{\pi}$. Furthermore, P_{π} is the transition probability matrix assuming policy π is followed, and $\vec{r}_{\pi}$ and $\vec{v}^{\pi}$ are vectors with dimensions equal to the number of states given policy π is followed. Rearranging the terms further by multiplying both sides of the final equality by the inverse of the transition probability matrix, P_{π}^{-1}, yields the following:

$$\vec{v}^{\pi} = (I - \lambda P_{\pi} \vec{v}^{\pi})^{-1} \vec{r}_{\pi}, \tag{4.26}$$

where I is the $|S| \times |S|$ identity matrix with ones along the diagonal and zeros for all other entries.

4.7.2 Optimality Equations for Infinite-horizon MDPs

Recall that the optimality equations for a finite-horizon problem were defined as:

$$v_t(s_t) = \max_{a_t \in A} \left\{ r_t(s_t, a_t) + \lambda \sum_{s_{t+1} \in S} p(s_{t+1} | s_t, a_t) v_{t+1}(s_{t+1}) \right\}, \forall s_t \in S, t = 1, \ldots, T-1, \tag{4.27}$$

$$v_T(s_T) = R(s_T), \forall s_T \in S. \tag{4.28}$$

In the infinite-horizon case in which all the model parameters are stationary, the optimal value function converges to a single value for each state. Substituting $v(s)$ for $v_t(s)$ in the optimality equations leads to the following optimality equations for the infinite-horizon case:

$$v(s) = \max_{a \in A} \left\{ r(s, a) + \lambda \sum_{s' \in S} p(s' | s, a) v(s) \right\}, \forall s \in S. \tag{4.29}$$

Alternatively, the equations can be expressed in vector form as follows:

$$\vec{v} = \max_{\pi \in \Pi} \{ \vec{r}_{\pi} + \lambda P_{\pi} \vec{v} \},$$

where vector $\vec{v}$ has elements, $v(s)$, for the optimal value function for all states, reward vector $\vec{r}_{\pi}$ and transition probability matrix P_{π} are defined by policy π and, as before, Π denotes the set of all possible policies. The optimality equations can be expressed more compactly by defining an *operator*, L, to represent the operation $\max_{\pi \in \Pi}\{\cdot\}$, which results in the following compact expression for the optimality equations:

$$\vec{v} = L\vec{v}. \tag{4.30}$$

With this new set of optimality equations defined for an infinite-horizon MDP, the focus shifts to finding an optimal policy that solves the equations. If L were simply a matrix, then the equations in (4.30) would be a system of linear equations, for which many standard linear algebra methods can be applied. However, the nature of the operator L, specifically the maximization element, introduces additional complexity.

The discussion above of the optimality equations for the infinite-horizon MDP does not provide a rigorous proof that a stationary optimal policy exists. This is addressed shortly. However, first, this is motivated numerically by observing the limiting behavior as the time horizon length, T, tends to infinity for a stationary finite-horizon MDP. The following example illustrates that increasing T has a diminishing effect on the optimal value function at the beginning of the horizon. This is due to the increasing discounting of future rewards over time. As $T \to \infty$, the MDP tends towards an infinite-horizon MDP. However, the rate of convergence depends on the discount factor λ, as is shown below. The remainder of this section gives an example showing how a finite-horizon MDP converges to an infinite-horizon MDP as the horizon increases. Then, two well-known algorithms, *value iteration* and *policy iteration*, that solve Equations 4.30 are described.

Python Example 4.8: Convergence from finite- to an infinite-horizon MDP

Consider the following finite-horizon MDP to optimize a policy for charging a battery backup. Each day, the battery is in one of two states, full (F) or empty (E), and the owner has two available decisions: charge (C) or defer (D). The decision to charge always leaves the battery fully charged. The decision to defer charging leaves a battery empty; furthermore, deferring charging a fully charged battery risks discharge.

Decision epochs: $t = 0, 1, 2, \ldots, T$

States: At each epoch, the battery is full or empty, $S = \{F, E\}$

Decisions: At the start of each epoch, the owner charges the battery or defers charging, $A = \{C, D\}$

Transition probabilities: State transitions depend on which decision is chosen:

$$P(F|F, D) = 0.8, \quad P(E|F, D) = 0.2,$$

$$P(F|E, D) = 0, \quad P(E|E, D) = 1,$$

$$P(F|F, C) = 1, \quad P(E|F, C) = 0,$$

$$P(F|E, C) = 1, \quad P(E|E, C) = 0.$$

Rewards: Charging the battery if it is full has a cost, and a cost is accrued to defer charging:

$$R(F, C) = -1,$$

$$R(F, D) = 0,$$

$$R(E, C) = -2,$$

$$R(E, D) = -3,$$

Assuming the discount factor is $\lambda = 0.97$, the optimality equations are these:

$$v_t(s_t) = \max_{a \in A} \left\{ r(s_t, a_t) + 0.97 \sum_{s'_t \in S} P(s'|s, a) v_t(s'_t) \right\}, \forall s_t \in S, t = 0, \ldots, T-1$$

and the boundary condition at the end of the time horizon is this:

$$v_T(s_T) = 0, \forall s_T \in S.$$

The MDP for a 300-day horizon can be solved using backward induction with the following Python code:

```
import numpy as np

# Input data
nstates = 2
ndecisions = 2
```

```
states = ['F', 'E']          #States: Full (F) and Empty (E)
decisions = ['C', 'D']       #Decisions: Charge (C) and Defer (D)
T = 300                      #Time horizon
Lambda = 0.9                 #Discount factor

# Transition probability matrix (2, 2, 2):  T[s, d, s'] = P(s' | s, d)
P = np.array([
    [[1.0, 0.0], [0.8, 0.2]],   # From F: [C, D]
    [[1.0, 0.0], [0.0, 1.0]],   # From E: [C, D]
])

# Reward function (2, 2): rewards[s, d] = R(s, d)
rewards = np.array([
    [-1, 0],    # F: [C, D]
    [-2, -3],   # E: [C, D]
])

# Initialize
value = np.zeros((T + 1, nstates, ndecisions))  #Value-to-go: value[t, s, d]
v = np.zeros((T + 1, nstates)) #Optimal value function
optimal_policy = np.zeros((T + 1, nstates), dtype=int) # Optimal policy

# Solve MDP to find optimal policy
for t in range(T - 1, -1, -1):
    for s in range(nstates):
        for d in range(ndecisions):


            value[t, s, d] = rewards[s, d] + Lambda * np.dot(P[s, d], v[t + 1])
        optimal_policy[t, s] = np.argmax(value[t, s])
        v[t, s] = value[t, s, optimal_policy[t, s]]

# Print the results for first epoch
print("Optimal Value Function at epoch 1:")
for s in range(nstates):
    best_d = optimal_policy[1, s]
    print(f"  State {states[s]}:  {value[1, s, optimal_policy[1, s]]}")

print("\nOptimal Policy:")
for s in range(nstates):
    print(f"  State {states[s]}: optimal policy = {decisions[optimal_policy[1, s]]}")
```

Figure 4.6 illustrates the convergence of the finite horizon MDP as the horizon length increases. Smaller values of λ result in faster convergence because the accumulated effect of discounting results in the expected rewards in future periods converging to zero. For the case of $\lambda = 0.9$, which is revisited in the next section, the results converge to the optimal value function, $\vec{v} = (-3.051, -4.746)$, as T grows large, as illustrated for state F in Figure 4.6. The optimal policy is to charge in the E state and defer charging in state F.

4.7.3 Solving the Optimality Equations for Infinite-horizon MDPs

When solving the optimality equations of an infinite-horizon MDP, $\vec{v} = L\vec{v}$, the standard backward induction approach previously used no longer works because no finite starting point defines the end-of-horizon boundary condition. In fact, infinite-horizon MDPs do not have a boundary condition. Algorithms for solving infinite-horizon MDPs are typically

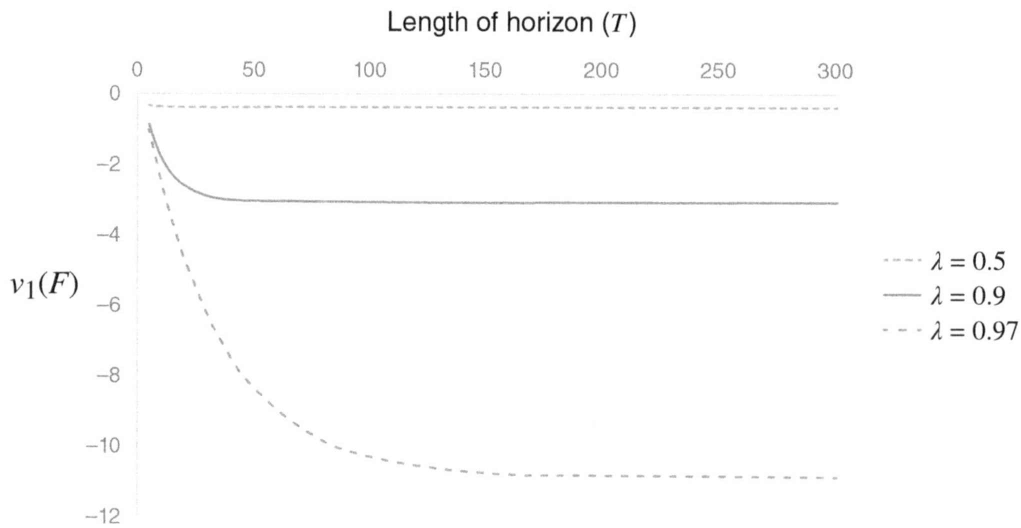

Figure 4.6 Plot of the optimal value function at epoch 1 for state F in the battery charging MDP showing convergence behavior for several choices of λ.

iterative the two alternative methods, known as *value iteration* and *policy iteration*, are introduced to solve the optimality equations.

4.7.3.1 Value Iteration

Value iteration is one of the best-known algorithms for solving infinite-horizon MDPs. Its basic approach is to start with some initial estimate of the optimal value function, $\vec{v}^0$, and apply the operator L to the initial vector to create a new estimate. Repeating this process results in a sequence of iterates, $\vec{v}^0, \vec{v}^1, \vec{v}^2, \ldots$, that converge to the optimal solution, $\vec{v}^*$. The method is described in Algorithm 4.3.

Algorithm 4.3 Value Iteration for Infinite-horizon MDPs

1: Input: MDP data elements: states, decisions, transition probability matrix, rewards, discount factor
2: Initiation: Select some initial vector $\vec{v}^0$ and set the iteration index, $n = 0$.
3: Generate new value function estimate: $\vec{v}^{n+1} = L\vec{v}^n$.
4: Check solution accuracy: If $||\vec{v}^{n+1} - \vec{v}^n|| < \dfrac{\epsilon(1-\lambda)}{2\lambda}$ continue to Step 5. Otherwise, set $n = n + 1$ and return to Step 3.
5: Return: ϵ-optimal policy.

In Algorithm 4.3, a new parameter, ϵ, defines the acceptable error tolerance on the optimal value function vector. The smaller the value of ϵ, the closer the final iterate will be to the optimal value function $\vec{v}$. Note that this algorithm differs from backward induction, which requires a finite number of steps to reach the exact optimal solution. In value iteration, the algorithm generates an ϵ-optimal policy, and the error converges to zero as the number of iterations tends to infinity.

Before explaining why the value iteration algorithm works, the battery-charging MDP of Example 4.8 is used to confirm that the solution equals the limiting value of the finite MDP as $T \to \infty$ as illustrated in Example 4.8, Figure 4.6.

Python Example 4.9: Value iteration

Consider the infinite-horizon version of the battery-charging MDP from Example 4.8. The key difference is that the number of decision epochs is now unbounded: $t = 1, 2, 3, \ldots, \infty$. Assume $\lambda = 0.9$ and set the initial value function estimate as follows:

$$\vec{v}^0 = \begin{bmatrix} 0 \\ 0 \end{bmatrix}.$$

Iterating using Step 3 of Algorithm 4.3 yields the following estimates for the first 50 iterations:

$$\vec{v}^{\,1} = \begin{bmatrix} 0 \\ 0 \end{bmatrix}, \ \vec{v}^{\,2} = \begin{bmatrix} 0 \\ -2 \end{bmatrix}, \ \vec{v}^{\,3} = \begin{bmatrix} -0.36 \\ -2.0 \end{bmatrix}, \ \vec{v}^{\,4} = \begin{bmatrix} -0.619 \\ -2.324 \end{bmatrix}, \ldots, \vec{v}^{\,50} = \begin{bmatrix} -3.034 \\ -4.729 \end{bmatrix}.$$

These initial iterations indicate that the iterates are converging to a limiting value. The following Python code implements Algorithm 4.3 and solves the infinite-horizon version of the battery-charging MDP with an error tolerance of $\epsilon = 0.001$.

```
import numpy as np

# Input data
nstates = 2
ndecisions = 2
TPM = np.array([
    [[0.8, 0.2], [1.0, 0.0]],
    [[0.0, 1.0], [1.0, 0.0]]
])
rewards = np.array([
    [0, -1],
    [-3, -2]
])
Lambda = 0.9
epsilon = 0.001

#Initialize variables
v = np.zeros(nstates) #optimal value function
n = 0 # Iteration index for value iteration

# Perform value iteration for an infinite horizon MDP.
while True:
    new_v = np.zeros(nstates)
    value = np.zeros((nstates, ndecisions))
    for s in range(nstates):
        for d in range(ndecisions):
            value[s, d] = rewards[s, d] + Lambda * np.dot(TPM[s, d, :], \
                                                                    v)
        new_v[s] = np.max(value[s, :]) #Choose optimal decision
    if np.max(np.abs(new_v-v)) \
        < epsilon*(1-Lambda)/(2*Lambda):


        break
    v = new_v
    n += 1

# Get epsilon optimal policy from the value function
optimal_policy = np.zeros(nstates, dtype=int)

for s in range(nstates):
    for d in range(ndecisions):
      value[s, d] = rewards[s, d] + Lambda * np.dot(TPM[s, d, :], v)
    optimal_policy[s] = np.argmax(value[s, :])
```

```
print("Number of iterations:", n)
print("Optimal Policy:", optimal_policy)
print("Optimal Value Function:", v)
print("Error Tolerance:", epsilon)
```

Running the above code yields the following output for the battery-charging example:

```
Number of iterations: 83
Optimal Policy: [0 1]
Optimal Value Function: [-3.05031656 -4.74523182]
Error Tolerance: 0.001
```

and the result has converged in 83 iterations to a predefined error tolerance of 0.001 according to the stopping criteria described in Theorem 4.7 below. The results match the limiting behavior of the finite-horizon MDP as the horizon grows large, as illustrated numerically in Figure 4.6.

4.7.3.2 The Reasons Value Iteration Works

To understand the value-iteration algorithm and why it converges to an optimal solution, the concept of a vector norm is needed. There are many different types of vector norms, typically denoted by L_n, which, when applied to a vector, can be expressed as follows:

$$||\vec{x}||_n = (x_1^n + x_2^n + ... + x_d^n)^{1/n}$$

where d is the dimension of the vector. The best-known is the L_2 norm, also called the Euclidean norm L, which is the square root of the sum of the squares of the vector elements. For purposes of this book, the L_∞ norm will serve well, which is defined as follows.

Definition: Define the *norm* of a vector, $\vec{v}$, as

$$||\vec{v}|| = \max_{s\in S} |v(s)|.$$

The norm defined above is known as the L_∞ norm because it is the limit as $n \to \infty$, and it serves well for some of the analysis to follow. For simplicity, it will just be referred to as the norm in the remainder of the book.

It will also be helpful to define the norm of a matrix, P, as

$$||P|| = \max\{||P\vec{v}|| : ||\vec{v}|| \leq 1\}.$$

When P is a transition-probability matrix, $||P|| = 1$, as all the elements of a row must sum to 1. For a linear transformation in which a matrix is multiplied by a vector, it is bounded if some finite number K exists such that

$$||P\vec{v}|| \leq K||\vec{v}||.$$

These definitions of vector and matrix norm can be used to understand why value iteration converges.

Theorem 4.4 *[5]: If there is a vector $\vec{v}$ that solves $\vec{v} = L\vec{v}$, then $\vec{v}$ is the expected discounted future rewards under an optimal policy π^*.*

The interested reader is referred to Appendix A.3 for a complete proof of Theorem 4.4, the basic idea of which is described here in two parts:

1. Prove if $\vec{v} \geq L\vec{v}$ then $\vec{v} \geq E_\omega^{\pi^*}[W(\omega)]$
2. Prove if $\vec{v} \leq L\vec{v}$ then $\vec{v} \leq E_\omega^{\pi^*}[W(\omega)]$

It follows that if inequalities (1) and (2) hold and if $\vec{v} = L\vec{v}$, then $\vec{v}$ is the expected discounted future reward achieved by the optimal policy, π^*.

From a practical perspective, Theorem 4.4 establishes that a solution to Equations 4.30 gives the optimal value function and, hence, an optimal policy for an infinite-horizon MDP. This result establishes that developing an algorithm to solve Equations 4.30 yields the optimal value function. To derive the algorithm, the notion of mathematical operators with the *contraction mapping* property is introduced.

Definition: Operator L is a *contraction mapping* if there exists a $0 < \lambda < 1$ such that

$$||L\vec{v} - L\vec{u}|| \leq \lambda||\vec{v} - \vec{u}||$$

for all $\vec{u}$ and $\vec{v}$.

This definition implies that applying the contraction mapping operator to two the difference of vectors reduces the distance between them, as measured by the vector norm defined above. The following theorem confirms that the operator, L, from Equations 4.30 is a contraction mapping.

Theorem 4.5 *[5]:* *If* $0 < \lambda < 1$, *then L is a contraction mapping.*

The proof of this property is given in Appendix A.3. Theorem 4.5 implies the following important theorems, which validate the optimality equations and explain why value iteration works.

Theorem 4.6 *[5]:* *For an arbitrary vector* v^0 *the sequence* $\{\vec{v}^0, \vec{v}^1, ..., \vec{v}^n\}$ *defined by* $\vec{v}^{n+1} = L\vec{v}^n$ *converges to the optimal value function,* $\vec{v}$, *as* $n \to \infty$, *where* $\vec{v} = L\vec{v}$.

See Appendix A.3 for the proof of Theorem 4.6. This theorem guarantees that the solution to the infinite horizon optimality equations is the optimal value function. Therefore, an algorithm that can find a solution to $\vec{v} = L\vec{v}$ produces the optimal value function from which the optimal policy can be extracted.

The next theorem shows that repeated application of the operator, L, on an initial arbitrary vector, creates a sequence of vectors that converge to the optimal value function as the number of iterations tends to infinity. Furthermore, Theorem 4.7 gives an error bound on the value-iteration algorithm.

Theorem 4.7 *[5]:* *Let the sequence* $\{\vec{v}^0, \vec{v}^1, \dots, \vec{v}^n\}$ *satisfy* $\vec{v}^{n+1} = L\vec{v}^n$, *then:*

$$||\vec{v}^{n+1} - \vec{v}|| \leq \frac{\epsilon}{2} \quad \text{whenever} \quad ||\vec{v}^{n+1} - \vec{v}^n|| < \frac{\epsilon(1-\lambda)}{2\lambda}.$$

See Appendix A.3 for the proof. The theorem asserts the validity of the stopping condition in value iteration, Algorithm 4.3. The stopping condition in Step 4 of Algorithm 4.3 satisfies the condition that the current iterates error is not more than ϵ, a user-selected worst-case error tolerance.

In summary, value iteration is a simple iterative approach to solving infinite-horizon MDPs to a specified tolerance guaranteed to converge to within ϵ of the optimal value function provided $\lambda < 1$.

Next, consider the second approach, called *policy iteration*, for solving infinite horizon MDPs. It is also iterative but yields an exact optimal solution within a finite number of iterations. Unlike value iteration, which seeks to find the optimal value function (from which the optimal policy can be derived), policy iteration iteratively improves policies until an optimal policy is found (from which the optimal value function can be determined via policy evaluation).

4.7.3.3 Policy Iteration

Policy iteration involves selecting an initial, arbitrary policy, evaluating it, and then improving the policy based on the evaluation results. If the new policy is optimal, then the method stops. Otherwise, it returns to the evaluation step and finds a new policy. The algorithm is described in Algorithm 4.4.

When the set of policies is finite, which is true for all problems considered here due to the assumptions of finite states and finite decisions, policy iteration generates a finite sequence of policies, $\pi^0, \pi^1, \pi^2, ...$, and value functions, $\vec{v}^0, \vec{v}^1, \vec{v}^2, ...$, which converge to the optimal value function vector, $\vec{v}$, and the optimal policy, π^*. The following example, based on the simple two-state model shown in Figure 4.7, illustrates the application of the policy iteration Algorithm 4.4.

Algorithm 4.4 Policy Iteration for Infinite-horizon MDPs

1: Input: MDP data elements: states, decisions, transition probability matrix, rewards, discount factor
2: Initiation: Select an initial policy, π^0 and set the iteration index, $n = 0$.
3: Policy Evaluation: Find $\vec{v}^n$ by solving $(I - \lambda P_{\pi^n})\vec{v} = \vec{r}_{\pi^n}$.
4: Policy Improvement: Choose π^{n+1} to satisfy

$$\pi^{n+1} \in \arg\max_{\pi\in\Pi}\{r_\pi + \lambda P_\pi \vec{v}^n\},$$

setting $\pi^{n+1} = \pi^n$ if possible.
5: If $\pi^{n+1} = \pi^n$, stop and set $\pi^* = \pi^n$. Otherwise, set $n = n + 1$ and return to step 3.
6: Return: Policy π^*.

Example 4.10: *Policy iteration*

Consider the MDP represented by the diagram in Figure 4.7. The decisions and transition probabilities are denoted in the figure, and the rewards are $r(s_1, a_{1,1}) = 8$, $r(s_1, a_{1,2}) = 12$, $r(s_2, a_{2,1}) = 1$. Assume the discount rate is $\lambda = 0.9$. Use policy iteration to find the optimal value function for each state and the optimal policy. Assuming $\pi^0 = (a_{1,1}, a_{2,1})$ is chosen as the initial policy, the first step is to evaluate the policy to determine the value function by solving the following linear system in Step 3 of Algorithm 4.4, as follows:

$$\left(\begin{pmatrix}1 & 0\\ 0 & 1\end{pmatrix} - 0.9\begin{pmatrix}0.2 & 0.8\\ 0 & 1\end{pmatrix}\right)\vec{v} = \begin{pmatrix}8\\ 1\end{pmatrix} \Rightarrow \vec{v} = \begin{pmatrix}18.54\\ 10\end{pmatrix}.$$

Now try to improve the decision in state s_1 using Step 4 of Algorithm 4.4 (state s_2 has only one decision and does not need to be evaluated):

$$\pi^1(s_1) \in \arg\max\left\{8 + 0.9(0.2v(s_1) + 0.8v(s_2)),\ 12 + 0.9v(s_2)\right\} = \arg\max\{18.54, 21\} = a_{1,2}.$$

So, $\pi^1 = (a_{1,2}, a_{2,1})$. Next, check the stopping criteria in Step 5. Since $\pi^1 \neq \pi^0$, it is not satisfied, and the algorithm returns to Step 3 and evaluates the new policy by solving the following system of linear equations:

$$\left(\begin{pmatrix}1 & 0\\ 0 & 1\end{pmatrix} - 0.9\begin{pmatrix}0 & 1\\ 0 & 1\end{pmatrix}\right)\vec{v} = \begin{pmatrix}12\\ 1\end{pmatrix} \Rightarrow \vec{v} = \begin{pmatrix}21\\ 10\end{pmatrix}.$$

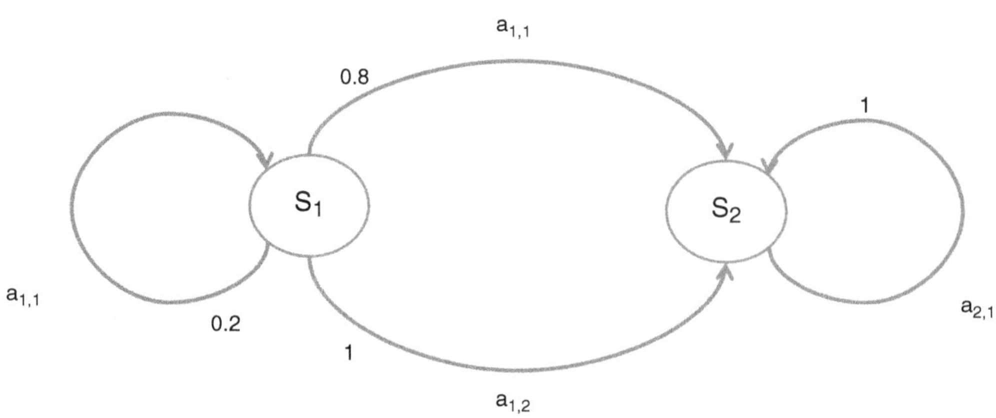

Figure 4.7 Illustration of a two-state infinite horizon Markov chain to be solved with policy iteration (PI).

The policy improvement step yields:

$$\pi^2(s_1) = \arg\max\left\{8 + 0.9(0.2v(s_1) + 0.8v(s_2)),\ 12 + 0.9v(s_2)\right\} = \arg\max\{18.98, 21\} = a_{1,2}.$$

Step 4 yields the same policy, i.e., $\pi^1 = \pi^2$, so the stopping criterion is satisfied, and the optimal policy is $\pi^* = (a_{1,2}, a_{2,1})$. The next example illustrates the use of policy iteration implemented in Python for a larger example.

Python Example 4.11 : Machine maintenance

Consider a manufacturing machine with a function that decays over time and requires maintenance. At each of an infinite number of decision epochs, the system is in a state $s \in S = \{\text{Excellent, Good, Fair, Poor, Broken}\}$ with decision choices $a \in A = \{\text{Run, Fix, New}\}$. Decision Run allows production but risks further deterioration; Fix incurs a cost but tends to improve functionality. Decision *New* is expensive but returns the machine to excellent condition with new parts. The reward function is

$$r(s,a) = \begin{cases} \text{profit} - c(a), & s \in \{\text{Excellent, Good, Fair, Poor}\}, \\ \text{loss} - c(a), & s = \text{Broken}. \end{cases}$$

Where $c(a)$ is the cost of decision a: $c(\text{Run}) = 0,\ c(\text{Maintain}) = 3,\ c(\text{Replace}) = 9$. The machine generates an incremental profit of 5 in all states except Broken, and a loss of -10 in the Broken state. $P(a)$ denotes the transition probability matrix for decision a, which can be written as follows:

$$P(\text{Run}) = \begin{pmatrix} 0.7 & 0.3 & 0 & 0 & 0 \\ 0 & 0.7 & 0.3 & 0 & 0 \\ 0 & 0 & 0.7 & 0.3 & 0 \\ 0 & 0 & 0 & 0.7 & 0.3 \\ 0 & 0 & 0 & 0 & 1 \end{pmatrix},\ P(\text{Fix}) = \begin{pmatrix} 0.9 & 0.1 & 0 & 0 & 0 \\ 0.6 & 0.3 & 0.1 & 0 & 0 \\ 0 & 0.6 & 0.3 & 0.1 & 0 \\ 0 & 0 & 0.6 & 0.3 & 0.1 \\ 0 & 0 & 0 & 0 & 1 \end{pmatrix},\ P(\text{New}) = \begin{pmatrix} 1 & 0 & 0 & 0 & 0 \\ 1 & 0 & 0 & 0 & 0 \\ 1 & 0 & 0 & 0 & 0 \\ 1 & 0 & 0 & 0 & 0 \\ 1 & 0 & 0 & 0 & 0 \end{pmatrix}.$$

Assume $\lambda = 0.99$ and the initial policy is:

$$\pi^0(Excellent) = New,\quad \pi^0(Good) = New,\quad \pi^0(Fair) = New,\quad \pi^0(Poor) = New,\quad \pi^0(Broken) = New.$$

The following Python code computes the optimal policy using policy iteration.

```
import numpy as np

#Input data
states = ["Excellent", "Good", "Fair", "Poor", "Broken"]
decisions = ["Run", "Fix", "New"]
Lambda = 0.99  # discount factor
profit = 5     # reward in non-Broken states
loss = -10     # reward in Broken state
cost = {"Run": 0, "Fix": 3, "New": 9}
MAX_ITS = 100 #max allowable iterations of policy iteration
# Transition probability matrices P(a)
P = {
    "Run": np.array([
        [0.7, 0.3, 0.0, 0.0, 0.0],
        [0.0, 0.7, 0.3, 0.0, 0.0],
        [0.0, 0.0, 0.7, 0.3, 0.0],
        [0.0, 0.0, 0.0, 0.7, 0.3],
        [0.0, 0.0, 0.0, 0.0, 1.0],
```

```
        ]),
        "New": np.array([
            [1.0, 0.0, 0.0, 0.0, 0.0],
            [1.0, 0.0, 0.0, 0.0, 0.0],
            [1.0, 0.0, 0.0, 0.0, 0.0],
            [1.0, 0.0, 0.0, 0.0, 0.0],
            [1.0, 0.0, 0.0, 0.0, 0.0],
        ]),
        "Fix": np.array([
            [0.9, 0.1, 0.0, 0.0, 0.0],
            [0.6, 0.3, 0.1, 0.0, 0.0],
            [0.0, 0.6, 0.3, 0.1, 0.0],
            [0.0, 0.0, 0.6, 0.3, 0.1],


            [0.0, 0.0, 0.0, 0.0, 1.0],
        ]),
    }
    # Initial policy
    policy_0 = ["New", "New", "New", "New", "New"]

    def build_reward_matrix(states, decisions, profit, loss, cost):
        r = np.zeros((len(states), len(decisions)))
        for i, s in enumerate(states):
            base = profit if s != "Broken" else loss
            for j, a in enumerate(decisions):
                r[i, j] = base - cost[a]
        return r

    def policy_evaluation(policy, P, r, Lambda, decisions):
        nS = len(policy)
        P_pi = np.zeros((nS, nS))
        r_pi = np.zeros(nS)
        decision_to_index = {a: j for j, a in enumerate(decisions)}
        for i, a in enumerate(policy):
            P_pi[i, :] = P[a][i, :]
            r_pi[i] = r[i, decision_to_index[a]]
        A = np.eye(nS) - Lambda * P_pi
        v = np.linalg.solve(A, r_pi)
        return v

    def policy_improvement(policy, v, P, r, Lambda, states, decisions, tol=1e-12):
        decision_to_index = {a: j for j, a in enumerate(decisions)}
        policy_new = []
        stop = True  # assume no change unless one is found
        for i, s in enumerate(states):
            value = []
            for a in decisions:
                q = r[i, decision_to_index[a]] + Lambda * P[a][i, :].dot(v)


                value.append(q)
```

```
        value = np.array(value)
        best_policy = decisions[int(np.argmax(value))]
        policy_new.append(best_policy)
        if best_policy != policy[i]:   # if no decision changes stop
            stop = False
    return policy_new, stop

def policy_iteration(states,decisions,P,r,Lambda,policy_0,max_iters=MAX_ITS):
    policy = list(policy_0)
    for n in range(max_iters):
        v = policy_evaluation(policy, P, r, Lambda, decisions)
        policy_new,Stop = policy_improvement(policy,v,P,r,Lambda,states,decisions)
        if Stop:
            print(f"Optimal Policy: {policy}")
            return policy, v
        else:
            print(f"Iteration {n}: policy = {policy}")
        policy = policy_new
    print("Policy iteration did not converge within max_iters.")

#Solve infinite horizon MDP with policy iteration
r = build_reward_matrix(states, decisions, profit, loss, cost)
policy_optimal,v=policy_iteration(states,decisions,P,r,Lambda,policy_0)
```

The results from running the Python code are as follows, where the state ordering is the same as in the problem description above:

```
Iteration 0: pi = ['New', 'New', 'New', 'New', 'New']
Iteration 1: pi = ['Run', 'Run', 'Run', 'Run', 'New']
Iteration 2: pi = ['Run', 'Fix', 'New', 'New', 'New']
Iteration 3: pi = ['Run', 'Fix', 'Run', 'New', 'New']
Optimal Policy: ['Run', 'Run', 'Run', 'New', 'New']
```

The optimal policy was reached in four iterations. The initial policy, π^0, was intentionally set to be poor, yet policy iteration quickly found the optimal policy. For small problems such as this, the policy can be chosen arbitrarily. However, in practice, for large problems, it is typically beneficial to start with a good policy, informed by the decision-maker's understanding of the problem, to reduce the number of iterations necessary to reach the optimal policy.

4.7.3.4 The Reason Policy Iteration Works

The reason policy iteration converges to an optimal solution is summarized in the following theorem, attributed to Howard [6].

Theorem 4.8 (*[6]*): *The policy-iteration algorithm terminates in a finite number of iterations, with a solution of the optimality equations and an optimal policy, π^*.*

See Appendix A.3 for the proof.

The concepts behind the proof of Theorem 4.8 can be restated in words as follows:

- If an improvement to a policy is possible, then policy iteration will find it.
- If no improvement is possible, then the algorithm terminates with an optimal solution to $\vec{v} = L\vec{v}$.

The formal proof considers both assertions. The first step is to show that the policy improvement step of Algorithm 4.4 is guaranteed to find a better policy if one exists. The second step follows from the stopping criterion in line 5. Thus, the proof follows from the fact that Algorithm 4.4 finds a finite improvement (if such a policy exists) and the fact that there is a finite number of possible policies (because of the assumption of states and decisions).

A comparison of value iteration and policy iteration reveals two fundamental differences between Algorithms 4.3 and 4.4. First, Algorithm 4.3 (value iteration) generates ϵ-optimal solutions, whereas Algorithm 4.4 (policy iteration) generates an exact optimal policy. Second, policy iteration requires solving a linear system at each iteration. This computational step imposes an additional burden on policy iteration at each step of the algorithm. The best algorithm for a particular application depends on several factors, including the size of the MDP, the available computation time, and the relative importance of an exact solution (policy iteration) versus a near-optimal solution with a defined error tolerance (value iteration).

For very large-scale infinite-horizon MDPs, more advanced methods than these two have been developed to achieve computational efficiencies, such as *modified policy iteration*, which seeks to accelerate Algorithm 4.4 by using inexact evaluations in Step 3 obtained using value-iteration steps, thus avoiding solving a linear system of equations at each iteration [7]. The modified policy iteration algorithm can be tuned by defining ϵ as a time-dependent parameter that decreases over time; thus, the evaluation tolerance is relaxed in early iterations, when the policies may be far from optimal, and then decreased as the policy approaches optimality.

4.8 Concluding Remarks

This chapter introduced MDPs and used several small examples to build intuition about the algorithms. The assumptions of finite additive rewards and discrete state and decision sets for the Markov chain, combined with the principle of optimality, led to an efficient backward-induction algorithm that can solve large-scale problems. These algorithms can be implemented computationally using Python or other programming languages. Of course, assuming that all problems can be modeled accurately using Markov chains is unreasonable. The next chapter considers an alternative approach to MDPs based on a special class of mathematical programs called *multi-stage stochastic programs*, which relax some of the assumptions and limitations of MDPs, albeit at the cost of greater computational effort.

Various parts of this chapter, including the models, algorithms, and examples, were motivated by original sources that provide greater mathematical detail on the theory and methods of MDPs, including Ross [8], Bertsekas [9], and Puterman [1]. A more recent book by Powell [10] takes an application-focused perspective, illustrating the use of related methods through examples motivated by real-world problems in sequential decision-making. Puterman [1] provides a rigorous treatment of infinite-horizon MDPs, extending value and policy iteration to include advanced methods that enhance computational efficiency in certain situations, such as *modified policy iteration*. Puterman also includes several examples, including the production-planning example in this chapter (Section 4.5).

The theorems presented in this chapter, in their most general form, are typically attributed to the original work of Bellman, much of which is covered in the compilation of his work in his 1957 book, *Dynamic Programming*, which gives a thorough treatment of the theoretical foundations of Markov decision processes and methods like value iteration. Policy iteration is often attributed to Howard [6] and to Denardo [11]. The treatment in this chapter follows the theorems and proofs in Puterman [1], with many simplifications enabled by the assumption of finite state and decision spaces.

Estimating transition probabilities was also briefly discussed in this chapter, with an example based on maximum-likelihood estimation of the proportion of transitions among health states for patients with diabetes. This example was motivated by the work published in Zhang et al. [2]. In practice, challenges can arise from systemic bias. For instance, in the diabetes example, some data may be missing for patients who do not undergo testing in accordance with the recommended guidelines. In some cases, ignoring missing data like this could lead to significant bias in the estimated transition probabilities if the data used to create the model differ from the population for which the model is intended. Such systematic challenges, although outside the scope of this book, are nevertheless important considerations for any models used in decision-making. The book by [4] is an introductory source on the challenges of real-world data.

Some software packages are available for solving MDPs. The *MDPToolbox* package, for example, is available in many computing environments, including Python. It provides solvers and methods for discrete-time MDPs, including *backward induction* for finite-horizon problems and value iteration and policy iteration for infinite-horizon problems [12].

4.9 Practice Exercises

1. Consider the production planning example of Section 4.5. How many possible policies are there? How does the number grow as a function of the number of states, decisions, and decision epochs?

2. Write Python code to solve Example 4.5 and test the code by confirming the optimal solution to the small example solved. Next, solve a 12-epoch MDP and answer the following question: how do the optimal cost and policy change as variance in the demand distribution increases while the mean stays constant? What can be concluded about the variance in the demand? Use graphs or charts to illustrate your analysis and support the claims.

3. For the stochastic shortest path problem of Example 4.3, determine the complete set of possible paths that could be taken under the optimal policy and determine the probability of each path. Compute the mean and variance of path length under the optimal policy. Can you find a policy with lower variance?

4. Degenerative disk disease is a common cause of back pain. This condition's degeneration process can be modeled by a set of discrete states $S = \{\text{Excellent}, \text{ Good}, \text{ Bad}, \text{ Failed}\}$, indexed by $s = 1,\ 2,\ 3,\ 4$, respectively. In any state, the available decisions are $A = \{R, D\}$, where R denotes repairing the disk via surgery, and D represents deferring surgery for one epoch. The surgery, regardless of the initial state, restores the disk to the Excellent state with probability q and fails with probability $1 - q$. The patient receives a reward $r(s, D) = 1 - \rho(s)$ for each epoch surgery is deferred and $r(s, R) = 1 - c$ during an epoch in which surgery is performed. The decision horizon spans T decision epochs, starting in the Excellent state. A terminal reward $R(s) = \mu(s)$ is received in the last epoch. A Markov chain defines the progression among states without surgery. Given the problem parameters $q = 0.9$, $\rho(E) = 0$, $\rho(G) = 0.1$, $\rho(B) = 0.2$, $\rho(F) = 0.5$, $c = 0.7$, $\mu(E) = 4$, $\mu(G) = 3$, $\mu(B) = 2$, $\mu(F) = 1$ for all states s, $T = 20$ and the following transition probability matrix that defines the transition among states in the absence of surgery, find the optimal policy.

$$P = \begin{bmatrix} 0.9 & 0.1 & 0 & 0 \\ 0 & 0.8 & 0.2 & 0 \\ 0 & 0 & 0.7 & 0.3 \\ 0 & 0 & 0 & 1 \end{bmatrix}.$$

Implement the backward induction algorithm for this model to find an optimal policy.

5. A campaign manager for a political candidate has a budget of $\$x$ to allocate to advertising her candidate in each of T sequential winner-take-all regional elections. She allocates to regions sequentially, starting with Region 1 and ending with Region T, and she observes the results of each allocation before proceeding to the next. If she allocates $\$x_i$ to region i, her candidate wins that region with probability $p(x_i)$ and gets all d_i delegates. Otherwise, her candidate gets 0 delegates for the region. Her candidate must win at least D delegates to be nominated. Provide a complete description of the dynamic program, assuming her goal is to maximize the probability of winning. State any assumptions made for each case. Explain the optimal value function at each stage for the two dynamic programs. Solve the model for the following specific case: budget = \$30 million, $T = 5$, $D = 25$, and the following delegates per region (Table 4.3):

Table 4.3 Number of delegates by region.

Region	Delegates
1	5
2	5
3	10
4	5
5	15
6	10

The probability distribution of winning each region if $\$x$ is spent is as follows:

$$p(x) = \begin{cases} 0.125 & x = \$5 \text{ million}, \\ 0.25 & x = \$10 \text{ million}, \\ 0.625 & x = \$15 \text{ million}. \end{cases}$$

Compute the optimal policy and provide a critical analysis of the model and the resulting policy.

6. Derive the expression for policy π_2 in the two-state example given in Equation 4.21.

7. For the two-state infinite-horizon MDP in Example 4.10, complete the following:
   a. Use Python to solve this infinite-horizon MDP using value iteration, with an initial estimate of the value function of , $v(s_1) = v(s_2) = 0$, an error tolerance of $\epsilon = 0.01$, and a discount factor of $\lambda = 0.8$
   b. How many iterations were necessary to guarantee that the optimal value function is within the ϵ tolerance? At what iteration was the ϵ-optimal policy determined?
   c. Describe how the answers to part (b) are affected by changes in the discount factor, λ.

8. Compare and contrast the computational effort required for value iteration vs. policy iteration. Is one better than the other?

9. An inventory storage system has an end-of-period capacity of 3 units for a particular part. During a period in which production takes place, a fixed setup cost of \$4 is incurred. A \$1 holding cost is assessed against each unit of a period's ending inventory. Also, a variable per-unit production cost of \$1 per unit is incurred. During each period, demand is equally likely to be 1 or 2 units. All demand must be met on time, and the discount factor is $\lambda = 0.8$. The goal is to find a production policy that minimizes expected discounted costs over an infinite horizon. Implement value iteration and policy iteration to solve this problem. How does the choice of λ influence the number of iterations for each algorithm to converge?

10. A financial call option provides the holder the right to purchase an asset (e.g., a stock) at a fixed price called the strike price, w, over a fixed number of days, T. Each day, if the stock price, p, is above w, then the holder can choose to exercise the option (where exercise means that they purchase and immediately sell the stock) and get a return of $p - w$ or keep the option. The option may be exercised only once and must be exercised within T days. If the stock price is below w then there is no value to exercising the option. Each day the price increases by $u\%$ with probability q and decreases by $d\%$ with probability $1 - q$. Assume the objective is to maximize expected return from the option.

a. Formulate a dynamic program to determine the optimal exercise policy, stating any assumptions made in the model.
b. Prove there are nonincreasing numbers $p_1 \geq p_2 \geq \cdots \geq p_T$ such that at day t the option should be exercised if and only if $p \geq p_t$.
c. Suppose IBM stock can be purchased for \$50/share today. The stock price is currently \$50, and each day it increases by 5% with a probability of 0.55 and decreases by 10% with a probability of 0.45. The holder can exercise the option on any of the next five days. What is the optimal exercise policy? If someone offers to purchase the option, what is the minimum price the holder should accept if the holder is risk-neutral?

11. Consider a small hospital with a 5-bed intensive care unit (ICU) and an attending physician who must decide whether to admit arriving patients. In each decision epoch, at most one arrival occurs as follows: Emergency with probability $p_E = 0.2$, non-emergency with $p_L = 0.5$, no arrival $p_0 = 0.3$. Each occupied bed independently discharges a patient with probability $p_D = 0.3$ each epoch. Emergency patients must be admitted if there is an unoccupied bed. Emergency patients who are not admitted cannot queue; they are diverted to another hospital with a penalty of $c_D = 0.8$ if no bed is available. The net reward per occupied bed per epoch is $r_B = 0.2$. A consultant to the hospital's operations team wishes to formulate and solve an infinite-horizon MDP with a discount factor of $\lambda = 0.95$ to assist the attending physician with admission decisions.

12. Prove the inverse of the matrix in Equation 4.26 for policy iteration exists and is well-defined.

13. Consider the following card game. A dealer begins with a shuffled deck with m red cards and n black cards. Each turn, the dealer flips a card. The player wins \$1 if a red card is turned and loses \$1 if a black card is turned. The player can stop playing anytime. What is the optimal policy that maximizes the expected winnings?
a. Formulate this as a dynamic program with state space defined by the number of red and black cards, (R, B), remaining in the deck at each turn. Provide a thorough explanation of all elements of the dynamic program, including the optimality equations.
b. Assume it is optimal to play in some state (R, B). Prove or disprove, by counterexample, the following propositions:
  i. It is optimal to play in state $(R + 1, B)$
  ii. It is optimal to play in state $(R - 1, B)$
  iii. It is optimal to play in state $(R, B + 1)$
  iv. It is optimal to play in state $(R, B - 1)$
c. Solve the model for the case of 5 red cards and 3 black cards. Provide the optimal policy and the expected winnings for the policy.

14. In a campus parking garage, there is a single electric vehicle (EV) charging station. There are two types of EVs: (1) *Slow-charging vehicles* that require 3 hours to charge and (2) *Fast-charging vehicles* that require 1 hour to charge. Users pay an hourly charging fee of \$5/hour. At the beginning of each hour, a sign is posted indicating what kind of vehicles are currently being accepted: "Accepting fast-charging and slow-charging vehicles," "Accepting fast-charging vehicles only," "Accepting slow-charging vehicles only," and "Not accepting vehicles." If a vehicle is accepted, the charging station becomes busy until that vehicle finishes charging. While the station is busy, all arriving cars are turned away. The following acceptance decisions are made:

$$1{:}00\text{ pm},\ 2{:}00\text{ pm},\ \ldots,\ 9{:}00\text{ pm},\ 10{:}00\text{ pm}.$$

Any vehicle accepted at 10:00 pm pays the full amount and finishes charging after closing. The charging station is available at 6:00 am each day. At the beginning of every hour:

$$\Pr(\text{“slow-charging car arrives”}) = p_s = 0.20, \qquad \Pr(\text{“fast-charging car arrives”}) = p_f = 0.40.$$

Arrivals of slow-charging and fast-charging vehicles occur independently. If both arrive at the same time, the fast-charging vehicle gets the spot, and the other is turned away.
a. Formulate an MDP describing states, decisions, rewards, transition probabilities, and the optimality equations (state any assumptions). Describe the optimal acceptance schedule you obtain from the optimal policy. What is the expected revenue of this schedule?

b. An advertisement on social media would increase the probability of a slow-charging vehicle arrival from $p_s = 0.20$ to $p_s = 0.40$ for the day. What is the most that should be paid for this advertisement? If the advertisement is made, does the acceptance schedule change? If so, how?

c. Explain what happens to the number of states as the number of charging stations grows from 1 to 2 to n.

References

1 Puterman, M. L. (1994). *Markov Decision Processes: Discrete Stochastic Dynamic Programming*. John Wiley & Sons. (Reprinted 2014.)

2 Zhang, Y., McCoy, R. G., Mason, J. E., Smith, S. A., Shah, N. D. & Denton, B. T. (2014). Second-line agents for glycemic control for type 2 diabetes: Are newer agents better? *Diabetes Care*, *37* (5), 1338–1345.

3 Wasserman, L. (2004). *All of Statistics: A Concise Course in Statistical Inference*. Springer.

4 Hernán, M. A. & Robins, J. M. (2020). *Causal Inference: What If*. Chapman & Hall/CRC. Available at: https://www.hsph.harvard.edu/miguelhernan/causal-inference-book/.

5 Bellman, R. (1957). *Dynamic Programming*. Princeton University Press.

6 Howard, R. A. (1960). *Dynamic Programming and Markov Processes*. MIT Press / Technology Press.

7 Puterman, M. L. & Shin, M. C. (1978). Modified policy iteration algorithms for discounted Markov decision problems. *Management Science*, *24* (11), 1127–1137.

8 Ross, S. M. (1983). *Introduction to Stochastic Dynamic Programming*. Academic Press. (Reprinted 2014.)

9 Bertsekas, D. P. (2012). *Dynamic Programming and Optimal Control* 4th. ed. Athena Scientific. (Original work published 1987.)

10 Powell, W. B. (2022). *Sequential Decision Analytics and Modeling: Modeling with Python*. Now Publishers.

11 Denardo, E. V. (1967). Contraction mappings in the theory underlying dynamic programming. *SIAM Review*, *9* (2), 165–177.

12 Chadès, I., Chapron, G., Cros, M.-J., Garcia, F. & Sabbadin, R. (2014). MDPtoolbox: A multi-platform toolbox to solve stochastic dynamic programming problems. *Ecography*, *37* (9), 916–920.

5

Constrained Optimization Models

5.1 Introduction

This chapter introduces a fundamentally different type of decision-making model. These problems were historically called *mathematical programs*, but in this book they will be termed *optimization models* to align with modern terminology. However, it should be noted that the DP and MDP models in Chapters 3 and 4 also seek optimal solutions and, colloquially, may be referred to as optimization models in other contexts. The formulation and notation of optimization models differ greatly from those of DPs and MDPs. As a result, the methods for finding optimal solutions vary widely. As will become apparent shortly, each type of formulation, DP/MDP models and optimization models, has pros and cons, and each makes an important contribution to the toolbox for solving sequential decision-making problems.

Chapter 3 covers many decision-making problems that can be formulated as DP models. One such problem, the resource allocation problem, is a special case of the *knapsack problem*, where the decision-maker is tasked with selecting which items, of varying sizes and rewards, to add to a one-dimensional container with limited space, such that the rewards are maximized and the space constraint is not violated. This choice of starting point is intended to provide a gentle introduction to how optimization models are formulated, assuming the reader is familiar with the conceptual idea of resource allocation problems already from Section 3.5 of Chapter 3.

More formally, the knapsack problem calls for a decision-maker to select items of varying sizes, $c_1, c_2, \dots, c_N$, and rewards, $r_1, r_2, \dots, r_N$, to maximize total rewards, subject to a constraint on the total size of items, i.e., the capacity of the knapsack, C. The optimization model formulation is as follows:

$$\max \sum_{i=1}^{N} r_i a_i, \tag{5.1}$$

$$\text{subject to: } \sum_{i=1}^{N} c_i a_i \leq C, \tag{5.2}$$

$$0 \leq a_i \leq 1, \forall i, \tag{5.3}$$

where a_i for $i = 1, \dots, N$ are referred to as *decision variables*. This model is known as the *continuous knapsack problem* when no restrictions are imposed on the decisions being integers, and, thus, items can be selected in whole or in part. The optimal solution to the continuous knapsack problem is to sequence items in increasing order of the ratio of rewards to size, r_i/c_i, and sequentially add items until the last one, which uses all remaining space, for which only a fraction of the item may be selected. Thus, it is considered an easy problem to solve.

An alternative formulation includes the binary constraint on the decision variables, $a_i \in \{0, 1\}$, which requires an "all-or-nothing" decision when selecting items. Alternatively, the decisions could be integer-valued, mirroring the resource allocation model in Chapter 3, in which an integer number of each project (or item) can be selected. (This latter formulation assumes that projects can be repeated and multiple instances of each item are available.) The two alternatives, together with their additional constraints, transform the problem from a continuous linear program into a binary or integer program. In some cases, seemingly small changes like this can have a profound impact on solution efficiency and even on the types of methods used to solve models. A detailed exploration of the vast array of methods is beyond the scope of this book. Instead, this chapter focuses on formulating the models and solving them using optimization software integrated with Python. As shown below, many problems solved using DP or MDP models can also be formulated as optimization problems. In fact, optimization models can solve the same problems as DP models, but their formulations and solution methods differ. One difference is that optimization models, unlike DP models, do not rely on a specific definition of the state describing the system at a point in time. Rather, they are built upon three main elements:

1. *Decision variables:* These variables represent the decisions available to the decision-maker. They may be continuous or discrete, and they may be indexed by time or other factors that define the decision (e.g., for the knapsack problem above, the decision variables are indexed by the items to be selected).
2. *Objective function:* This function defines the criterion for determining how one choice of decision compares to another. The goal may be to maximize or minimize the objective function, with *optimal solutions* defined as feasible points that achieve the best objective function value.
3. *Constraints:* These are expressed as equalities (=) or inequalities ($\leq, \geq$) with a function of the decision variables on the left-hand side of the constraint and a scalar on the right-hand side (e.g., $x_1 + 2x_2 + 1.5x_3 \leq 3.2$). The constraints in the model serve to define the feasible region for the decision variables.

Constrained optimization models are widely varied. This chapter focuses on those that relate directly to sequential decision-making and, within that context, provides introductory coverage suitable for learning how to apply these models. The models covered below belong to the following categories:

Linear Programs: These models have a linear objective function and linear constraints with respect to the decision variables. All of the decision variables are continuous-valued, possibly constrained to some range of relevant values (e.g., $x \geq 0$ or $a \leq x \leq b$).

Integer Programs: These models include variables that are restricted to taking on integer values (e.g., $x \in \{0, 1\}, x \in \{-2, -1, 0, 1, 2\}$).

Mixed-Integer Programs: These models include continuous and integer-valued decision variables.

Stochastic Programs: These models can be any of the above types, but with some of the model parameters (objective function coefficients, constraint coefficients, and right-hand sides of constraints) being random variables.

Several factors influence the difficulty of solving constrained optimization problems, including the number of decision variables, the number of constraints, and the type of decision variables. Linear programs are generally much easier to solve than integer or mixed-integer programs of similar size because the search for optimal solutions is more complex when variables are restricted to discrete values. Stochastic programs, parallel to the extension of deterministic DP models to the stochastic MDP models of Chapter 4, add complexity but also functionality (responsive to randomness and uncertainty). Indeed, stochastic programs, especially stochastic integer or mixed-integer programs, are among the most challenging to solve due to the large model sizes that must account for the random outcomes of model parameters.

To illustrate the use of an optimization model, consider the following example from Chapter 2 that was solved using DP.

Example 5.1: *Investment selection*

Consider 10 alternative investments that a decision-maker may select, in whole or in part. The costs of the investments in increments of \$1,000, from least to most expensive, are $c_1 = 1.5$, $c_2 = 1.8$, $c_3 = 2.1$, $c_4 = 3.0$, $c_5 = 3.6$, $c_6 = 4.4$, $c_7 = 5.3$, $c_8 = 5.9$, $c_9 = 6.0$, and $c_{10} = 6.6$. Furthermore, the expected returns from the investments one year later are $r_1 = 2.8$, $r_2 = 2.9$, $r_3 = 2.5$, $r_4 = 4.5$, $r_5 = 4.0$, $r_6 = 4.6$, $r_7 = 6.5$, $r_8 = 9.1$, $r_9 = 7.2$, $r_{10} = 6.8$. Assuming the decision-maker has a budget of \$12,000 to invest, what is the optimal choice of investments?

To answer this question, the data provided are used to create an instance of the model in Equations 5.1–5.3, which yields the following:

$$\max \quad 2.8a_1 + 2.9a_2 + 2.5a_3 + \cdots + 9.1a_8 + 7.2a_9 + 6.8a_{10}, \tag{5.4}$$

$$\text{subject to: } 1.5a_1 + 1.8a_2 + 2.1a_3 + \cdots + 5.91a_8 + 6.0a_9 + 6.6a_{10} \leq 12000, \tag{5.5}$$

$$0 \leq a_i \leq 1, \forall i = 1, \ldots, 10. \tag{5.6}$$

As mentioned above, the optimal solution is obtained by sorting the investments (items) from highest to lowest based on the ratio of investment returns (rewards) to costs (item sizes). Computing the ratios of rewards to investment costs for the 10 investments yields:

$$\text{Ratio}_1 = 1.87,$$

$$\text{Ratio}_2 = 1.61,$$

$$\text{Ratio}_3 = 1.19,$$

$$\text{Ratio}_4 = 1.50,$$

$$\text{Ratio}_5 = 1.11,$$

$$\text{Ratio}_6 = 1.05,$$

$$\text{Ratio}_7 = 1.23,$$

$$\text{Ratio}_8 = 1.54,$$

$$\text{Ratio}_9 = 1.20,$$

$$\text{Ratio}_{10} = 1.03.$$

Ordering from highest to lowest ratio reveals that the optimal investment selection is in the order 1, 2, 8, 4; however, the total cost of including all items is $12,200, which exceeds the budget constraint. Thus, optimally, the reduction should target the least favorable investment, item 4, reducing it to 0.93 (93% of item 4), yielding a feasible optimal solution.

Constrained optimization models rarely have solutions as easily obtainable as those of the continuous knapsack problem. They are typically much more complex to solve and require advanced algorithms, the most well-known of which is the *simplex method* [1], which has been thoroughly studied. Nowadays, many highly effective algorithms are implemented in various optimization software. Since they come in varying forms and levels of difficulty, the chapter begins with an introduction to linear programming.

5.2 A Visual Introduction to Linear Programming Models

This section begins with a conceptual description of linear programming models to facilitate an intuitive understanding of the fundamental principles for a small (2-dimensional) problem. The fundamental knowledge is subsequently extended to specific approaches for solving constrained optimization models using software tools.

The easiest to visualize and the most straightforward to analyze constrained optimization model is a linear program with two decision variables. Because it entails only two decision variables, this model is easy to visualize and illustrates several properties of linear programs that extend to larger problems. A common example arises when two activities coincide, corresponding to two decision variables, each of which requires a shared resource to be completed. If the shared resource is limited, this can result in interdependent decision variables, as both decisions affect the level of available shared resources.

Consider the example of a simple capacity-planning model for a small surgery practice used for two types of outpatient surgery on a particular day: (1) urology and (2) ophthalmology. Both types of surgery require an operating room (OR) and a (shared) post-anesthesia care unit (PACU), where patients can recover after surgery. ORs are allocated to specific surgical procedures, but PACU capacity is shared across them. As a result, the level of PACU capacity available impacts both types of surgery.

With the above assumptions in mind, the model can be defined as follows. Let decision variables x_1 and x_2 denote the daily number of urology and ophthalmology surgeries. Assume each surgery is associated with some relative reward, $r_1 = 3$ and $r_2 = 5$ for urology and ophthalmology, respectively. Moreover, assume urology surgeries each require 1 unit of OR capacity and 1.5 units of PACU capacity, while ophthalmology surgeries require 2 units of OR capacity and 1 unit of PACU capacity. Finally, assume the practice has 4 units of urology OR capacity, 12 units of ophthalmology capacity, and 9 units of PACU capacity. The constrained optimization model with the data specified is written as the following linear program:

$$\max 3x_1 + 5x_2, \tag{5.7}$$

$$\text{subject to:} \quad x_1 \leq 4, \tag{5.8}$$

$$2x_2 \leq 12, \tag{5.9}$$

$$1.5x_1 + x_2 \leq 9, \tag{5.10}$$

$$x_1 \geq 0, x_2 \geq 0. \tag{5.11}$$

Constraints 5.8 and 5.9 enforce limited urology and ophthalmology OR capacity, respectively, and Constraint 5.10 enforces limited PACU capacity, which is shared by both surgery services. Finally, Constraint 5.11 ensures that the decision variables are nonnegative.

Because the above model has only two decision variables, it can be illustrated in Figure 5.1. In the figure, the two axes correspond to the two decision variables. Because the decision variables are nonnegative, only the upper-right quadrant is relevant. The remaining three constraints define the boundary of the feasible region. Because they are ($\leq$) inequalities, they each define a *half-space* in which a line, based on replacing the inequality ($\leq$) with equality ($=$) defines the boundary, and only points on one side or the other are feasible.

This simple example illustrates several properties of linear programs. First, only points in the grey-shaded region of Figure 5.1, i.e., points simultaneously satisfying all constraints, are feasible. Changing the right-hand side of a linear program constraint can increase or decrease the feasible region, depending on the direction of the change. For example, if the capacity of either the ORs or the PACU increases, the feasible region expands. Thus, the number of feasible solutions would increase, and the optimal solution to the new linear program would be at least as good as that of the original problem (since there are more choices available to the decision-maker). Alternatively, if the capacity of ORs or the PACU decreases, the feasible region decreases, and the resulting optimal solution to the new linear program can be no better than the original problem (because there are fewer choices available to the decision-maker).

The optimal solution to a small linear program like this can be found by visual inspection of the objective function in Equation 5.7. An *isoprofit* line is a line for which all points on the line have the same objective function value. It can be generated by setting the objective function equal to a constant (e.g., $3x_1 + 5x_2 = 8$). The resulting isoprofit line may have a segment that passes through the feasible region, depending on the value assigned to the right-hand side. If the line intersects the feasible region for a given objective function value, then any point on the line is feasible and achieves that value. Therefore, a decision-maker is indifferent to which feasible point on the line is selected.

Taking this one step further, as the objective function value increases, the isoprofit line shifts, revealing feasible solutions with higher objective function values. This is illustrated in Figure 5.2. Increasing the objective function value eventually causes the isoprofit line to touch the boundary of the feasible region at ($x_1 = 2,\ x_2 = 6$) which occurs for the isoprofit line: $3x_1 + 5x_2 = 36$. At this single point on the edge of the feasible region, any further increase yields an isoprofit line that does not intersect the feasible region; any decrease yields feasible solutions inferior to the optimal solution. Therefore, this isoprofit line defines a unique optimal solution for the surgery planning problem. Figure 5.2 shows several instances of isoprofit lines, including the one at the extreme most point of the feasible region, which is the optimal solution to the linear program, with decision variable choices of $x_1 = 2$, $x_2 = 6$, and an objective function value of 36.

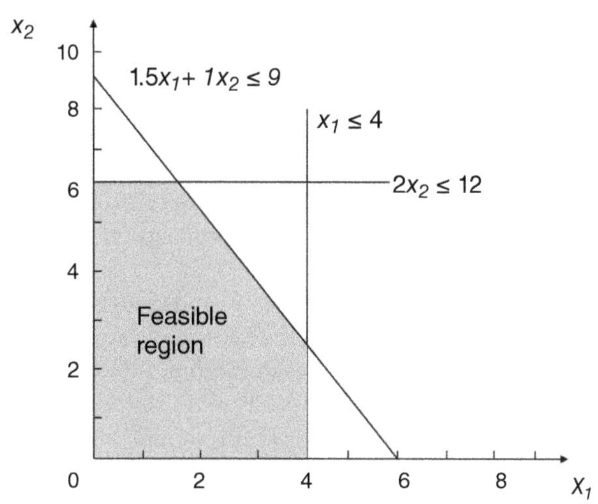

Figure 5.1 A visualization of a two-decision variable linear program for a surgery-planning model.

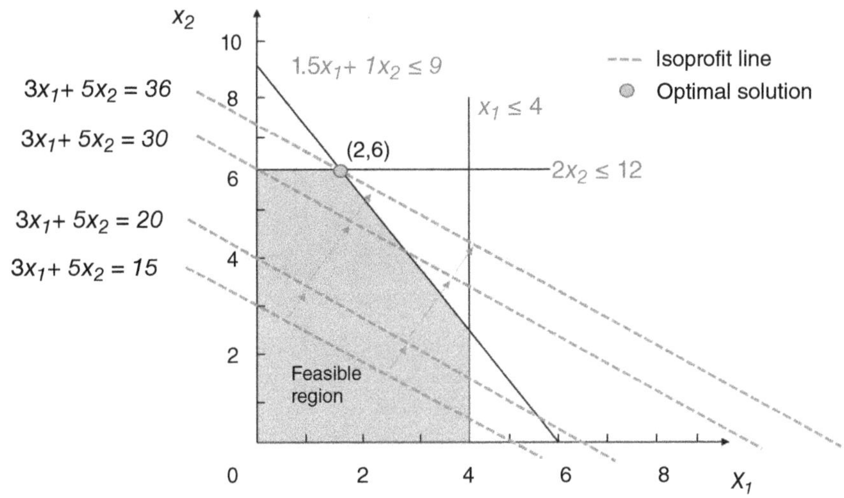

Figure 5.2 A visualization of isoprofit (dashed) lines and the optimal solution, $(x_1 = 2,\ x_2 = 6)$, for which the objective function is 36 for the two-decision variable linear program for surgery planning.

The surgery-planning example offers a conceptual representation of linear programs to foster a basic understanding, build intuition, and explain several important properties of linear programs in general, including the following:

1. *Infeasible linear program:* If the feasible region is empty, no optimal solution exists. This occurs when no point satisfies all of the constraints.
2. *Feasible optimal solution:* If the feasible region is non-empty, the optimal solution, if one exists, is somewhere along that region's boundary. Moreover, multiple optimal solutions are possible in some cases. For example, this would have occurred in Figure 5.2 if the isoprofit line were parallel to Constraint 5.10.
3. *Unbounded linear program:* If the feasible region is unbounded, then the objective value can conceivably increase without any restrictions, thus tending to infinity. This indicates that the linear program is poorly formulated and warrants reconsideration.

The above properties all generalize to linear programs with more than two decision variables. However, advanced algorithms are necessary to solve larger-scale problems. A treatment of these methods goes beyond the scope of this book; instead, the remainder of this chapter focuses on how to formulate models and solve them using optimization software. Many excellent sources cover solution methods for readers who want to delve deeper into the methodology. Some examples are given in the concluding remarks to this chapter.

5.3 The Shortest Path Problem Revisited

Consider the shortest path problem first introduced in Chapter 3 as a DP model, where the goal was to traverse a network from a start node to a finish node in a way that minimizes the total path length of the selected arcs that connect the nodes. To formulate the optimization model, define a set of nodes indexed by $i = 1, \ldots, N$, with the start node at 1 and the end node at N. The weight for the arc from i to j is denoted by w_{ij}, and $w_{ij} = \infty$ is used to indicate if there is no arc between nodes i and j. The weights can be organized into an $N \times N$ matrix as follows:

$$W = \begin{bmatrix} w_{11} & w_{12} & \cdots & w_{1N} \\ w_{21} & w_{22} & \cdots & w_{2N} \\ \vdots & \vdots & & \vdots \\ w_{N1} & w_{N2} & \cdots & w_{NN} \end{bmatrix}.$$

Note that elements on the diagonal, $w_{11} = w_{22} = \cdots = w_{NN} = 0$ because they are the distance from and to the same node in the network. Using the above notation, the shortest path optimization model can be written as follows:

$$\min \quad \sum_{i=1}^{N} \sum_{j=1, j \neq i}^{N} w_{ij} x_{ij}, \tag{5.12}$$

$$\text{subject to:} \quad \sum_{j=1, j \neq 1}^{N} x_{1j} - \sum_{i=1, i \neq 1}^{N} x_{i1} = 1 \tag{5.13}$$

$$\sum_{i=1, i \neq N}^{N} x_{iN} - \sum_{j=1, j \neq N}^{N} x_{Nj} = 1, \tag{5.14}$$

$$\sum_{j=1, j \neq i}^{N} x_{ij} - \sum_{k=1, k \neq i}^{N} x_{ki} = 0, \quad \forall i \in \{2, \dots, N-1\}, \tag{5.15}$$

$$x_{ij} \in \{0, 1\}, \quad \forall i, j \in \{1, \dots, N\}, i \neq j, \tag{5.16}$$

where x_{ij} is a binary decision variable indicating whether arc (i, j) is on the shortest path ($x_{ij} = 1$) or not ($x_{ij} = 0$). The first constraint 5.13, ensures exactly one outgoing arc from the start node 1 and no incoming arcs. The second constraint 5.14, ensures exactly one incoming arc to the finish node N and no outgoing arcs. The third set of constraints 5.15, ensures flow conservation at every node except the source and target, with each node having exactly one incoming arc and one outgoing arc. These constraints are enumerated over all nodes except node 1 and node N, as specified by the range on i in constraint set 5.15.

Notice that the optimization model formulation differs from the DP model formulation of the shortest path problem in Chapter 3. This alternative formulation of the shortest path problem is defined in terms of decision variables and constraints. Drawing parallels with the DP formulation, the objective function equates to Equation 5.12 with the total reward of the DP model. The nodes in the optimization model correspond to the states of the DP model; they represent locations where decisions are made. The decision variables x_{ij}, $\forall (i, j)$ are equivalent to the decisions in the DP model, where $x_{ij} = 1$ or 0 denotes the decision whether or not to traverse the arc from node i to node j, respectively. Finally, the constraints in Equations 5.13–5.16 collectively define the allowable values for the decision variables. The complete formulation, Equations 5.12–5.16, seeks the vector of decision variables that minimizes the objective function in Equation 5.12 while also satisfying all the constraints. Ultimately, both models—the DP model of Chapter 3 and the linear programming model—yield a shortest path.

Python Example 5.2: Shortest path problem

Consider the shortest path problem in Figure 5.3, for which the start node is labeled A and the end node is labeled D. The following Python code solves this problem as an optimization model, specifically a linear programming model. The code utilizes a commercial solver called *Gurobi*, which is freely available for educational and (non-commercial) academic research purposes. Similar code can be generated using other solvers, including *scipy.optimize*, a free, open-source Python package.

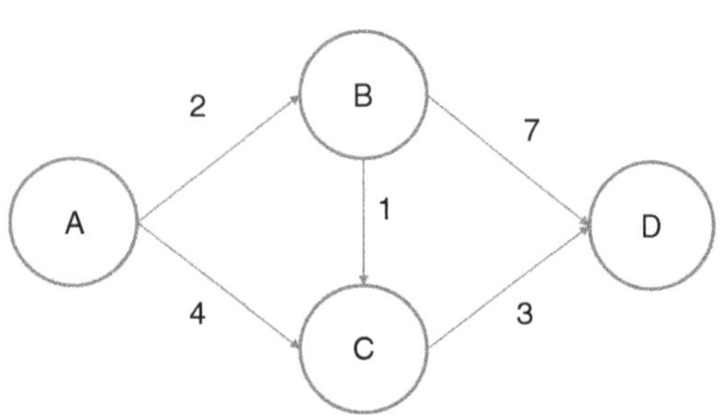

Figure 5.3 A simple shortest path problem with start node A and finish node D.

```
import gurobipy as gp
from gurobipy import GRB

# Input data
vertices = ['A', 'B', 'C', 'D'] # Start vertex A and end vertex D
edges = {('A','B'): 2, ('A','C'): 4, ('B','C'): 1, ('B','D'): 7, ('C','D'): 3}

# Create a Gurobi linear programming model
model = gp.Model("shortest_path")

# Vector of decision variables: Add variables for edges
decisions = model.addVars(edges, vtype=GRB.BINARY, name="edge")

# Create objective function to minimize the total cost of selected edges
model.setObjective(gp.quicksum(decisions[i, j]*edges[i, j]
                               for (i, j) in edges), GRB.MINIMIZE)

# Add constraint for start vertex A (ensure sum of outgoing edges is 1)
model.addConstr(gp.quicksum(decisions['A',j]
                            for j in vertices if ('A',j) in edges) == 1,"start")

# Add constraint for end vertex D (ensure sum of incoming edges is 1)
model.addConstr(gp.quicksum(decisions[i,'D']
                            for i in vertices if (i,'D') in edges) == 1, "end")

# Add flow balance constraints for all vertices except start and end
for vertex in vertices[1:-1]: # Vertices B and C in this case
    model.addConstr(
    gp.quicksum(decisions[j, vertex] for j in vertices if (j, vertex) in edges) ==
    gp.quicksum(decisions[vertex, k] for k in vertices if (vertex, k) in edges),
    f"flow_balance_{vertex}")

# Optimize the model
model.optimize()

# Print results
if model.status == GRB.OPTIMAL:
    print("Total Cost of Shortest Path:", model.objVal)
    for (i, j), var in decisions.items():
        if var.x > 0.999: # Interpret values close to 1 as 1
            print(f"Path includes edge {i} to {j} with weight {edges[(i, j)]}")
```

Executing the above Python code with Gurobi yields the following optimal solution:

```
Optimal solution found (tolerance 1.00e-04)
Best objective 6.000000000000e+00, best bound 6.000000000000e+00, gap 0.0000%
Total Cost of Shortest Path: 6.0
Path includes arc from A to B with weight 2
Path includes arc from B to C with weight 1
Path includes arc from C to D with weight 3
```

The shortest path length is 6.0, and the *gap* of 0% in the output above indicates the solver successfully found an optimal solution. The gap refers to the *optimality gap*, which is an upper bound on the relative error in the obtained solution. In this case, the error is 0 within an error tolerance of 10^{-4} under the Gurobi default parameters. For much larger problems, models can be difficult to solve so accurately, and the optimality gap gives guidance on the worst-case bound. Not that a non-zero optimality gap does not guarantee the solution found is not optimal; it only asserts the worst-case error for the solution.

Example 5.3: *Production lot-sizing*

Consider the production lot-sizing problem, first introduced in Chapter 3, in which the goal is to determine how much of a product to produce in each period to satisfy a customer demand schedule. The decision variables are z_t, whether or not to produce any product at epoch t, x_t, the amount to produce at epoch t, and the amount of inventory, I_t, to hold over from epoch t to $t+1$ for all epochs $t = 0, \dots, T$. The fixed cost of production setup is c_t^S, the variable (per unit) cost of production is c_t^V, and the inventory holding cost is c^H. Finally, constraints determine the feasible decisions that satisfy demand d_t, $t = 0, \dots, T$. Putting this all together leads to the following optimization model:

$$\min \quad \sum_{t=1}^{T} c_t^S z_t + c_t^V x_t + c_t^H I_t, \tag{5.17}$$

$$\text{subject to:} \quad I_{t-1} + x_t - I_t = d_t, \quad t = 1, \dots, T, \tag{5.18}$$

$$I_0 = 0, \tag{5.19}$$

$$x_t \le M z_t, \quad t = 1, \dots, T, \tag{5.20}$$

$$x_t, I_t \ge 0, \quad t = 1, \dots, T, \quad z_t \in \{0, 1\} \quad t = 1, \dots, T. \tag{5.21}$$

Constraints 5.18 require for each period that the inventory at the end of the period plus the production during the period is equal to the period's demand plus the initial inventory carried over to the next period. This is referred to as an *inventory balance* constraint. Constraint 5.19 initializes the inventory at the beginning of the planning horizon. Constraints 5.20 enforce the fixed cost aspect of production, which can only occur if binary decision variable $z_t = 1$, in which case production setup cost c_t^S is incurred; otherwise, if $z_t = 0$, production is forced to 0 and there is no setup cost incurred. Note that parameter M in Constraints 5.20 must be large enough that production x_t is not limited if the decision is made to produce in period t; a valid choice of M is the sum of all demands, since optimal production will certainly not be higher than that. This is an example of a *logical constraint*, which commonly arises when there are binary (e.g., on/off) decisions, such as production setup, and additional decision variables (e.g., production amount x_t) that are affected. Finally, Constraints 5.21 require that production quantities and inventory levels be non-negative.

5.4 Relationship Between Dynamic Programs and Linear Programs

The above examples show that some deterministic DP models considered in Chapter 3 can be formulated as optimization problems. This raises the question: how often is this true? To answer this, consider the optimality equations for the general DP model discussed in Chapter 3:

$$v_t(s_t) = \max_{a_t \in A} \{r_t(s_t, a_t) + \lambda v_{t+1}(s_{t+1})\}, \forall s_t \in S, t = 0, \dots, T.$$

$$v_T(s_T) = r_T(s_T), \forall s_T \in S.$$

A linear programming model can be formulated to solve this problem as follows:

$$\min \sum_{t=0}^{T} \sum_{s=1}^{N} \alpha(s_t, t) v_t(s_t), \tag{5.22}$$

$$\text{subject to:} \quad v_t(s_t) - \lambda \sum_{s_t=1}^{N} v_{t+1}(s_{t+1}) \ge r_t(s_t, a_t), \quad \forall a_t \in A, \forall s_t \in S, t = 0, \dots, T-1, \tag{5.23}$$

$$v_T(s_T) \ge R_T(s_T), \quad \forall s_T \in S. \tag{5.24}$$

In this linear programming model, the optimal value functions from the DP formulation, $v_t(s)$, are the decision variables. This may seem counterintuitive at first because decision variables parallel the decisions prescribed by policies in Chapters 3 and 4. However, in the context of linear programs, Equations 5.22–5.24 provide a means to search for the optimal solution to the optimality equations, which compute the optimal value function. The objective function coefficients, $\alpha(s, t)$, are arbitrary strictly positive parameters. The constraints are defined by linearizing the optimality equations and by recognizing that an equation of the form $f = \max\{e_1, e_2, \dots, e_k\}$ can be expressed as follows:

$$\begin{aligned} \min \quad & f \\ \text{subject to:} \quad & f \geq e_1 \\ & f \geq e_2 \\ & \vdots \\ & f \geq e_k, \end{aligned} \tag{5.25}$$

Minimizing this objective will automatically select the largest of $e_1, e_2, \dots, e_k$ to be the value of f.

The above formulation shows that any deterministic DP model is equivalent to a constrained optimization model. The formulation of MDPs as constrained optimization models is straightforward, requiring only a modification of Constraints 5.22–5.24. For a finite horizon MDP, the constraints are constructed using Equations 4.9–4.10 in the same manner as the deterministic DP formulation. Similarly, for an infinite horizon MDP, Equations 4.29 are used to generate the formulation as follows:

$$\min \sum_{\forall s \in S} \alpha(s) v(s), \tag{5.26}$$

$$\text{subject to:} \quad v(s) - \lambda \sum_{\forall s' \in S} p(s' | s, a) v(s') \geq r(s, a), \quad \forall a \in A, \forall s \in S, \tag{5.27}$$

This linear programming model for the infinite-horizon MDP simplifies considerably compared to the finite-horizon counterpart of Equations 5.22–5.24, because time is absent in a stationary model like this, and there is no boundary condition. Thus, the number of decision variables and constraints is much smaller. Just as in the case of a deterministic DP model formulation, the objective function coefficients, $\alpha(s)$, $\forall s \in S$, must be strictly positive, and the exact choice is theoretically arbitrary (although extremely large or highly variable choices are best avoided due to the potential for them to create numerical instability for solvers like Gurobi).

In the remainder of this chapter, deterministic constrained optimization models are extended to *stochastic programs*, which incorporate uncertainty. This extension is analogous to extending deterministic DP models to MDP models, as discussed in Chapters 3 and 4.

Python Example 5.4: Linear program formulation of an MDP

Consider Example 4.9 of Chapter 4 based on the model pictured in Figure 4.7. This small two-state example yields a linear program with two decision variables, one for each state, and three constraints, two corresponding to the two decisions for s_1 and one corresponding to the single decision for state s_2. The constraints in the formulation are

$$v(s_1) \geq 8 + \lambda(0.2v(s_1) + 0.8v(s_2)), \quad v(s_1) \geq 12 + \lambda v(s_2), \quad v(s_2) \geq 1 + \lambda v(s_2),$$

collecting the terms for $v(s_1)$ and $v(s_2)$ and substituting $\lambda = 0.9$ yields the following linear program:

$$\min v(s_1) + v(s_2), \tag{5.28}$$

$$\text{subject to:} \quad 0.82v(s_1) - 0.72v(s_2) \geq 8, \tag{5.29}$$

$$v(s_1) - 0.9v(s_2) \geq 12, \tag{5.30}$$

$$0.1v(s_2) \geq 1. \tag{5.31}$$

In this example, the objective function coefficients were selected to be $\alpha(s_1) = \alpha(s_2) = 1$. As expected, solving this linear program yields exactly the same solution as policy iteration in Chapter 4: $v(s_1) = 21,\ v(s_2) = 10$. With the optimal value function obtained from solving the linear program, the optimal policy is easily obtained using

$$\pi \in \arg\max_{\pi \in \Pi}\{\vec{r}_\pi + \lambda P_\pi \vec{v}\},$$

which in this case yields

$$\pi(s_1) = \arg\max\left\{8 + 0.9(0.2v(s_1) + 0.8v(s_2)), 12 + 0.9v(s_2)\right\} = \arg\max\{18.98, 21\} = a_{1,2}.$$

for state s_1. Since there was only a single decision in state s_2, no evaluation is necessary.

In summary, linear programs constitute a third approach to solve MDPs, in addition to the value- and policy-iteration algorithms of Chapter 4. Generally, solving the linear programming counterpart to a DP or MDP is less computationally efficient than the other methods. Nevertheless, the linear programming formulation provides insight into the relationships among the different types of models. Moreover, the linear programming formulation is a powerful tool for approximating optimal policies in very large-scale models, as will become clear in Chapter 10.

5.5 Two-stage Stochastic Linear Programs

The most basic stochastic program for sequential decision-making is the *two-stage stochastic linear program* (2-SLP). The fundamental elements of 2-SLP models are the following:

First-Stage Decision Variables: One or more first-stage decisions, denoted by a vector $x \in \Re^n$ of n decision variables. These decisions are made before learning the outcomes of any of the random variables.

Random Variables: One or more random variables, with outcomes $\omega \in \Omega$ where ω indexes a vector of outcomes from a collection of random variables, $z \in \Re^m$, referred to as a *scenario*. Note that this is somewhat analogous to the notion of sample paths discussed in Chapter 4, in the sense that it is a sequence of samples of random variables, but the samples are not states, and no decisions are included in a scenario. In fact, scenarios are independent of the decisions in the standard stochastic programming model formulations discussed here.

Second-Stage Decision Variables: A second-stage recourse decision vector $y(\omega) \in \Re^m$ for all $\omega \in \Omega$. These decisions are made sometime after observing the outcomes of the random variables. The exact time point at which the random outcomes become known to the decision-maker is not specified in the mathematical formulation; however, the time between stages is an important consideration when creating a model, as it ensures that the sequence of events (first-stage decisions, random variable outcomes, second-stage decisions) is well-defined. The sequence of events can be summarized mathematically as follows:

$$\text{choose } \vec{x} \rightarrow \text{observe } \vec{z}(\omega) \rightarrow \text{choose recourse } \vec{y}(\vec{z}(\omega), \vec{x}),$$

where the arguments for the vector of second-stage decision variables, $\vec{y}$, indicate they are made with knowledge of the first-stage decisions and the random outcomes, $\vec{x}$ and $\vec{z}(\omega)$, respectively. Often, for simplicity, the dependency of $\vec{y}$ on $\vec{x}$ and $\vec{z}$ is suppressed by writing $\vec{y}(\omega)$ with the understanding that the second-stage has an implicit dependence on the first-stage decision vector, $\vec{x}$, and the random variables $\vec{z}$.

Figure 5.4 illustrates the decision process using a *scenario tree*, with an initial node defining the vector first-stage decisions, $\vec{x}$, and a leaf node for each outcome of the random variables, indexed by $\omega = 1, \ldots, |\Omega|$. The second-stage variables, $\vec{y}(\omega)$ for all ω, are determined by a second-stage optimization model. Overall, the complete 2-SLP formulation can be written as follows:

$$\begin{aligned}
\min\ \vec{c}\cdot\vec{x} + \quad & \min E_{\vec{z}(\omega)}[\min \vec{q}(\omega)\cdot\vec{y}(\omega)] \\
\text{subject to:} \quad & A\vec{x} = \vec{b} \\
& T(\omega)\vec{x} + W\vec{y}(\omega) = \vec{h}(\vec{z}(\omega)),\ \omega \in \Omega \\
& \vec{x} \geq 0,\ \vec{y}(\omega) \geq 0,\ \omega \in \Omega.
\end{aligned} \tag{5.32}$$

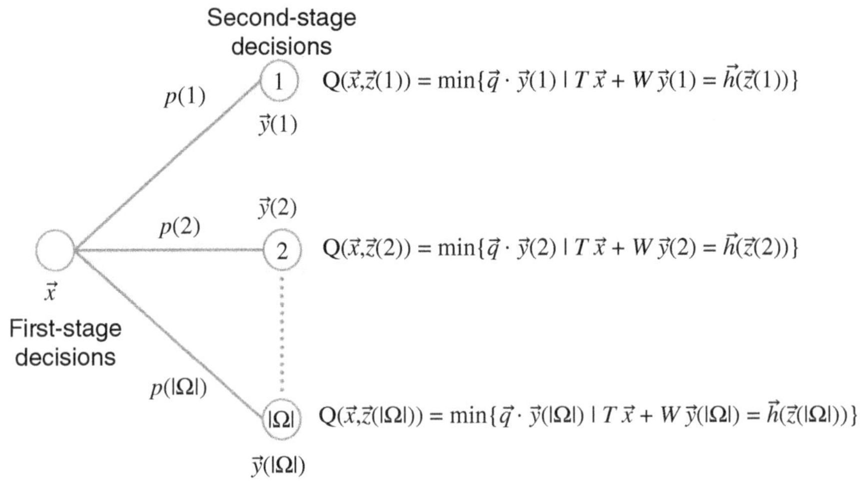

Figure 5.4 The decision process for a two-stage stochastic program from the first-stage decisions in the leftmost node to the second-stage decisions on the right, each with an objective function defined by $Q(\vec{x}, \vec{z}(\omega))$.

The objective function has two terms. The first term is the *first-stage cost*, given by the dot product of the vectors $\vec{c}$ and $\vec{x}$. The second term is the second-stage expected cost, referred to as the *recourse function*, which is determined by a second layer of optimization in which the product's second-stage costs and decision variables are minimized. Matrix $W(\omega)$ defines the constraint coefficients of second-stage variables for scenario ω, often called the *recourse matrix*, and $T(\omega)$ is a matrix of constraint coefficients for the first-stage variables for scenario ω, often called the *technology matrix*. The right-hand side of the inequality $h(z(\omega))$ is a vector of random elements that could be some function of random variables, $z(\omega)$.

To expand on the basic understanding of stochastic programs presented above and to reinforce the formulation and advantages of such models, the sections below present a series of progressively more complex examples, starting with the simplest two-stage problem that can be modeled as a 2-SLP.

5.5.1 The Newsvendor Problem

The *newsvendor problem* aims to find the optimal number of units of a perishable product (e.g., newspapers, which carry no value after the publication date), denoted by a single decision variable x, to purchase to meet random customer demand, with outcomes $z(\omega)$, $\forall\omega \in \Omega$ and probabilities $p(\omega)$, assuming a per-unit cost c^u for having less supply than demand (*underage*) and a per-unit cost c^o for having more supply than demand (*overage*).

Using the above problem definition, the newsvendor model can be written as follows:

$$\min_x\{c^u E_{z(\omega)}[\max\{z(\omega) - x, 0\}] + c^o E_{z(\omega)}[\max\{x - z(\omega), 0\}]\},$$

where the first term is the cost of underage (when $z > x$) and the second is the cost of overage (when $z < x$). This can be reformulated using the above notation as follows:

$$\min\{c^u \sum_{\omega\in\Omega} p(\omega)y^u(\omega) + c^o \sum_{\omega\in\Omega} p(\omega)y^o(\omega)\}$$

$$\text{subject to:} \quad y^u(\omega) - y^o(\omega) = z(\omega) - x, \forall\omega \in \Omega$$

$$y^u(\omega), y^o(\omega) \geq 0, \forall\omega \in \Omega.$$

This is the simplest 2-SLP model and can help illustrate several important points. First, although the model addresses a stochastic problem, it is a linear program. 2-SLP models are often referred to as *two-stage recourse problems*, indicating the presence of second-stage decision variables that provide recourse following the first-stage decision, x, and the outcome of a random variable, z. The newsvendor model is said to have *simple recourse* because when given x and z, the optimal recourse decision is obvious.

Python Example 5.5: Vaccine purchase planning

A pharmacy manager responsible for purchasing COVID-19 vaccine doses for a regional hospital system must place an order in the spring to ensure receipt of the necessary doses for administration in the fall. The demand for vaccinations is uncertain, but historical data suggest the following distribution:

$$z = \begin{cases} 1,500 & \text{with probability } 0.4, \\ 2,200 & \text{with probability } 0.5, \\ 2,800 & \text{with probability } 0.1. \end{cases}$$

The manager estimates the harm per patient of not having enough doses as four times worse than that of too many. Therefore, the manager chooses $c^u = 0.8$ and $c^o = 0.2$, yielding the following 2-SLP:

$$\begin{aligned} & \min\{0.8(0.4y^u(1) + 0.5y^u(2) + 0.1y^u(3)) + \\ & \quad 0.2(0.4y^o(1) + 0.5y^o(2) + 0.1y^o(3))\} \\ \text{subject to:} \quad & y^u(1) - y^o(1) = z(1) - x \\ & y^u(2) - y^o(2) = z(2) - x \\ & y^u(3) - y^o(3) = z(3) - x \\ & y^u(1), y^u(2), y^u(3), y^o(1), y^o(2), y^o(3) \geq 0 \end{aligned}$$

The following Python code solves the "newsvendor" model of this form (here applied to vaccine procurement) by solving the linear program corresponding to the 2-SLP model.

```
import gurobipy as gp
from gurobipy import GRB

# Input data
c_u = 0.8 # cost coefficient for underage, y_u
c_o = 0.2 # cost coefficient for overage, y_o
p = [0.4, 0.5, 0.1] # probability vector for demand
z = [1500, 2200, 2800] # z values for demand for each omega
Omega = range(len(p)) # Index set for all demand outcomes omega

# Create a new Gurobi model
model = gp.Model("Newsvendor Model")

# Decision variables
decisions = [('x',)] + \
            [('y_u', omega) for omega in Omega] + \
            [('y_o', omega) for omega in Omega]
decisions = model.addVars(decisions, vtype=GRB.CONTINUOUS, name="decisions")

# Objective function
obj = c_u*gp.quicksum(p[omega]*decisions[('y_u', omega)] for omega in Omega) + \
    c_o*gp.quicksum(p[omega] * decisions[('y_o', omega)] for omega in Omega)
model.setObjective(obj, GRB.MINIMIZE)
```

```
# Constraints
for omega in Omega:
    model.addConstr(decisions[('y_u', omega)] - decisions[('y_o', omega)] == \
                    z[omega] - decisions[('x',)], f"balance_{omega}")

# Optimize model
model.optimize()

# Print results
if model.status == GRB.OPTIMAL:
    solution = {
        'Objective Value': model.objVal,
        'x': decisions[('x',)].x,
        'y_u': {omega: decisions[('y_u', omega)].x for omega in Omega},
        'y_o': {omega: decisions[('y_o', omega)].x for omega in Omega}
    }

    print("Optimal Objective Value:", solution['Objective Value'])
    print("Optimal x:", solution['x'])
    print("Optimal y_u:", solution['y_u'])
    print("Optimal y_o:", solution['y_o'])

else:
    print("No solution found.")
```

Executing the above Python code with the Gurobi solver yields the following solution:

```
Optimal objective  1.040000000e+02
Optimal Objective Value: 104.00000000000003
Optimal x: 2200.0
Optimal y_u: {0: 0.0, 1: 0.0, 2: 600.0}
Optimal y_o: {0: 700.0, 1: 0.0, 2: 0.0}
```

Thus, the optimal vaccine purchase quantity for this example is $x^* = 2,200$ vaccine doses, which has an expected cost of 104. This minimizes total expected cost and yields three possible outcomes according to the demand probability distribution: $y^o(1) = 700$ and $y^u(\omega) = 0$ for scenario 1, with probability 0.4; $y^o(2) = 0$ and $y^u(2) = 0$ for scenario 2, with probability 0.5; and $y^o(3) = 0$ and $y^u(3) = 600$ for scenario 3, with probability 0.1. Therefore, the most likely outcome is that demand is met with no overage or underage. With probability 0.4, there is overage, and with probability 0.1, there is underage. The overage and underage outcomes are undesirable, but the manager can be reassured that, on average, the decision is optimal given the uncertainty in vaccine demand and the estimated cost penalties and demand scenarios.

5.5.2 Value of the Stochastic Solution

The above example illustrates the application of the newsvendor model to optimize the purchase of perishable inventory, as COVID-19 vaccine doses are typically seasonal (as with flu vaccines). A variation of the newsvendor model replaces the random variable z with its mean. This is known as the *mean value problem* (MV), which is formulated as follows:

$$\text{MV} = \min_x \{c^u \max\{E_z[z] - x, 0\} + c^o \max\{x - E_z[z], 0\}\}.$$

In general, the MV problem is easier to solve than its 2-SLP counterpart because the random variables are replaced by their means, yielding a smaller, more straightforward deterministic model. In this example, the optimal solution is trivial, with $x = E_z[z]$ achieving the minimum objective function value of MV = 0. However, the solution may be suboptimal because the MV model ignores uncertainty. The quality of the MV solution can be measured by evaluating it as a feasible solution to the 2-SLP by computing the expectation of the MV solution (EMV) as follows:

$$\text{EMV} = c^u E_z[\max\{z - E_z[z], 0\}] + c^o E_z[\max\{E_z[z] - z, 0\}].$$

For the vaccine-purchase stochastic program, setting $x = p(1)z(1) + p(2)z(2) + p(3)z(3) = 1,980$ results in EMV = 192, which is substantially higher than the optimal solution to the 2-SLP of 104. Since this is a minimization problem, the MV problem can yield a solution no better than the 2-SLP by definition of optimality.

The MV problem is essential in defining a benchmark for the expected value of solving the more difficult 2-SLP, called the *Value of the Stochastic Solution* (VSS). For the above example, the VSS is calculated as follows:

$$\text{VSS} = \text{EMV} - \text{SP} = 88,$$

where SP denotes the optimal objective function value for the 2-SLP. VSS is more commonly expressed in relative terms as a percentage:

$$\text{Relative VSS\%} = \frac{\text{EMV} - \text{SP}}{\text{SP}} \times 100\% = 213.6\%.$$

Thus, the VSS quantifies the value of optimizing vaccine dose selection in the presence of uncertain demand. Note that VSS = EMV − SP because the model is a minimization problem; thus, by definition, EMV will be no smaller than *SP*. For a maximization problem, VSS = SP − EMV. Given that the optimal solution to the 2-SLP can be no worse than any other solution, including the solution to the MV problem, there is the following guarantee:

$$\text{VSS} \geq 0.$$

5.5.3 Scheduling Arrivals to a Stochastic Server

Some important scheduling problems can be formulated as 2-SLPs. As a specific example, consider appointment scheduling at a stochastic server. This arises in many contexts, such as scheduling patient arrivals at a doctor's office, clients at legal services, or vehicles at a service center. The key decisions are when to schedule customer arrivals, assuming that appointment times cannot be easily changed and that service time is a random variable, thereby creating the possibility of customer waiting and server idling.

Let a_i denote the appointment time assigned by a decision-maker to $i = 1, ..., n$ customers. The first appointment will be assigned to the server's first available time, which is defined as time zero in the model formulation. The appointments can then be written as $a_1 = 0, a_2 = a_1 + r_1, ..., a_n = a_{n-1} + r_{n-1}$ where r_i is the amount of time allotted for customer i's service. As noted above, the service times are random variables denoted by $Z_1, Z_2, \dots, Z_n$. In this model formulation, the first-stage decision variables, $r_i, i = 1, ..., n$, are the primary consideration of the decision-maker. Second-stage decision variables $w_i(\omega), s_i(\omega), i = 1, ..., n, \forall \omega \in \Omega$ are enumerated over all scenarios defining the possible service time outcomes. The second stage variables define the customer waiting and idling using the constraints in Equations 5.40–5.43, which enforce the following piecewise linear equations:

$$w_i(\omega) = \max(0, w_{i-1}(\omega) + z_{i-1}(\omega) - r_i), \quad s_i = \max(0, r_i - w_{i-1}(\omega) - z_{i-1}(\omega)), i = 2, ..., n, \forall \omega \in \Omega. \tag{5.33}$$

Note that by assumption $s_1(\omega) = w_1(\omega) = 0$, $\forall \omega \in \Omega$ because the first customer is assumed to be punctual and finds the server free at time zero. The 2-SLP formulation uses the reformulation approach of Equation 5.25 to express the above equations as linear constraints by splitting each equation into two constraints as follows:

$$w_i(\omega) \geq w_{i-1}(\omega) + z_{i-1}(\omega) - r_i, \quad i = 2, ..., n, \tag{5.34}$$

$$s_i(\omega) \geq r_i - w_{i-1}(\omega) - z_{i-1}(\omega), \quad i = 2, ..., n, \tag{5.35}$$

$$w_i(\omega), s_i(\omega) \geq 0, \quad i = 1, ..., n, \forall \omega \in \Omega. \tag{5.36}$$

This set of linear inequalities can be further simplified by recognizing that the absolute value of the right-hand side (rhs) of the constraints in Equations 5.34–5.35 are the same, i.e., $w_i(\omega) \geq \text{rhs}$ and $s_i(\omega) \geq -\text{rhs}$. Both waiting and idling time must be non-negative, implying either $w_i(\omega) > 0$ or $s_i(\omega) > 0$ but not both, unless rhs is precisely zero. Thus, if rhs > 0, there is waiting time but no idle time; if rhs < 0, there is idle time but no waiting time. As a result, the constraints can be expressed more compactly as $w_i - s_i = \text{rhs}$. Written in full, this is

$$w_i(\omega) - s_i(\omega) = w_{i-1}(\omega) + z_{i-1}(\omega) - r_i, \quad i = 2, \ldots, n, \forall \omega \in \Omega, \tag{5.37}$$

$$w_1(\omega), s_1(\omega) = 0, \forall \omega \in \Omega, \tag{5.38}$$

which corresponds to the constraints in Equations 5.40–5.43. Next, the 2-SLP model can be written in full as follows:

Letting $w_i(\omega)$ and $s_i(\omega)$ denote the waiting time for customer i and the server idle time prior to customer i under scenario ω, the following 2-SLP model formulation selects $r_1, r_2, \ldots, r_{n-1}$ to minimize the expected cost of waiting and idling as follows.

$$\min E_\omega \left[\sum_{i=2}^{n} c_i^w w_i(\omega) + c_i^s s_i(\omega) \right], \tag{5.39}$$

subject to:

$$w_2(\omega) - s_2(\omega) = Z_1(\omega) - r_1, \forall \omega \in \Omega, \tag{5.40}$$

$$w_3(\omega) - w_2(\omega) - s_3(\omega) = Z_2(\omega) - r_2, \forall \omega \in \Omega, \tag{5.41}$$

$$\vdots$$

$$w_n(\omega) - w_{n-1}(\omega) - s_n(\omega) = Z_{n-1}(\omega) - r_{n-1}, \forall \omega \in \Omega, \tag{5.42}$$

$$r_i \geq 0, i = 1, \ldots, n-1, \; w_i(\omega) \geq 0, \; s_i(\omega) \geq 0 \; \forall i = 1, \ldots, n, \forall \omega \in \Omega. \tag{5.43}$$

The following example illustrates how to solve the 2-SLP for customer arrival scheduling using Python and Gurobi.

Python Example 5.6: Value of the stochastic solution for scheduling deliveries

Consider the scheduling of raw material deliveries (e.g., transmissions) to a loading dock at an automobile manufacturer. Appointment times for truck arrivals must be scheduled in advance; however, the time required to unload each vehicle is uncertain. Thus, the manager who sets the schedule has to assign arrival times without knowing exactly how long it will take to service each truck. The objective is to minimize the costs of truck waiting and loading dock idle time. Unfortunately, these two goals are conflicting because the unload time is a random variable. The manager setting the schedule estimates that the per-unit time cost of idling the loading dock is four times that of a single truck waiting. Thus, the costs can be set to $c_w = 0.7$ and $c_s = 0.3$. Based on historical data, each truck appears to be distributed according to the following discrete uniform distribution:

$$z = \begin{cases} 1.5 & w.p.\ 0.25 \\ 2.0 & w.p.\ 0.25 \\ 2.5 & w.p.\ 0.25 \\ 3.0 & w.p.\ 0.25 \end{cases}$$

The following Python code solves the model to determine the optimal time allocation for each truck. This can be converted to a schedule by having the first truck arrive at the first available time at the dock and subsequently spacing truck arrivals by $r_1, r_2, \ldots, r_{n-1}$.

```
import gurobipy as gp
from gurobipy import GRB
import numpy as np
import itertools
```

```
# Input data
n=5 #of customers arriving
c_w = 0.7 # cost coefficient for customer waiting
c_s = 0.3 # cost coefficient for server idling

p=[0.25, 0.25, 0.25, 0.25] #probability of service times
z=[1.5, 2.0, 2.5, 3.0] #service times

# Construct all len(z)^n scenarios by generating all combinations of indices
comb_indices = list(itertools.product(range(len(z)), repeat=n))

# Initialize an array to hold the combinations and their probabilities
# The array will have len(comb_indices) rows and len(z) + 1 columns (+1 for probability)
scenarios = np.zeros((len(comb_indices), n + 1))

# Populate the array with scenarios
for i, comb in enumerate(comb_indices):
    # For each combination, calculate the product of the probabilities
    prob = np.prod([p[idx] for idx in comb])

    # Set the elements and their probability in the result array
    scenarios[i, :-1] = [z[idx] for idx in comb] # Set elements from z based on indices
    scenarios[i, -1] = prob # Set the probability

Omega = np.zeros((len(scenarios), n)) # 2D array of all possible scenarios of length n

# Create a new Gurobi model
model = gp.Model("Appointment Scheduling Model")

# Second stage decision variables for customer waiting (w) and server idling (s)
decisions = [('y_w', i, omega) for i in range(n) for omega in range(len(Omega))] + \
            [('y_s', i, omega) for i in range(n) for omega in range(len(Omega))] + \
            [('r', i ) for i in range(n - 1)]
decisions = model.addVars(decisions, vtype=GRB.CONTINUOUS, name="decisions")

# Set the objective function
obj = gp.quicksum(scenarios[omega, n] * c_w * decisions['y_w', i, omega] + \
                  scenarios[omega,n] * c_s*decisions['y_s', i, omega] for i \
                      in range(n) for omega in range(len(Omega)))
model.setObjective(obj, GRB.MINIMIZE) # Set to minimize objective

# Constraints
for omega in range(len(Omega)):
    model.addConstr(decisions['y_w', 1, omega] - decisions['y_s', 1, omega] == \
                    scenarios[omega, 0] - decisions['r', 0], f"balance_{0}_{omega}")
    for i in range(1, n - 1):
        model.addConstr(- decisions['y_w', i, omega] + \
                        decisions['y_w', i + 1, omega] - decisions['y_s', i + 1, omega] \
                    == scenarios[omega, i] - decisions['r', i], f"balance_{i}_{omega}")

# Solve the 2-SLP appointment scheduling model
model.optimize()
```

```
# Print results
if model.status == GRB.OPTIMAL:
    solution = {
            'Objective Value': model.objVal,
            'r': [decisions[('r', i)].X for i in range(n - 1)],
        }
    print("Optimal Objective Value:", solution['Objective Value'])
    print("Optimal r:", solution['r'])
else:
    print("No solution found.")
```

Executing the above Python code with Gurobi yields the following solution for the truck arrivals (as well as additional information about the algorithm performance, which is suppressed here for brevity):

```
Solved in 1092 iterations and 0.05 seconds (0.04 work units)
Optimal objective  8.437500000e-01
Optimal Objective Value: 0.8437499999999658
Optimal r: [2.5, 3.0, 3.0, 2.5]
```

Thus, the scheduled arrival times for the trucks are $a_1 = 0$ (by assumption), $a_2 = 2.5$, $a_3 = 5.5$, $a_4 = 8.5$, $a_5 = 11.0$. Moreover, the expected value of the objective function for this optimal choice is 0.844. Similar to the vaccine purchasing example above, this problem allows for many possible decisions (e.g., arrival times) that introduce varying degrees of idling and waiting; however, the manager can be confident that, on average, this is the best solution (assuming they trust the data).

Example 5.7: *Multi-stage financial planning*

Suppose a parent wants to save for their child's college education with the plan to use the funds 15 years from now. They currently have \$$b$ to invest in a to-be-determined allocation between two mutual funds: one comprising a diversified portfolio of stocks ($i = 1$) and another comprising bonds ($i = 2$). They plan to rebalance their portfolio every 5 years; thus, the model has three stages corresponding to years 0 (Stage 1), 5 (Stage 2), and 10 (Stage 3). Let s_t index the two possible market outcomes—low yield (L) and high yield (H)—at each stage, for each fund type (stocks or bonds), resulting in 8 possible scenarios comprising all combinations of low and high yield at each of the three decision stages, denoted as $\Omega = \{\text{LLL, LLH, LHL, ..., HHH}\}$. At stage 1, there are two decision variables, denoted by $x(1)$ and $x(2)$, where the argument indexes the investment type. At stage 2, there are 4 decision variables for the two investments and two market outcomes (L or H) observed up to that stage. The stage 2 variables are denoted as $x(1, \text{L})$, $x(1, \text{H})$, $x(2, \text{L})$, and $x(2, \text{H})$. The number of decision variables at each subsequent stage doubles because the number of possible market histories doubles. Assume each scenario is equally likely, i.e., $p(s_1, s_2, s_3) = 0.125, \forall s_1, s_2, s_3$. Let $z(i, t, s_t)$ denote the return for investment i, at stage t, given market condition s_t. The stage t decision variables are denoted by $x(i, s_1, ..., s_{t-1})$, which is the amount of investment i at stage t given market history $\{s_1, ..., s_{t-1}\}$. The final stage decision variables (stage 4) are denoted $w(s_1, s_2, s_3)$ and $y(s_1, s_2, s_3)$, which are the *shortage* and *surplus* relative to the goal at the end of the 15-year horizon, respectively. Assume that exceeding the target G results in an income of G plus $q\%$ of the surplus, denoted by $y(s_1, s_2, s_3)$. Conversely, not meeting the goal requires borrowing to cover the shortage at a relative cost of $r\%$, $w(s_1, s_2, s_3)$. Using this notation, the model formulation is as follows:

$$\max \sum_{s_3} \sum_{s_2} \sum_{s_1} p(s_1, s_2, s_3)(-rw(s_1, s_2, s_3) + qy(s_1, s_2, s_3)), \tag{5.44}$$

$$\text{subject to: } \sum_{i=1}^{2} x(i) = b, \tag{5.45}$$

$$\sum_{i=1}^{2} x(i, s_1) - \sum_{i=1}^{2} z(i, s_1)x(i) = 0, \ \forall s_1, \tag{5.46}$$

$$\sum_{i=1}^{2} x(i, s_1, s_2) - \sum_{i=1}^{2} z(i, s_1, s_2)x(i, s_1) = 0 \ \forall s_1, s_2, \tag{5.47}$$

$$\sum_{i=1}^{2} z(i, s_1, s_2, s_3)x(i, s_1, s_2) - y(s_1, s_2, s_3) + w(s_1, s_2, s_3) = G \ \forall s_1, s_2, s_3, \tag{5.48}$$

$$x(i, s_1, \ldots, s_{t-1}) \geq 0, y(s_1, s_2, s_3) \geq 0, w(s_1, s_2, s_3) \geq 0, s_i \in \{\text{L}, \text{H}\}, t = 1, 2, 3. \tag{5.49}$$

Constraint 5.45 restricts the investments to a maximum of b in the first stage. Constraints 5.46 require that the total investments in stage 2 equal the total return from stage 1 investments. Similarly, constraints 5.47 require the total investments in stage 3 to equal the total return from stage 2 investments. Constraints 5.48 assess the surplus or shortage of the investment return at stage 3 relative to the target G.

Python Example 5.8: Multi-stage financial planning

Consider the following specific example based on the formulation in Equations 5.44–5.49. Assume an initial investment of $b = \$15,000$, a goal of $G = \$25,000$ at Year 15, a surplus rate of $q = 1.12$, and a shortage rate of $r = 1.15$. Further, assume that uncertainty in the markets can be approximated as two possible scenarios following each stage:

Scenario 1: Stocks return 1.25 times the initial investment, and bonds return 1.14 times the initial investment ("High [H]" scenario),

Scenario 2: Stocks return 1.06 on the initial investment, and bonds return 1.12 on the initial investment ("Low [L]" scenario).

The following Python code solves the resulting three-stage stochastic program.

```
import gurobipy as gp
from gurobipy import GRB

# Input data
b = 15000 # Initial investment
G = 25000 # Target goal
q = 1.12 # Surplus rate
r = 1.15 # Shortage rate
prob = 0.125 # Probability of each scenario
scenarios = ['L', 'H']
z = {('H',1): 1.25, ('L',1): 1.06, ('H',2): 1.14, ('L',2): 1.12} # Returns

# Create a new Gurobi model
model = gp.Model("Financial Planning")

# Decision variables
decisions = [('x', i, 1) for i in [1, 2]] # Stage 1, x: stocks (i=1),bonds (i=2)
decisions += [('x',i,2,s1) for s1 in scenarios for i in [1,2]] # Stage 2 variables
decisions += [('x', i, 3, s1, s2) for s1 in scenarios for s2 \
              in scenarios for i in [1, 2]] # Stage 3 variables
```

```
decisions += [(var, s1, s2, s3) # Stage 4,w: shortage, y: surplus
            for s1 in scenarios
            for s2 in scenarios
            for s3 in scenarios
            for var in ['w', 'y']]
decisions = model.addVars(decisions, name="decisions")

# Objective function
obj = gp.quicksum(prob * (- r * decisions['w', s1, s2, s3] + q * \
                          decisions['y', s1, s2, s3])
                    for s1 in scenarios for s2 in scenarios for s3 in scenarios)
model.setObjective(obj, GRB.MAXIMIZE)

# Constraints
model.addConstr(gp.quicksum( decisions['x', i, 1] for i in \
                      [1, 2]) == b,"Initial_Investment") # Investment constraint

for s1 in scenarios:  # Portfolio rebalancing constraints for stage 2
    model.addConstr(gp.quicksum(- z[(s1, i)] *  decisions['x', i, 1] +\
                                decisions['x', i, 2, s1]
                    for i in [1, 2]) == 0, f"Rebalance_2_{s1}")

for s1 in scenarios: # Portfolio rebalancing constraints for stage 3
    for s2 in scenarios:
        model.addConstr(gp.quicksum(- z[(s2, i)] *  decisions['x', i, 2, s1] +\
                                    decisions['x', i, 3, s1, s2]
                        for i in [1, 2]) == 0, f"Rebalance_3_{s1}{s2}")
# Final stage constraints
for s1 in scenarios:
   for s2 in scenarios:
      for s3 in scenarios:
        model.addConstr(gp.quicksum(z[(s3, i)]*decisions['x',i,3,s1,s2] \
                for i in [1, 2]) - decisions['y', s1, s2, s3] + \
                        decisions['w', s1, s2, s3] == G, f"Goal_{s1}{s2}{s3}")

model.optimize()

# Print results
if model.status == GRB.OPTIMAL:
    for v in model.getVars():
        if v.X > 1e-6:
            name = v.VarName.replace("decisions[", "").rstrip("]")
            parts = [p.strip().strip("'") for p in name.split(",")]
            print(f"{parts[0]}({','.join(parts[1:])}): {v.X}")
else:
    print("Optimal solution not found.")
```

Running the Python code yields the following optimal solution to this multi-stage stochastic linear program. Note that decision variables with a value of zero are suppressed for brevity.

```
Optimal objective -2.187332687e+03
x(1,1): 15000.0
x(1,2,L): 15900.0
x(1,2,H): 18750.0
x(1,3,L,L): 16854.0
x(1,3,L,H): 19875.0
x(1,3,H,L): 19875.0
x(1,3,H,H): 23437.5
w(L,L,L): 7134.76
w(L,L,H): 3932.5
w(L,H,L): 3932.5
w(L,H,H): 156.25
w(H,L,L): 3932.5
w(H,L,H): 156.25
w(H,H,L): 156.25
y(H,H,H): 4296.875
```

Based on the solution, several observations can be made. First, the optimal investment strategy is expected to yield an average shortfall of \$2,187. However, most people exhibit some risk aversion about personal finances and may want to know the worst possible outcome, which corresponds to the LLL scenario, which has a shortfall of \$7, 135, which would occur with probability 0.125, resulting in the decision-maker needing to borrow funds to reach the goal. Third, the surplus at the end of 15 years could be as high as \$4, 296, corresponding to the HHH scenario, again with probability 0.125. In all but this one scenario (HHH), the decision-maker will have a shortfall and therefore may wish to try to increase the initial investment, b, to a level sufficient to cover the shortfall under some or all scenarios.

The above example also illustrates a fundamental property of multi-stage decision-making models. The initial investment choice in the first stage is the most crucial one. The decision-maker needs it to make the *here-and-now* decision about what to do at the start of the decision horizon. In this case, the model recommends investing the entire initial amount in investment 2, thus selecting bonds over stocks. The recommendations for future stages, 5 and 10 years into the future, are not actionable at the start of the planning horizon. They represent the intended recourse in future stages based on the current available information and the first-stage decision. However, these future recourse decisions are important components of the model, providing some representation of how today's decisions influence the future. In reality, the two-scenario model is a substantial simplification of reality, and the decision-maker should expect to rerun the model with updated data in 5 years, which will almost certainly not correspond exactly to one of the two future scenarios. Thus, such models are run repeatedly to update the here-and-now decisions over the decision horizon.

5.6 Expected Value of Perfect Information (EVPI)

Recall that EVPI, encountered in Chapter 2, measures how much value could be gained from predicting future outcomes. In the context of stochastic programs, EVPI is the difference between the expected value of solving each scenario subproblem under the relaxed condition that the first-stage decisions, x, can take on different values for each subproblem. This has been referred to as the *Wait and See* (WS) solution because the decision-maker can wait to observe the random variable outcomes before committing to the choice of $\vec{x}$. To represent this, separate $\vec{x}$ into multiple vectors, one for each complete scenario, $x(\omega)$, $\forall \omega \in \Omega$. For each scenario, define the scenario model as $\mathrm{SM}(\omega)$ as follows:

$$\mathrm{SM}(\omega) = \min_{\vec{x}}\{\vec{c}\cdot\vec{x} + \min_{\vec{y}(\omega)}\{\vec{q}(\omega)\cdot\vec{y}(\omega) | W\vec{y}(\omega) = \vec{h}(\omega) - T(\omega)\vec{x}, \vec{y}(\omega) \geq 0\} | A\vec{x} = \vec{b}, \vec{x} \geq 0\}.$$

With this the WS solution can be defined as $E_{\omega}[SM(\omega)]$ where the expectation is over all scenarios. The definition of EVPI, assuming a maximization problem, follows:

$$\mathrm{EVPI} = \mathrm{WS} - \mathrm{SP}. \tag{5.50}$$

Similar to VSS, EVPI is always nonnegative since the availability of additional information, i.e., the outcomes of the random variables, cannot lead to a worse solution. Also, similar to VSS, the definition of EVPI depends on whether the stochastic program is a minimization or a maximization problem (for a min problem, EVPI = SP − WS).

Example 5.9: *Management of energy systems*

Consider a set of power-generating units $i \in I$ and decision epochs $t = 0, \ldots, T$. The *stochastic unit commitment problem* (SUCP) aims to minimize the expected total operating cost over the planning horizon, considering a finite set of scenarios Ω, each with probability $p(\omega)$, indexed by $\omega \in \Omega$. Each scenario represents a possible realization of consumer demand. Operators must create a plan well in advance, which is complicated by demand uncertainty in each period. If the plan results in a power shortage, the excess demand is met by using more expensive short-term thermal power units. The objective function and supporting notation for decision variables and model parameters are as follows:

$$\min_{\vec{u}} \sum_{t=0}^{T} \sum_{i \in I} c_{it}^{\text{start}} u_{it} + \min_{\vec{x},\vec{e}} \sum_{\omega \in \Omega} p(\omega) \left(\sum_{t=0}^{T} \sum_{i \in I} \left(c_{it}^{\text{var}} x_{it}(\omega) + c_t^{\text{short}} e_t(\omega) \right) \right), \tag{5.51}$$

where the decision variables in the model are defined as follows:

- u_{it} is the first-stage binary decision variable that indicates if unit i is operational during period t ($u_{it} = 1$) or not ($u_{it} = 0$)
- $x_{it}(\omega)$ is the second-stage continuous decision variable for the amount of power generated by unit i in period t under scenario ω
- $e_t(\omega)$ is the second-stage decision variable for the short-term power generation needed to meet excess demand during period t under scenario ω

and the model parameters are:

- c_{it}^{start} is the fixed start-up cost of unit i in period t
- c_{it}^{var} is the variable generation cost for unit i in period t
- c_t^{short} is the short-term variable generation cost to meet excess demand in period t
- M_{it} is the maximum possible power generation for unit i at period t
- $d_t(\omega)$ is the demand in period t for scenario ω

Lastly, the SUCP model has the following power-balance constraints:

$$\sum_{i \in I} x_{it}(\omega) + e_t(\omega) = d_t(\omega), \quad t = 0, \ldots, T, \forall \omega \in \Omega, \tag{5.52}$$

$$x_{it}(\omega) \leq u_{it} M_{it}, \forall i \in I, t = 0, \ldots, T, \forall \omega \in \Omega, \tag{5.53}$$

$$u_{it} \in \{0, 1\}, \forall i \in I, t = 0, \ldots, T, \ x_{it}(\omega), e_t(\omega) \geq 0, \forall i \in I, t = 0, \ldots, T, \forall \omega \in \Omega. \tag{5.54}$$

Overall, the decision process represented by the 2-SLP model is one in which the decision-maker makes first-stage (binary) decisions u_{it} for all units i and epochs t, determining each unit's availability in each period. These first-stage decisions are made well in advance—operators must plan for personnel and operational resources—before knowing which demand scenario will be realized. In the second stage, the decision-maker determines how much to produce from each unit, x_{it}, for all units i and epochs t, subject to constraints on total production in each period. Any shortfall in supply must be addressed with costly short-term generation to meet any excess demand, denoted by e_t for all epochs t. The constraints in Equation 5.53 are another example of a logical constraint, in this case representing an on/off decision for each unit at each epoch. The parameters M_{it} need to be sufficiently large to not constrain unit i energy generation in epoch t if $u_{it} = 1$. This could be achieved by choosing $M_{it} = \max_{\omega}\{d_{it}(\omega)\}$.

Python Example 5.10: Expected value of perfect information for the stochastic unit commitment problem

Consider the following Python code that solves the SUCP formulation above and provides a small example with 2 units, 3 decision epochs, 2 scenarios (demand levels: high or low), and specific choices of other model parameters. Assume the costs are in units of $1,000.

```
import gurobipy as gp
from gurobipy import GRB

# Input data
I = [1, 2] # Set of power-generating units
T = [1, 2, 3] # Set of time periods
Omega = ['High', 'Low'] # Scenarios
p = {'High': 0.6, 'Low': 0.4} # Scenario probabilities
c_start = {(1, 1): 10, (1, 2): 10, (1, 3): 10,
           (2, 1): 12, (2, 2): 12, (2, 3): 12} # Start-up costs
c_var = {(1, 1): 20, (1, 2): 20, (1, 3): 20,
         (2, 1): 22, (2, 2): 22, (2, 3): 22} # Variable generation costs
c_short = {1: 50, 2: 50, 3: 50} # Short-term generation costs
M = {1: 100, 2: 80} # Maximum generation capacity
d = {(1, 'High'): 150, (2, 'High'): 100, (3, 'High'): 200,
     (1, 'Low'): 100, (2, 'Low'): 80, (3, 'Low'): 120} # Demand scenarios

# Create a new Gurobi model
model = gp.Model("SUCP")

# Decision variables
decisions = {}
decisions.update(model.addVars((('u', i, t) for i in I for t in T), \
                               vtype=GRB.BINARY, name="u"))
decisions.update(model.addVars((('x', i, t, omega) for i in I for t in T for \
                         omega in Omega), vtype=GRB.CONTINUOUS, name="x",lb=0))
decisions.update(model.addVars((('e', t, omega) for t in T for omega in Omega),\
                               vtype=GRB.CONTINUOUS, name="e", lb=0))

# Objective function
obj = (gp.quicksum(c_start[i, t]*decisions['u',i,t] for i in I for t in T) +\
      gp.quicksum(p[omega]*(gp.quicksum(c_var[i,t]*decisions['x',i,t,omega] for\
                      i in I for t in T) + gp.quicksum(c_short[t] * \
                      decisions['e',t,omega] for t in T)) for omega in Omega))
model.setObjective(obj, GRB.MINIMIZE)

# Power balance constraints
model.addConstrs((gp.quicksum(decisions['x', i, t, omega] for i in I) + \
                  decisions['e', t, omega] == d[t, omega]
                  for t in T for omega in Omega), "PowerBalance")

# Generation limits
model.addConstrs((decisions['x', i, t, omega] <= decisions['u', i, t] * \
             M[i] for i in I for t in T for omega in Omega), "GenerationLimits")

# Solve model
model.optimize()

# Print results
if model.status == GRB.OPTIMAL:
    for i in I:
        for t in T:
```

```
            if decisions['u', i, t].X > 0.5: # Operational units
                print(f"Unit {i} is operational at time {t}.")
                for omega in Omega:
                    print(f" Generation under scenario {omega}: \
                          {decisions['x', i, t, omega].X:.2f}")
    for t in T:
        for omega in Omega:
            if decisions['e', t, omega].X > 0.01: # Short-term generation
                print(f"Short-term generation at time {t} under scenario \
                      {omega}: {decisions['e', t, omega].X:.2f}")
```

Running the Python code yields the following optimal solution to this 2-SLP:

```
Optimal solution found (tolerance 1.00e-04)
Best objective 8.386000000000e+03, best bound 8.386000000000e+03, gap 0.0000%
Unit 1 is operational at time 1.
  Generation under scenario High: 100.00
  Generation under scenario Low: 100.00
Unit 1 is operational at time 2.
  Generation under scenario High: 100.00
  Generation under scenario Low: 80.00
Unit 1 is operational at time 3.
  Generation under scenario High: 100.00
  Generation under scenario Low: 100.00
Unit 2 is operational at time 1.
  Generation under scenario High: 50.00
  Generation under scenario Low: 0.00
Unit 2 is operational at time 3.
  Generation under scenario High: 80.00
  Generation under scenario Low: 20.00
Short-term generation at time 3 under scenario High: 20.00
```

According to the solution, unit 1 operates throughout all periods, with power generation varying with the demand scenario. Unit 2 is operational during periods 1 and 3. Finally, more expensive short-term generation is necessary under the "high" scenario during period 3.

To compute the EVPI, use Equation 5.50 and the above solution, $SP = \$8,386$. For WS, solve the SUCP for the High (1) and Low (2) scenario, which yields two solutions:

$$\text{SM}(1) = \$9,814, \tag{5.55}$$

$$\text{SM}(2) = \$6,082. \tag{5.56}$$

Thus, given this is a minimization problem, $\text{EVPI} = \$8,386 - 0.6(\$9,814) - 0.4(\$6,082) = \$64.8$, which quantifies the expected value of having perfect information about which scenario will occur. In this case, the relative benefit of predicting demand in advance is very low.

Naturally, the above formulation of the SUCP is a simplified version of what is found in practice. A more realistic model would be an M-SLP that makes decisions about which units to operate—both startup and shutdown—with appropriate lead time for implementing these major decisions. More realistic models would also include minimum up- and downtime for power-generating units, maintenance schedules, and ramp constraints on how much production can change from one period to the next. Nevertheless, the above model captures the essence of the SUCP.

5.7 Concluding Remarks

This chapter summarizes approaches for formulating sequential decision-making problems of various types, including deterministic linear and mixed-integer programs, two-stage stochastic programs, and multi-stage stochastic programs. A much more detailed treatment of this content can be found in the references below, which provide thorough coverage of theory and methodology for solving problems of this type. For a concrete introduction to linear and integer programming, Lee [2] is a free, open-source book suited for advanced undergraduate or introductory graduate-level courses. Additional resources, such as Wolsey [3] and Wolsey and Nemhauser [4], provide in-depth coverage of integer programming.

Stochastic programs can grow very large when uncertainty is introduced through scenarios. For large models with many random variables, it may be necessary to introduce a very large number of scenarios to achieve a good approximation. Each scenario results in many new decision variables and constraints, making the problem harder to solve. This is an example of the *curse of dimensionality,* which refers to the exponential growth in model size as the number of dimensions (random variables in this case) grows. Many approaches have been developed to address this challenge. One of the most well-known approaches for 2-SLPs is the *L-shaped* method, which exploits the fact that stochastic programming models decompose into scenario subproblems when the first-stage decisions, x, are fixed. Since the subproblems are much smaller than the complete model, they can be solved quickly and provide information about the performance of the first-stage decisions, allowing them to be improved and the subproblems to be solved again, and so on. Iterations between the first and second stages continue until the stopping criterion based on a user-specified error tolerance is met. A related approach for M-SLPs, *nested decomposition*, applies a similar strategy in an iterative framework. A thorough introduction to these and many other methods for solving 2-SLPs and M-SLPs can be found in Birge and Louveaux [5] and other sources, including those listed below [6–9].

This chapter used Python 3.11.5 [10] and Gurobi Optimizer 10.0.3 to solve example problems [11]. Many other commercial and open-source solvers can solve large-scale model instances arising in practice. An example of a free, open-source solver is COIN-OR [12], which includes implementations of many well-known algorithms for solving optimization problems, as well as several advanced, specialized methods suitable for solving large model instances.

5.8 Practice Exercises

1. Solve the investment selection problem of Example 5.1 for the case in which the decision variables, a_i, $i = 1, \dots, 10$ are constrained to be binary, $\{0,1\}$. How does it compare to the solution to the continuous version of this problem?

2. An industrial engineering student is planning to register for five courses next semester and is choosing among the following courses: Linear Algebra, Manufacturing, Human Factors, Computer Programming, Business Management, Economic Analysis, Stochastic Models, Deterministic Models, and Scheduling. If the student registers for Manufacturing, Human Factors, or Computer Programming, they must also register for the corresponding lab (a lab counts as a full course). She must register for at least one modeling course (Stochastic Models or Deterministic Models) and one management course (Business Management or Economic Analysis). Linear Algebra and Stochastic Models are both challenging, and she prefers not to take them concurrently. The Manufacturing lab is offered during the same time as Deterministic Models, and the Computer Programming lab is offered at the same time as Economic Analysis. She ranked the courses according to the preference score below.

Course	Preference score
Linear Algebra	9
Manufacturing	4
Manufacturing Lab	6
Human Factors	7
Human Factors Lab	5
Computer Programming	8
Computer Programming Lab	6
Business Management	7
Economic Analysis	9
Stochastic Models	6
Deterministic Models	10
Scheduling	7

Formulate and solve an integer program to find a feasible schedule that maximizes the student's preferences for next semester.

3. Build your intuition about the solution generated by the newsvendor model for the vaccine purchasing problem of Example 5.5 by doing two sets of experiments: (1) vary the ratio of the objective function coefficients, c^u, and c^o; (2) vary the probability distribution for the random variable, z. For both experiments, use figures to illustrate the effects of the changes. Summarize your findings.

4. A firm producing a single product has three manufacturing plants and four customers. The three plants will produce 3,000, 5,000, and 5,000 units during the next period. The firm has committed to sell 4,000 units to customer 1, 3,000 units to customer 2, and at least 3,000 units to customer 3. Both customers, 3 and 4, also want to buy as many of the remaining units as possible. The profit of shipping a unit from plant i to customer j is given in Table 5.1.

Table 5.1 Profit per unit from plant to customer.

From	To customer			
	1	2	3	4
Plant 1	65	63	62	64
Plant 2	68	67	65	62
Plant 3	63	60	59	60

Formulate a constrained optimization model to maximize the company's profit.

5. For the multi-stage financial planning problem, use the Python code provided to determine the approximate level of initial investment b needed to achieve the desired goal with a probability of at least 80%.

6. Use the vaccine purchasing model of Example 5.5 to answer the following question: Can the MV problem ever provide an optimal solution to the 2-SLP? If so, for what choice of 2-SLP model parameters is this true?

7. For the multi-stage financial planning problem, Example 5.10, compute the EVPI and VSS. Explain what these values tell the decision-maker about the problem.

8. Formulate and solve the linear program formulation for the infinite-horizon MDP for the battery charging model of Chapter 4 as a linear program.

9. Prove that EVPI and VSS will always be non-negative.

10. Consider the two-stage infinite horizon MDP example of Section 5.4. Create a graph of $v(s_1)$ vs. $v(s_2)$ and include all constraints to illustrate the feasible region. Use an isoprofit line with $\alpha(s_1) = \alpha(s_2)$ to identify the optimal feasible solution. Vary the choices of $\alpha(s_1), \alpha(s_2)$ to show that any choices such that $\alpha(s_1) > 0$ and $\alpha(s_2) > 0$ yield an optimal solution to the linear program.

11. A public health clinic must decide today how many COVID and Flu vaccine doses to procure for the current season. Two managers make the first-stage ordering decisions. They share a budget and cold-storage capacity. In each demand scenario s, doses are administered until inventory is depleted or all demand is met. Unmet demand has an estimated opportunity cost of −\$5 for Flu and −\$7 for COVID per patient. The costs per dose for COVID and Flu are \$25 and \$15, respectively. The cold-storage volume required per dose is 5 cm^3, regardless of the vaccine type. The total budget is \$120,000 and total storage is 25, 000 cm^3. The clinic receives reimbursement of \$40 for each flu vaccine dose administered and \$30 for each COVID vaccine dose administered. Demand is uncertain, with the following scenarios (d_C, d_F) and probabilities $\{p_s\}$ are:

Scenario s	p_s	d_{Cs} **(COVID)**	d_{Fs} **(Flu)**
1 (Mild season)	0.25	2000	3000
2 (COVID surge)	0.35	5000	3500
3 (Flu surge)	0.25	2500	5500
4 (Both surge)	0.15	5500	6000

Formulate and solve the stochastic program for this problem to find the optimal solution. How, if at all, do space and budget limitations affect the ability of the clinic to meet demand? If cold-storage space could be expanded by 20% at a cost of \$19,000, should the clinic allocate this funding from their budget or not?

References

1 Dantzig, G. B. (1951). Maximization of a linear function of variables subject to linear inequalities. In Koopmans, T. C. (Eds.), *Activity Analysis of Production and Allocation, Cowles Commission Monograph* 13 (pp. 339–347). New York: John Wiley & Sons.

2 Lee, J. (2013–2022). *A First Course in Linear Optimization* 4th Version 4.07). Reex Press. Available at: https://github.com/jon77lee/JLee_LinearOptimizationBook/blob/master/JLee.4.07.zip.

3 Wolsey, L. A. (2020). *Integer Programming*. John Wiley & Sons.

4 Wolsey, L. A. & Nemhauser, G. L. (2014). *Integer and Combinatorial Optimization*. John Wiley & Sons.

5 Birge, J. R. & Louveaux, F. (2011). *Introduction to Stochastic Programming*. Springer.

6 Haneveld, W. K. K. & Van der Vlerk, M. H. (2020). *Stochastic Programming*. Springer.

7 Sen, S. & Higle, J. L. (1999). An introductory tutorial on stochastic linear programming models. *Interfaces, 29* (2), 33–61.

8 Prékopa, A. (2013). *Stochastic Programming*. Springer.

9 Kall, P. & Stein, W. W. (1994). *Stochastic Programming*. John Wiley & Sons.

10 Anaconda, Inc. (2023). Python 3.11.5 (Anaconda distribution; main, Sep 11 2023; MSC v.1916 64-bit (AMD64)).

11 Gurobi Optimization, LLC. (2026). *Gurobi Optimizer Reference Manual*. Available at: https://www.gurobi.com. Accessed January 29, 2026.

12 COIN-OR: Computational Infrastructure for Operations Research. (2024). Available at: https://www.coin-or.org/. Accessed January 29, 2025.

6

Monte Carlo Simulation

6.1 Introduction

This chapter covers *Monte Carlo simulation*, a powerful statistical method that addresses important considerations for the sequential decision-making models discussed in previous chapters. It is widely used across various branches of scientific research due to its demonstrated effectiveness in analyzing complex problems. In this chapter, the focus is on applying Monte Carlo simulation for two important purposes in sequential decision-making problems: the *curse of dimensionality* and *sensitivity analysis*.

The *curse of dimensionality* refers to the fact that the size of mathematical models can quickly become prohibitive as the number of dimensions in the state space (MDPs) or scenario space (stochastic programs) grows. So far, this book's treatment of sequential decision-making methods has focused on finding optimal policies using algorithms that can solve models exactly. However, as will become clear, achieving exactitude in many situations is too time-consuming or even impossible due to the model's size. This can occur for various reasons, depending on the model type. For example, in the MDP models of Chapter 4, the number of states may be so large as to make it infeasible to solve the optimality equations. This occurs when the system state is defined by multiple factors, leading to exponential growth in the number of possible states. In the case of the stochastic programming models in Chapter 5, the scenario trees grow exponentially with respect to the number of stages, leading to optimization models that cannot be solved by the best available optimization software.

The second problem, *sensitivity analysis*, arises due to potential inaccuracies in a model's input parameters. Regardless of a mathematical model's size, errors in input parameter estimates can lead to model solutions that provide poor recommendations to decision-makers. This uncertainty necessitates conducting a sensitivity analysis to evaluate the robustness of the model solution and understand the risks associated with input parameter errors. This is one of the important mechanisms for evaluating the risk of using models to support decision-making.

This chapter begins with examples to instill a conceptual understanding of the curse of dimensionality, then delves into Monte Carlo sampling itself and its applications to sequential decision-making, including its strength in providing insightful sensitivity analysis. This introduction sets the stage for future chapters, all of which employ Monte Carlo sampling.

6.2 Curse of Dimensionality

Many problems lead to difficult or impossible-to-solve models due to exponential growth in the number of dimensions governing model size. This occurs in a DP setting when multiple factors must be considered in defining the state. For example, the production planning problems discussed in Chapters 3 and 4 assumed a single product, whereas most companies sell a portfolio of products, sometimes numbering in the thousands. If the products are completely independent, the problem can be decomposed into single-product problems and solved efficiently (e.g., using the Wagner-Whitin algorithm of Chapter 3). Alternatively, the products may be linked for reasons such as the following:

- Some products may be substituted for others. For example, if a particular color is unavailable when purchasing a smartphone, the customer may accept a different in-stock color. When purchasing a laptop, if a customer's desired specifications are unavailable, they may settle for a comparable model (e.g., a slower or faster processor speed, or more or less memory). This is true of many products, including clothing, food, bicycles, automobiles, and sports equipment.
- Companies that manage a product portfolio often have limited storage space for inventory. Thus, the inventory level for other products may influence the decision to produce a certain product, and vice versa.

- When products utilize the same manufacturing line or tools in production, the available tool capacity can limit production throughput. For instance, automobile manufacturers use *paint shops* to paint car panels. Typically, a large batch of a single color is processed consecutively before changing colors due to the changeover cost (e.g., flushing the lines to prepare for a new color).

All three examples above share the common feature of demonstrating dependencies among products. When products are interdependent, sequential decisions cannot be decomposed into single-product problems without potentially sacrificing solution quality. Thus, for example, in a company with n products that uses an MDP model to manage production and inventory control, it must consider all products collectively. If there is a maximum inventory allocation for each product (e.g., 100 units), then the total number of states is 100^n, making the problem unsolvable for all but the smallest examples.

Unfortunately, the curse of dimensionality is not confined to production planning problems. It arises commonly in applications in which multiple dimensions (sometimes called *factors*) must be considered when defining the state space. For example, in a problem involving medical treatment decisions, the state space may need to be defined by age, blood pressure, cholesterol, blood glucose, and liver enzymes. Similarly, in a financial setting, portfolio decisions may need to account for the prices of many assets. Yet another example is transportation systems, where the state determines the locations of multiple vehicles within a geographic network to optimize customer service decisions, such as for on-demand shuttle services.

In stochastic programs, as discussed in Chapter 5, although the definition of a state is not pertinent to formulating or solving the resulting models, similar computational challenges arise because the scenario tree that captures the uncertainty of the random variables can grow exponentially. In the case of production planning for multiple products, for instance, the scenarios reflect the many possible combinations of random demand realizations. Moreover, in multi-stage problems, the total number of scenarios also grows exponentially with time due to the absence of the Markov property inherent to MDPs. Considering the problem of multi-stage financial planning, if the number of possible states of the market were expanded to 10, for example, to more accurately represent possible market conditions, and the 15-year decision horizon were broken into annual stages, the scenario tree would have 10^{15} unique scenarios.

In short, the curse of dimensionality is a significant issue that frequently occurs in real-world situations. As will become clear, Monte Carlo simulation is a valuable tool for addressing this hazard in various decision-making contexts.

6.3 Monte Carlo Sampling

Random sampling is a common approach to estimating the mean of random variables. Consider, for example, flipping a fair coin with an equal probability of landing on either side. Assigning one side the value 1 and the other side the value −1, the mean can be computed to be $0.5 \times 1 + 0.5 \times -1 = 0$ based on probability theory. An alternative approach is to sample the outcomes of the random variable multiple times to estimate the mean, which is appropriate if a decision-maker is concerned that the coin may be biased. Consider Table 6.1, which has the results of an experiment based on averaging randomly sampled coin flips.

Table 6.1 Estimated mean from simulated coin flips with rewards of 1 and −1 for the two different sides.

Number of coin flips	Estimated mean
10	−0.4
30	0.333
100	0.04
1,000	0.05
10,000	−0.0088
1,000,000	−0.00063

If repeated, this set of experiments would yield different results due to sampling variability. However, one of the most fundamental results in statistics is the *Law of Large Numbers*, which guarantees that as the number of samples grows larger, the sample mean converges to the true mean. Furthermore, this convergence is expressed mathematically as follows: the amount of variation is inversely proportional to the square root of the sample size. As expected, as seen in Table 6.1, the estimate of the mean is converging to zero, the true expectation of the random variable in this example.

6.3.1 Pseudo Random Number Generators

The above coin-flip example is only feasible if the experiments are simulated computationally. Doing so requires using a pseudo-random number generator that produces outcomes of a random variable (e.g., a coin flip). The qualifier "pseudo" reflects that it is impossible to generate truly random variables computationally. Rather, pseudo-random number generators employ algorithms that create a deterministic sequence of numbers that appear random. The success of pseudo-random number generators can be evaluated by showing that they pass statistical tests, making them nearly indistinguishable from real-world random outcomes. Random number generators are commonplace in programming languages, interpreters, spreadsheets, and other software systems. The following example illustrates the use of a pseudo-random number generator in Python.

Python Example 6.1: Coin-flip simulation

Consider the above coin-flipping example again. The following Python code uses a standard random number generator to simulate coin flips and compute a statistical estimate of the mean. Moreover, it shows how Python can be used to estimate confidence intervals (CIs) for the mean, as reported in Table 6.2. The example below generates a confidence interval for each experiment to illustrate how natural variation changes with sample size.

Table 6.2 Coin-flip simulation results including 95% CIs.

Number of coin flips	Mean	95% confidence interval
10	0.2	(−0.440, 0.840)
30	−0.4	(−0.734, −0.066)
100	0.02	(−0.177, 0.217)
1000	−0.058	(−0.120, 0.004)
10000	−0.0174	(−0.037, 0.0022)
100000	−0.00204	(−0.00824, 0.00416)
1000000	−2.8e−05	(−0.00199, 0.00193)

```
import numpy as np
import scipy.stats as stats

# Input data
nsamples = 100 # User-defined variable N = # of coin flips
np.random.seed(42) #Fix seed for repeatability

# Simulate coin flips and estimate mean value and 95% Confidence
sample = np.random.choice([1, -1], size=nsamples)
mean = np.mean(sample) # Compute mean of outcomes

std_dev = np.std(sample, ddof=1) # Compute the standard deviation
std_error = std_dev / np.sqrt(nsamples)

CI = stats.norm.interval(0.95,loc=mean,scale=std_error) # 95% confidence interval

# Print results
print(f"N: {nsamples}")
print(f"Mean Estimate: {mean}")
print(f"95% Confidence Interval: {CI}")
```

Executing this code with $N = 100$ samples yields the following:

```
N: 100
Mean Estimate: -0.04
95% Confidence Interval: (-0.23682614198191507, 0.15682614198191505)
```

As observed previously, based on the Law of Large Numbers, the sample mean approaches the true mean of zero as the sample size, N, increases. The Python code also generates 95% confidence intervals for each experiment. These are computed using the fact that estimates of the mean converge to a normal distribution as the number of samples grows larger, a result motivated by the *Central Limit Theorem* (CLT), another important theorem underpinning probability and statistics. As the number of samples grows, estimates approach a normal distribution regardless of the underlying random variable, which aids in constructing CIs.

The 95% CIs are interpreted to mean that if a series of independent experiments were performed with a certain choice of N, say $N = 1,000$, then as the number of experiments grows larger, the proportion of CIs containing the true estimate of the mean converges to 95%. In other words, 95% of the time, the actual mean will be within the upper and lower limits given by the intervals generated from the independent experiments. Different CIs can be computed as needed, e.g., 90%, 99%, or 99.9%. If the confidence level is reduced, the interval decreases in size; conversely, if the confidence level is increased, the interval increases. Confidence levels are therefore important to consider in situations where the interval size (e.g., tolerance levels) may be paramount. A variation of adjusting the CI is to adjust the error margin and sampling number, which are discussed at the end of Section 6.4.

6.4 Using Monte Carlo Simulation to Estimate Expectations

A key role of Monte Carlo simulation in sequential decision-making is to estimate the expected value of complex functions of random variables. In theory, this can be done mathematically using calculus; however, in practice, the resulting integrals for evaluating the expectation cannot be derived exactly. For example, consider the case of a function $g(x) = e^{x^2}$ of a random variable, x, with a probability density function $f(x)$. The expectation can be written as follows:

$$E[g(x)] = \int_a^b e^{x^2} f(x)dx.$$

If x is uniformly distributed, $x \sim U(a, b)$, then the integral has no closed-form expression. However, Monte Carlo simulation can estimate this integral by sampling from $U(a, b)$ to get the following approximation:

$$E[f(x)] \approx \frac{1}{N} \sum_{k=1}^{N} e^{x_k^2}.$$

where $x_1, x_2, \ldots, x_N$ is a sequence of independent samples from the uniform distribution $U(a, b)$. According to the law of large numbers, as N grows larger, the approximation converges to the true expectation. This fact can be exploited to estimate the expected value of complex multivariable functions, including the optimal value function of MDPs and the recourse function of stochastic programs. As such, this can be an effective method for approximating the performance of policies (e.g., MDPs) or feasible solutions to constrained optimization problems (e.g., SPs) in complex sequential decision-making problems involving continuous random variables. In Chapter 8, Monte Carlo simulation will play an important role in *reinforcement learning*, a model-free approach to learning optimal policies for sequential decision-making problems. For now, the application of Monte Carlo sampling is illustrated for estimating multidimensional integrals to compute the expected value of a multidimensional function.

Python Example 6.2: Expected value of a multidimensional function of automobile revenue

Expanding on the above discussion, consider a three-dimensional function for the monthly revenue from new sales of an electric vehicle that comes in three colors (silver, black, and blue). Assume the automotive dealer has access to up to I_k units of each, and demand is D_k with $k = 1, 2, 3$, indexing the three colors, respectively. Assume that customers for silver vehicles ($k = 1$) are willing to purchase black ($k = 2$) at a 10% discount, and customers for black are willing to purchase blue ($k = 3$) at an 8% discount. The revenue can be expressed as follows:

$$\begin{aligned} R(D_1, D_2, D_3) = {} & r \times \min(D_1, I_1) + r \times \min(D_2, I_2) + r \times \min(D_3, I_3) \\ & + 0.9r \times \min(\max(I_2 - D_2, 0), \max(D_1 - I_1)) \\ & + 0.92r \times \min(\max(I_3 - D_3, 0), \max(D_2 - I_2, 0)), \end{aligned}$$

where $D_1 \sim \text{LogNormal}(a, b)$, $D_2 \sim \text{LogNormal}(c, d)$, and $D_3 \sim \text{LogNormal}(e, f)$. In theory, the expected revenue could be computed by evaluating the following three-dimensional integral:

$$E[R(D_1, D_2, D_3)] = \int_0^\infty \int_0^\infty \int_0^\infty R(D_1, D_2, D_3) f_1(D_1) f_2(D_2) f_3(D_3) dD_1 dD_2 dD_3,$$

where $f_1(D_1)$, $f_2(D_2)$, and $f_3(D_3)$ are probability density functions of demand for silver, black, and blue cars, respectively. Multidimensional integrals like this are difficult to evaluate because they lack a closed-form solution. Computational integration methods can be used, such as the *trapezoid rule*, often covered in introductory calculus classes; however, they are computationally intensive because they require estimating a discrete grid, which becomes intractable for all but very low-dimensional problems. This occurs because the number of points in the grid, and thus the computational effort, grows exponentially with respect to the number of dimensions. For instance, for a two-dimensional integral discretized in 10 points in each dimension, the grid size is $10 \times 10 = 100$; generalizing to n dimensions yields 10^n grid points, which quickly becomes computationally infeasible. Monte Carlo simulation, however, is relatively easy to implement, as illustrated in the following Python example. Furthermore, it scales to higher dimensions more easily than numerical integration methods, which quickly break down due to the curse of dimensionality.

```
import numpy as np
import scipy.stats as stats

# Input data
a, b, = 20, 3 # For D1 (silver) # Mean and standard deviation for demand
c, d = 15, 5 # For D2 (black)
e, f = 8, 4 # For D3 (blue)
r = 25000 # Base price for vehicles
I1, I2, I3 = 50, 40, 30 # Available inventory for silver, black, blue vehicles

# Monte Carlo sampling
nsamples = 100000 # Number of samples
np.random.seed(42) #Fix seed for repeatability

# Define the revenue function
def revenue(D1, D2, D3):
    revenue_silver = r * np.minimum(D1, I1)
    revenue_black = r * np.minimum(D2, I2)
    revenue_blue = r * np.minimum(D3, I3)
    shift_silver_black = 0.9*r*np.minimum(np.maximum(I2-D2,0),np.maximum(D1-I1,0))
    shift_black_blue = 0.92*r*np.minimum(np.maximum(I3-D3,0),np.maximum(D2-I2,0))
    total_revenue = (revenue_silver + revenue_black + revenue_blue
                    + shift_silver_black + shift_black_blue)
    return total_revenue

# Generate demand samples for each color
sample = np.random.normal(loc=[a, c, e], scale=[b, d, f], size=(nsamples, 3))

# Mean revenue estimate
revenues = np.array([revenue(*demand) for demand in sample])

# Compute the standard deviation and standard error
std_dev = np.std(revenues, ddof=1)
std_error = std_dev / np.sqrt(nsamples)

# Mean revenue and 95% confidence interval
mean = np.mean(revenues)
CI = stats.norm.interval(0.95, loc=mean, scale=std_error)
```

```
# Print results
print(f"Mean Revenue: ${mean:.2f}")
print(f"95% Confidence Interval: ${CI[0]:.2f} to ${CI[1]:.2f}")
```

Running the above Python code yields the following results:

MeanRevenue: $1,076,068.67

95% CI: $1,072,587.46 to $1,079,549.88.

The above type of result is often expressed more concisely in the form $\hat{\mu}$ [95%CI : L, U] where $\hat{\mu}$ is the mean estimate (superscript ^ is used to denote this is a statistical estimate) and L and U are the lower and upper bounds defining the confidence interval, respectively. In this example, it would be written as follows:

$$\$1,076,068.67 \quad [95\%\ \text{CI} : 1,072,587.46, \quad 1,079,549.88].$$

Rerunning the code will yield slightly different results because the random sampling is reinitialized with a new seed for the pseudorandom number generator. Many systems use a host computer's clock time, obtained from its central processing unit (CPU), to generate a seed, but this can be overridden to provide a user-defined seed, enabling reproducibility when the same seed is reused. This can be useful for debugging code that employs random samples.

In the above example, 10, 000 samples were used to estimate the mean revenue. By the looks of the CI, this was sufficiently accurate. (The interval covers $38 on a revenue of over $1 million.) Using a large number of samples is computationally feasible in this case because the function being evaluated is simple. In more complex situations, with many random variables or a difficult-to-evaluate function (e.g., the recourse function of a stochastic program from Chapter 5, which requires solving optimization models for each sample subproblem), a better approach is to estimate the necessary number of samples to achieve a desired level of accuracy. This can be done using the following formula:

$$n = \left(\frac{Z \times \sigma}{e}\right)^2, \tag{6.1}$$

where Z is the *z-score*, which quantifies the number of standard deviations for the chosen confidence level (e.g., for 95% confidence intervals $Z = 1.96$), σ is the standard deviation, and e is the acceptable margin of error. (Z-scores can be found in a look-up table from various statistics references or via predefined functions in programming languages.) In situations where σ is not known, some initial sampling may be necessary to estimate the standard deviation to be able to compute the number of samples needed. Typically, a minimum of 30 samples is recommended to estimate the standard deviation; however, larger sample sizes are recommended when the distribution is highly skewed. In the above example, sampling 100 times yields a variance of $\sigma = 189, 679$. If the decision-maker is indifferent to errors on the order of $1,000, then the number of samples can be computed as follows:

$$n = \left(\frac{1.96 \times 189,679}{1000}\right)^2 = 1,382.$$

Thus, for the automobile revenue problem, the Monte Carlo simulation could have limited the number of samples to a much smaller number to achieve the desired accuracy.

6.5 Sampling a Markov Chain

This section considers the second use of Monte Carlo simulation described in the chapter's introduction: sensitivity analysis. Consider a Markov chain with decision epochs $t = 0, \ldots, T$, states $S = \{S_1, S_2, \ldots, S_N\}$, and transition probability matrix P. To simulate a sequence of states, assume that the initial state is known or it is sampled using an initial probability distribution $q = (q(S_1), q(S_2), \ldots, q(S_N))$. The following steps will generate a random sample of states:

1. Sample a uniformly distributed random variable $rv \sim U(0, 1)$.
2. Determine the state at the first decision epoch $t = 1$: If $rv < q(S_1)$ assign state S_1, if $q(S_1) \le rv < q(S_1) + q(S_2)$ assign state S_2, and so on, until the last state, if $rv > \sum_{i=1}^{N-1} q(S_i)$ assign state S_N.
3. Assuming the sampled state at decision epoch $t = 1$ is S_k, sample a new random variable to determine the state at decision epoch $t = 2$ as follows: if $rv < P(S_1|S_k)$ assign state S_1, if $P(S_1|S_k) \le rv < P(S_1|S_k) + P(S_2|S_k)$ then assign state S_2, and so on, until at the last state, if $rv > \sum_{i=1}^{N-1} P(S_i|S_k)$ then assign state S_N.
4. Continue Step 3 for the remaining epochs $t = 3, \ldots, T$.

Repeating the above steps R times yields a set of sequences, $s_1^r, s_2^r, s_3^r, \ldots, s_T^r$ for $r = 1, 2, \ldots, R$, referred to as sample paths in Section 4.3. These can be used to generate a statistical estimate of the expected reward, as expressed in Equation 4.7 in Chapter 4:

$$\frac{1}{R}\sum_{r=1}^{R} W(r) \approx E_\omega^\pi[W(\omega)] .$$

Estimating the expected rewards with a statistical estimate based on R sample paths is desirable when evaluating policies, using backward induction is difficult or impossible, and there are too many sample paths to enumerate them completely.

Consider revising the above steps such that transition probabilities are influenced by the choice of decisions defined by a given policy, i.e., $p(s_{t+1}|s_t)$ becomes $p(s_{t+1}|s_t, a_t)$ where decision a_t is prescribed by a policy. If the policy is non-Markovian, then backward induction cannot be used to evaluate expected rewards. Such a policy could arise if decisions depend on the history rather than just the most recently visited state; in this case, Monte Carlo simulation may be the only practical way to estimate expected rewards for a policy.

Python Example 6.3: Simulation of a non-Markovian policy for a stock call option

Consider an *arithmetic call option*, a financial investment that pays off when the average closing stock price, defined by a Markov chain over a decision horizon of T days, is above a predetermined strike price, $\bar{p}$, and holds no value otherwise. The option can be executed at any decision epoch from $t = 0, \ldots, T$. Each day t', the investor reviews the price history from previous days $t = 0, \ldots, t' - 1$ and decides whether to exercise the option or defer execution by one more day. Executing the option yields a reward $\max\{(1/t')\sum_{t=0}^{t'} s_t - \bar{p}, 0\}$. Thus, a decision-maker is interested in the expected reward when the investor follows a given policy. For example, assume a policy recommends the decision-maker execute the option if the mean price is above $1.2 \times \bar{p}$, or if $t = T$ (by assumption). The history-dependent nature of the policy renders backward induction infeasible for computing the expected reward in this case; however, Monte Carlo simulation can readily estimate it. To illustrate this, assume a simple two-state Markov chain with a state set $S = \{I, D\}$ representing a 10% increase (I) or a 10% decrease in stock price, respectively, at a given decision epoch, with this transition probability matrix:

$$P = \begin{bmatrix} 0.80 & 0.20 \\ 0.25 & 0.75 \end{bmatrix}, \tag{6.2}$$

Finally, also assume a strike price of $\bar{p} = \$1,000$, an initial state I, an initial stock price of \$980, and a decision horizon of $T = 30$.

```
import numpy as np

# Input data
T = 30
strike_price = 1000
initial_price = 980
initial_state = "I"
nsamples = 10000
P = np.array([[0.80, 0.20],[0.25, 0.75]]) # Transition probability matrix
states = ["I", "D"]
state_effects = {"I": 1.10, "D": 0.90} #effects of state on price
rewards = []
```

```
# Markov chain simulation to find expected reward
np.random.seed(42) #Fix seed for repeatability
for _ in range(nsamples):
    current_state = initial_state # Initialize the starting state and price
    current_price = initial_price
    prices = [] # daily closing stock prices
    reward = 0
    for t in range(1, T + 1):
        prices.append(current_price) # Record the current price
        mean_price = np.mean(prices)  # Calculate mean price up to the current day
        # Decision rule to execute the option
        if mean_price > 1.2 * strike_price or (t == T and mean_price > strike_price):
            reward = max(mean_price - strike_price, 0)
            break
        # Transition to the next state
        if current_state == "I":
            next_state = np.random.choice(states, p=P[0])
        else:
            next_state = np.random.choice(states, p=P[1])
        current_state = next_state  # Update the state and stock price
        current_price = current_price * state_effects[current_state]
    rewards.append(reward)

mean = np.mean(rewards)

# Calculate the 95% confidence interval
std_error = np.std(rewards) / np.sqrt(nsamples)
z_score = 1.96 # Z-score for 95% confidence
CI = (mean - z_score * std_error, mean + z_score * std_error)

#Print results
print(f"Expected Reward: ${mean:.2f}")
print(f"95% Confidence Interval: (${CI[0]:.2f}, ${CI[1]:.2f})")
```

Running the above Python code yields the following estimate of the mean and 95% CI of the rewards:

Expected Reward: 162.55 [95%CI : \$160.38, \$164.72].

Of course, each time the code runs, it generates a new independent set of 10, 000 random samples of state transitions and expected rewards. Still, the confidence interval is relatively tight ($\pm 1.33\%$ with respect to the mean). As previously discussed, the accuracy of the estimates can be modified by adjusting the number of sampled sequences.

6.5.1 Using Monte Carlo Simulation to Analyze MDPs

An established approach to sensitivity analysis using Monte Carlo simulation is *probabilistic sensitivity analysis* (PSA) for MDPs. PSA seeks to evaluate the sensitivity of MDP solutions to variations in model parameters. Because most model parameters are subject to error, this step is important in assessing the sensitivity of policies to statistical variation. Algorithm 6.1 implements PSA to analyze the sensitivity of the total expected rewards for a given policy π.

Model parameters to consider in a sensitivity analysis of an MDP may include the reward function, $r(s, a)$, for all state and decision pairs, and the transition probability matrix P. The former is easy to sample if a probability distribution is specified. For example, a common choice is the normal distribution, $N(\mu, \sigma^2)$, when reward parameter estimates are obtained by sampling (e.g., customer surveys), as the mean is normally distributed according to the CLT. The *beta distribution*, $\text{Beta}(\alpha, \beta)$, is a popular choice for sampling transition probabilities in cases with two states because the beta distribution offers a straightforward way to sample random variates bounded below by zero and above by one. It has two shape parameters that provide flexibility in the distribution's form. It can be written as follows:

$$f(x; \alpha, \beta) = \frac{1}{\text{B}(\alpha, \beta)} x^{\alpha-1}(1-x)^{\beta-1},$$

Algorithm 6.1 Procedure for Probabilistic Sensitivity Analysis of an MDP

Generate K randomly sampled MDPs to evaluate a given policy π.
for $k = 1, \ldots, K$ **do**
    for $t = 1, \ldots, T-1$ **do**
        for $i = 1, \ldots, N$ **do**
            Generate $P_t^{(k),\pi(s_i)}$ by sampling the s_i^{th} row of the transition probability matrix, $P_t^{\pi(s_i)}$.
            Add $P_t^{(k),\pi(s_i)}$ to $MDP^{(k)}$.
        end for
    end for
end for
For $MDP^{(k)}$, $k = 1, \ldots, K$, compute the expected rewards by backward induction.
while $k \leq K$ **do**
    Compute $\vec{v}_1^{(k),\pi}(s_i)$ for all s_i.
end while
Examine the variation in $\vec{v}_1^{(k),\pi}(s_i)$ for all s_i.

where α and β are the shape parameters, and $B(\alpha, \beta)$ is a normalization constant. The shape of the beta distribution can be controlled by varying α and β. For example, setting $\alpha = \beta$ generates a symmetric distribution centered at 0.5. Setting $\alpha > \beta$ skews the distribution to the right, while setting $\alpha < \beta$ skews it to the left. A simple transformation can be used to reset the range from $[0, 1]$ to $[a, b]$ by defining a new random variable $Y = a + (b - a)X$, where $X \sim \beta(\alpha, \beta)$.

Situations with more than two states are often handled using a *Dirichlet distribution*, which reduces to a beta distribution when only two states are present. The Dirichlet distribution is a natural choice because all outcomes are bounded below by zero above by one, and all sum to one by assumption, thus maintaining the expected properties of rows of transition probability matrices. The Dirichlet distribution is denoted as $\text{Dirichlet}(\alpha)$, where α is a vector of distribution parameters with dimensionality equal to the number of states that can be reached from a given state. The probability density function for an N-dimensional distribution of random variates can be expressed as follows:

$$f(\mathbf{x}; \alpha) = \frac{1}{B(\alpha)} \prod_{i=1}^{N} x_i^{\alpha_i - 1},$$

where vector α defines the distribution shape, and $B(\alpha)$ is a normalization constant.

Python Example 6.4: PSA of machine maintenance

Consider a manufacturing tool that can be in one of four states, $S = \{\text{Excellent, Good, Damaged, Failed}\}$, indexed by $s = 1, 2, 3, 4$. In any state, the possible decisions are to repair the machine (R) or wait to repair it for at least one more epoch (W). Figure 6.1 illustrates the decay in the machine's function when repair is deferred. The repair decision returns the machine to excellent condition with probability q and fails to improve it with probability $1 - q$. The operating cost is $r(s, W) = -\exp(-\rho s_t)$ for each monthly epoch that maintenance is deferred and $r(s, R) = -c$ if the machine is repaired.

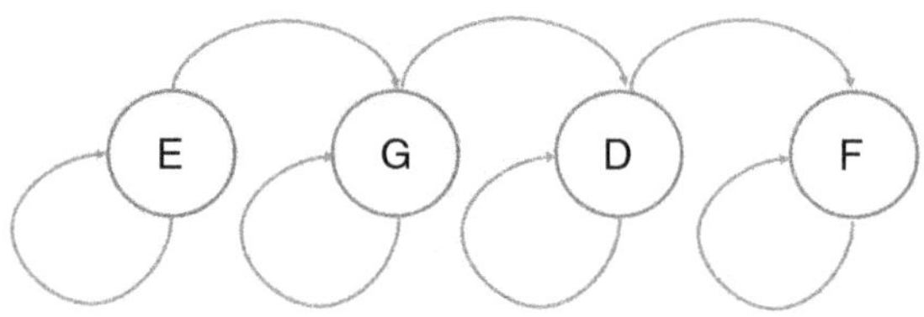

Figure 6.1 Illustration of the Markov chain for the machine repair MDP model.

The decision horizon is $T = 4$ epochs, and the terminal reward in the last period is $R(s) = \mu(s)$. In the absence of maintenance, the transition probability matrix is the following:

$$P = \begin{bmatrix} 0.7 & 0.3 & 0 & 0 \\ 0 & 0.5 & 0.5 & 0 \\ 0 & 0 & 0.4 & 0.6 \\ 0 & 0 & 0 & 1.0 \end{bmatrix}$$

and the reward function can be written as follows:

$$r_t(s_t, a_t) = \begin{cases} -\exp(-\rho\, s_t), & t = 1, \ldots, T-1,\ a_t = W, \\ -c, & t = 1, \ldots, T-1,\ a_t = R, \\ \mu(s_t), & t = T. \end{cases}$$

Assume the following model parameters: $q = 0.9,\ \rho = 1,\ c = 0.5,\ \mu(\text{E}) = 20,\ \mu(\text{G}) = 10,\ \mu(\text{D}) = 5,\ \mu(\text{F}) = 1$ for all $s, t = 0, 1, 2, 3$. Given these parameters, the solution to the MDP can be computed, which yields the following optimal policy:

s	$d_1^*(s)$	$d_2^*(s)$	$d_3^*(s)$
E	W	R	R
G	W	R	R
D	W	R	R
F	W	R	R

The optimal policy recommends deferring repair in all states in the first decision epoch and repairing all states in the remaining two decision epochs. A very conservative strategy that favors repair in any state for all but the first decision epoch. Naturally, the policy depends on the choice of model parameters, and a decision-maker may wish to better understand how these choices influence the optimal policy.

Given that the transition probability matrix, like most model parameters, is an estimate subject to error, evaluating the sensitivity of the value function to changes in the transition probabilities may be insightful. Assume that each transition probability could vary $\pm 20\%$ relative to the point estimate and that the distribution of the variation is symmetric around the original point estimate. Further assume that at each epoch, the state stays the same or shifts one state to the right (this implies the zero values are not subject to variation). Suppose the main concern is variation in transition probabilities for the "wait" decision W, in Equation 6.4, and further assume only the first three rows are subject to uncertainty, i.e., the "Failed" state is always absorbing. Sensitivity analysis can be performed by sampling each row of the transition probability matrix using the random variable $Y = a + (b - a)X$, where $X \sim \text{Beta}(1, 1)$, and $a = 0.8p$ and $b = 1.2p$, where p is a placeholder for the probability of self-transition, and $1 - p$ is the probability of moving to the next state to the right. Setting $\alpha = \beta = 1$ results in a symmetric distribution. The following Python code solves the MDP to find the optimal policy for the baseline parameter choices and then implements Algorithm 6.1. Figure 6.2 shows the resulting distribution of $v_1^{\pi^*}(E)$. Given the narrow range of x-axis values, the sensitivity analysis indicates that variation in transition probabilities is of little consequence in this case.

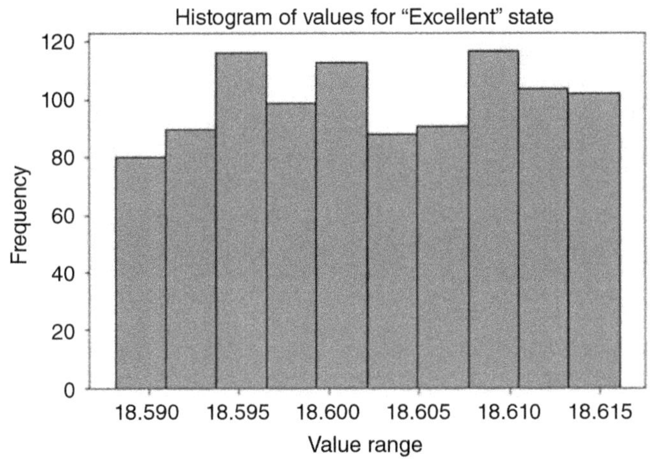

Figure 6.2 Histogram for the distribution of the value function for state "Excellent" at the first decision epoch based on 1,000 sampled transition probability matrices for the optimal policy of the machine repair MDP model, Example 6.4.

```
import numpy as np
import matplotlib.pyplot as plt

# Input data
T = 3
nstates = 4
ndecisions = 2
states = ['Excellent', 'Good', 'Damaged', 'Failed']
decisions = ['R', 'W']
q = 0.9
rho = 1
c = 0.5
mu = [20, 10, 5, 1] # Terminal rewards for Excellent, Good, Damaged, Failed
nsamples = 1000
# Transition probability matrix (4, 2, 4):  T[s, d, s'] = P(s' | s, d)
P = np.zeros((nstates, ndecisions, nstates))
P[:, 0, :] = np.array([ # d = 'R'
    [1.0, 0.0, 0.0, 0.0],
    [q, 1 - q, 0.0, 0.0],
    [q, 0.0, 1 - q, 0.0],
    [q, 0.0, 0.0, 1 - q]
])
P[:, 1, :] = np.array([ # d = 'W'
    [0.7, 0.3, 0.0, 0.0],
    [0.0, 0.5, 0.5, 0.0],
    [0.0, 0.0, 0.4, 0.6],
    [0.0, 0.0, 0.0, 1.0]
])

rewards = np.zeros((nstates, ndecisions))
for s in range(nstates):
```

```
        rewards[s, 0] = -c # r(s, 'R')
        rewards[s, 1] = -np.exp(- rho * (s + 1)) # r(s, 'W')
sample = np.random.beta(1,1,size=nsamples) # Generate samples for sensitivity analysis
v_sample = np.zeros((nsamples, nstates), dtype = float)

# Perform Monte Carlo simulations for sensitivity analysis
np.random.seed(42) #Fix seed for repeatability
for k in range(nsamples):
    value = np.zeros((T, nstates, ndecisions))  #Value-to-go function
    v = np.zeros((T + 1, nstates))  #Optimal value function
    optimal_policy = np.zeros((T, nstates), dtype=int)  # Policy
    v[T, :] = mu    # Boundary condition at epoch T
    P_W_sample = np.copy(P[:, 1, :]) #Sample rows of TPM[:, 1, :] matrix
    for i in range(nstates - 1):       # Generate k^th sample of P_W
        a = 0.8 * P[i, 1, i]
        b = 1.2 * P[i, 1, i]
        P_W_sample[i, i] = a + (b - a) * sample[k]
        P_W_sample[i, i + 1] = 1 - P_W_sample[i, i]
    P_W_sample[nstates - 1, nstates - 1] = 1 # State 'Failed' absorbing state
    for t in range(T - 1, -1, -1):  # Backward induction for current sample
        for s in range(nstates):
            value[t,s,0] = rewards[s,0] + np.dot(P[s,0,:],v[t+1,:])
            value[t,s,1] = rewards[s,1] + np.dot(P_W_sample[s],v[t+1,:])
            optimal_policy[t, s] = np.argmax(value[t, s])
            v[t, s] = value[t, s, optimal_policy[t, s]]
    v_sample[k, :] = v[0, :]

# Show a histogram of the distribution of v_1('Excellent')
v_sample_0 = v_sample[:, 0] #Extract all 'Excellent' state samples
plt.hist(v_sample_0, bins=10, edgecolor='black')
# Add titles and labels
plt.title("Histogram of Values for 'Excellent' State")
plt.xlabel("Value range")
plt.ylabel("Frequency")
plt.show()
```

6.6 Approximating Stochastic Programs with Monte Carlo Simulation

Chapter 5 covered scenario trees and the fundamental data structures that define uncertainty in stochastic programs. Scenario trees are often constructed using statistical sampling from a probability distribution. For example, in the 2-SLP model for scheduling arrivals at a stochastic server, the raw-material-delivery example assumed random variables with four possible outcomes. With five deliveries, the total number of scenarios was $4^5 = 2^{10}$, a large but manageable number with reasonable computing resources. However, what happens if the model must accommodate not just 5 deliveries but 20? This results in more than 2^{40} scenarios, which are too large to store in memory in most situations, let alone solve the massive 2-SLP model. Fortunately, Monte Carlo sampling can be used to evaluate the solution by simulating scenarios. The model created from sampled scenarios is the *sample average approximation* (SAA).

For 2-SLPs, an SAA enhancement uses Monte Carlo sampling of a discrete and finite set of scenarios as a means to either reduce the total number of scenarios to a manageable number (e.g., from 2^{40} in the example above to $1,000$ scenarios for the SAA) or convert a model with continuously distributed random variables (an infinite number of scenarios) into one with a discrete and finite set of scenarios. The 2-SLP model of Chapter 5, defined by Equation 5.32, can be approximated by randomly sampling a subset of scenarios to create the SAA, where each scenario $\omega \in \Omega$ is sampled with its associated probability $p(\omega)$ to generate $\tilde{\Omega} \subset \Omega$, where each of the scenarios in subset $\tilde{\Omega}$ has equal probability of $1/|\tilde{\Omega}|$.

Example 6.5: *Sample average approximation for stochastic server scheduling*

Consider Example 5.6 of Chapter 5 to generate the SAA for the stochastic server scheduling model of Chapter 5. In this case, assume that the random variables are independent, continuous, and uniformly distributed on $U(1.25,\ 3.25)$, with the sample mean of 4.5 as the original problem in Chapter 5. Since the number of scenarios is not finite, sampling is the only option. The vector $Z(\omega)$ with elements $Z_1(\omega),\ Z_2(\omega), \dots, Z_n(\omega)$, is randomly sampled to create $|\tilde{\Omega}|$ scenarios. Solving the following model gives a statistical estimate of the optimal solution as follows:

$$\min \frac{1}{|\tilde{\Omega}|} \sum_{\forall \omega \in \tilde{\Omega}} c_i^w w_i(\omega) + c_i^s s_i(\omega), \tag{6.3}$$

subject to:

$$w_2(\omega) - s_2(\omega) = Z_1(\omega) - r_1, \forall \omega \in \tilde{\Omega}, \tag{6.4}$$

$$w_3(\omega) - w_2(\omega) - s_3(\omega) = Z_2(\omega) - r_2, \forall \omega \in \tilde{\Omega}, \tag{6.5}$$

$$\vdots$$

$$w_n(\omega) - w_{n-1}(\omega) + s_n(\omega) = Z_{n-1}(\omega) - r_{n-1}, \forall \omega \in \tilde{\Omega}, \tag{6.6}$$

$$r_i \geq 0, i = 1, \dots, n-1,\ w_i(\omega) \geq 0,\ s_i(\omega) \geq 0\ \forall i = 2, \dots, n, \forall \omega \in \tilde{\Omega}. \tag{6.7}$$

Assume that $|\tilde{\Omega} = 100|$ is chosen. The natural question that arises is whether the resulting SAA is a good approximation. In other words, are 100 scenarios enough for an accurate estimate? One way to answer this question is to compute the optimal solution to the 2-SLP multiple times using independently sampled scenarios. For example, letting $k = 1, \dots, K$ define the total number of experiments (also called *replications*), each solving a $|\tilde{\Omega}|$ scenario model, from which the probability distribution can be estimated. Running the Python code in Chapter 5 for this problem $K = 30$ times, with $|\tilde{\Omega}| = 100$ scenarios for each model instance, yields a series of 30 independent estimates of the optimal objective function value, provided the scenarios are sampled independently. Table 6.3 gives the results of the first 10 experiments. The results provide a visual sense of the statistical variation in SAA solutions.

Using all 30 samples to compute the 95% confidence intervals yields the following estimate: 0.846 [95%CI : 0.834, 0.858]. This provides a measure of confidence in the use of 100 sampled scenario SAAs to solve the problem.

Table 6.3 Objective function values for the first 10 of 30 sample average approximations, each with 100 sampled scenarios, for the stochastic program for scheduling arrivals to a stochastic server covered in Chapter 5.

Sample number	Objective function value
1	0.895
2	0.918
3	0.822
4	0.852
5	0.865
6	0.794
7	0.834
8	0.840
9	0.859
10	0.868

6.6.1 Monte Carlo Sampling for Multi-stage Stochastic Programs

Many approaches have been proposed for generating scenario trees to solve M-SLPs. The large size of such problems makes them computationally challenging, with the scenario tree serving as the primary data structure that influences the overall size of the resulting model. Scenario-tree generation fundamentally involves a trade-off between two important considerations: (1) the accuracy with which the model represents reality and (2) the computational effort required to solve the model. Both are necessary factors to consider guaranteeing solution quality and computational feasibility.

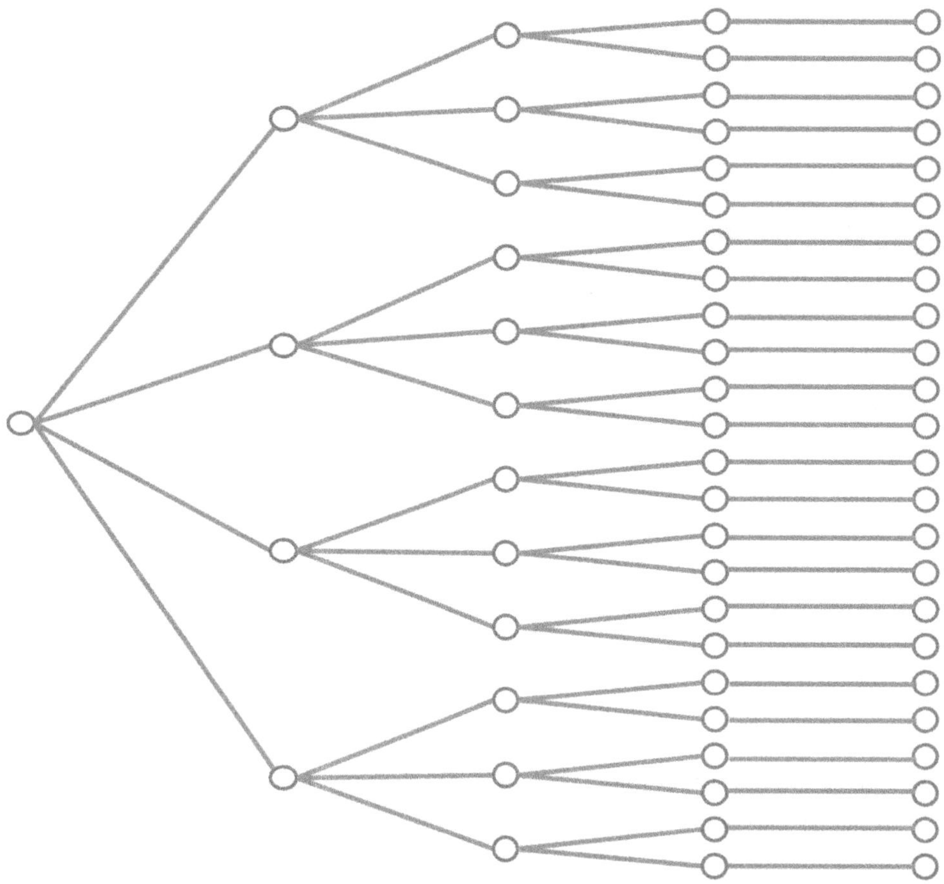

Figure 6.3 Illustration of a four-stage (4, 3, 2, 1) scenario tree with a total of 24 unique scenarios, with four branches in stage one, three branches in stage two, two branches in stage three, and one branch in stage four.

This section describes an approach for scenario-tree approximation that assumes an existing stochastic model that defines the evolution of the random variables. It is presumed that the scenario tree, as envisioned, is too large to solve the resulting stochastic program. Thus, the real scenario tree must be replaced by an approximation, which can be done using Monte Carlo sampling to reduce the tree size. Given that the stochastic process is known at each node of the tree, conditional probability distributions can be used to sample random variables at each stage, thereby constructing a scenario tree of the desired size and structure. Moreover, the sampling process and order are guided by the specific tree design, which often requires numerical experimentation to select an effective design that balances competing goals of computational effort and solution quality. For example, consider the scenario tree in Figure 6.3, which is a (4, 3, 2, 1) scenario tree in which the first stage admits four alternative outcomes, the second three outcomes, the third two such outcomes, and finally, one outcome for the last stage, for a total of 24 unique scenarios, which can be checked by counting the number of leaf nodes, i.e., the rightmost nodes in the tree. The choice reflects a common goal: to decrease the number of outcomes with respect to the stage, with greater emphasis on accuracy in the near-term portion of the horizon—since these decisions will be implemented first—versus later stages that provide an approximation to guide earlier-stage decisions.

Example 6.6: *Stock price scenario tree*

Consider a financial model with two independent stock prices, each represented by a binomial tree as illustrated in Figure 6.4. Let us assume there are four stages, and the decision-maker elects to have a (3, 2, 1, 1) scenario tree for SAA. Sampling at each stage is conditional on the prior stage stock prices, and each outcome is sampled according to its associated probability. Note that the tree is recombinant, similar to a binomial tree, because the outcomes for each stock are

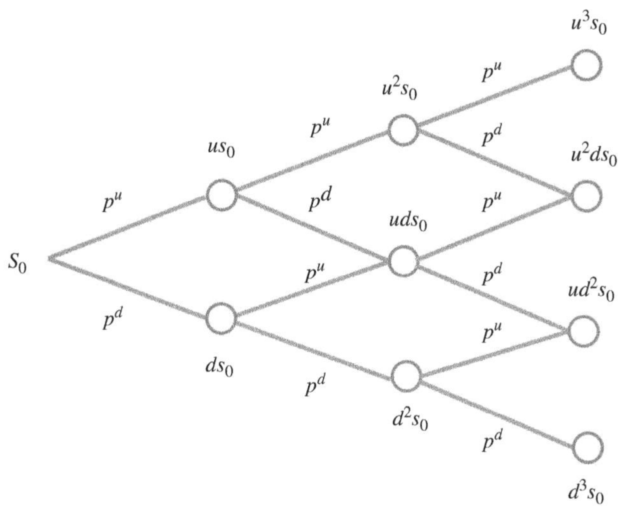

Figure 6.4 Illustration of the scenario tree for a single stock price model.

indifferent to permutations, i.e., for a particular stock, the price changes are independent of the sequence of changes. For example, an increase followed by a decrease in stock 1, denoted ud, is the same as a decrease followed by an increase, du.

Assume the following parameters for each of the stock price models:

$$s_t^i = \begin{cases} u^i s_{t-1}^i \text{ with probability } p_u^i, \\ d^i s_{t-1}^i \text{ with probability } p_d^i, \end{cases}$$

where $i = 1, 2$ indexes the two stocks, and $t = 0,\ 1,\ 2,\ 3$ indexes four stages. The stock prices are governed by p_u^i and p_d^i, the probabilities of an up and down tick in the stock price at each stage, respectively, and u^i and d^i, the up and down percentage changes, respectively. Table 6.4 gives the parameters for each model.

Table 6.4 Parameters for the stochastic model of stock prices for two independent stocks.

Stock	p_u^i	p_d^i	u^i	d^i	s_0^i
1	0.70	0.30	1.25	0.8	75
2	0.55	0.45	1.1	0.9	100

In the first stage, sample both stock prices independently to create pairs of stock prices. The sample outcomes are generated using a random-number generator, such as the two-stage example presented earlier in this chapter. This is done three times to create three such pairs representing the outcomes for the first stage. Next, for the second stage, two pairs are sampled. Finally, one pair is sampled for the third stage to complete a $(3,\ 2,\ 1)$ scenario tree. Figure 6.5 shows all possible scenarios with the sampled scenarios represented by darker lines.

6.7 Monte Carlo Sampling for MDPs

Monte Carlo sampling can approximate solutions to MDPs using a policy improvement algorithm. This applies to situations where the stochastic nature of a problem can be characterized by a Markov chain, but the exact model governing state transitions, i.e., the transition probabilities, is unknown to the decision-maker. Suppose the decision-maker can observe states sequentially over time, select decisions, and observe the outcomes. In that case, *reinforcement learning* methods, to be covered in Chapter 8, can be used to iteratively search for an optimal policy by way of the following main steps:

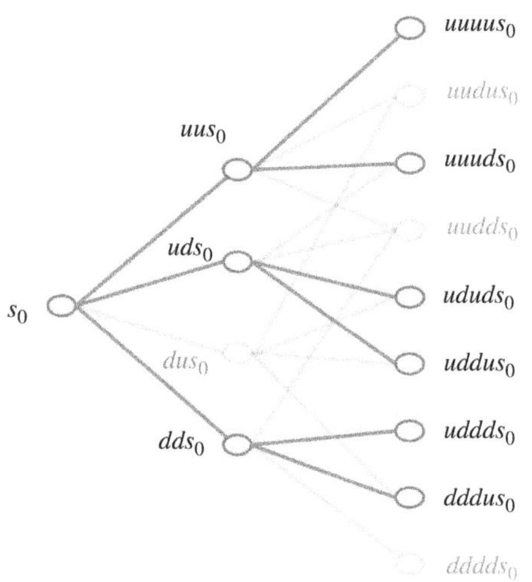

Figure 6.5 Illustration of the scenario tree for the first two stages of the two-stock price model. Dark lines indicate sampled scenarios, and light lines indicate unsampled scenarios. All sampled scenarios have equal probability.

Algorithm 6.2 Monte Carlo Policy Iteration

1: **Input:** initial policy π_0; initialize $t \leftarrow 0$; select stopping criterion

2: **while** stopping criterion not satisfied **do**
3:     **Policy evaluation:** $\hat{v} \leftarrow$ SIMULATE(π_t)
4:     **Policy improvement:** $\pi_{t+1} \leftarrow$ IMPROVEPOLICY($\hat{v}$)
5:     $t \leftarrow t + 1$
6: **end while**
7: **Output:** π_t

Chapter 8, gives an introduction to reinforcement learning, which provides a powerful portfolio of algorithms that utilize Monte Carlo sampling to approximate optimal policies using sequential-learning algorithms.

6.8 Concluding Remarks

Simulation is a broad and important topic, and this chapter covered only the most critical concepts needed to enable application to the analysis and solution of sequential decision-making problems. Many excellent, more in-depth resources on simulation exist, including books by Law and Kelton [1], Nelson [2], and Ross [3], among others.

Other methods exist that apply to the sensitivity analysis of MDPs. Smith [4] first reported methods for uniform sampling of polytopes (bounded regions defined by linear constraints). This is relevant to sampling transition probabilities because the estimates must lie in a given confidence region, i.e., an interval for each transition probability; they must belong to the *unit simplex*, a region defined by the requirement that all probabilities are between 0 and 1; and they must sum to 1. The use of hit-and-run methods for MDP sensitivity analysis was proposed by Zhang et al. [5].

The Monte Carlo sampling approach for SPs described in this chapter is effective when the SAA can be solved within a reasonable time limit, where "reasonable" is determined by the context of the problem and, more specifically, the time available to find a solution. However, this can be computationally intensive, requiring the solution of potentially many 2-SLP models (e.g., 30 experiments as in the example above) to evaluate confidence intervals before determining whether K is sufficiently large to achieve the desired accuracy. An approach described by Mak et al. [6] uses probabilistic bounds on SPs to compute confidence intervals for the optimality gap, thereby providing statistical bounds on the error in the SAA solution.

Extensive literature exists on scenario-tree generation for multi-stage stochastic programs. One well-known approach is based on matching the statistical properties of the underlying random variables, called *moment matching*, which attempts to design trees that have the same mean, variance, and possibly higher moments as the true random variables while maintaining a compact size that leads to a solvable model. The hope is that preserving these statistical properties will yield good approximations of the recourse function—defined as the conditional expectation over the scenario tree—and of the true optimal solution. Other approaches consider the effects of scenario tree design choices on measures such as EVPI and discrepancies between scenarios to avoid similar scenarios in the tree. Heitsch and Werner [7] reviewed methods proposed in the literature and presented and motivated heuristics for scenario generation.

6.9 Practice Exercises

1. A patient with an infectious disease progresses through a care pathway modeled as a Markov chain with four states: Mild, Moderate, Severe, Recovered. The one-week transition matrix is

$$P = \begin{pmatrix} 0.70 & 0.20 & 0.05 & 0.05 \\ 0.00 & 0.65 & 0.20 & 0.15 \\ 0.00 & 0.00 & 0.60 & 0.40 \\ 0.00 & 0.00 & 0.00 & 1.00 \end{pmatrix}.$$

Let N be the number of weeks until recovery, defined as the number of transitions until the chain first reaches state 4, starting from state 1. Use a Monte Carlo simulation to estimate $E[N]$. Evaluate different sample sizes $(10,\ 100,\ 1000)$, estimate the 95% CIs, and compare them to the exact probability.

2. Consider the following function $f(X) = \max(X, 4)$ of an exponentially distributed random variable $X \sim \exp(\lambda)$ with $\lambda = 1.0$. Use a Monte Carlo simulation to estimate $E[f(X)]$ using 1000 samples. Compute the 95% confidence interval on the expectation using 30 replications.

3. For the stochastic unit commitment problem, Example 5.9, of Chapter 5, formulate the SAA and solve it for the case of uniformly distributed demand, $d_t \sim U(50,\ 150)$, at each stage. Assume a $(4,\ 3,\ 2)$ scenario tree. Compute confidence intervals on the optimal solution to evaluate the accuracy of the SAA. Justify your choices of sample size and assumptions made.

4. Conduct a PSA of the optimal stopping time problem of Example 4.4, in Chapter 4, assuming the adverse event transition probabilities $p_t(E|s_t)$ are subject to variation of $\pm 25\%$. Plot the histogram for the optimal value function for each state.

5. Consider the following optimization problem:

$$\begin{aligned} &\min\{2x_1 + 3x_2 + E_{\tilde{\xi}}[\max\{\xi_1 - x_1, 0\} + \max\{\max\{\xi_1 - x_1, 0\} + \xi_2 - x_2, 0\}, \\ &+ \max\{x_1 - \xi_1, 0\} + \max\{x_2 - \xi_2, 0\}] \mid x_1 + x_2 \geq 1,\ \ x_1, x_2 \geq 0\}\} \end{aligned}$$

Assuming ξ_1 and ξ_2 have a discrete uniform distribution with values $\{1, 2, 3, 4, 5\}$ occurring with equal probability, show that it can be formulated as a two-stage stochastic linear program and compute the value of the stochastic solution (VSS) and the expected value of perfect information (EVPI).

6. Use a Monte Carlo simulation to illustrate the CLT. Given random variable $Y \sim \exp(\lambda)$. Let $\lambda = 1$ and sample $n \in \{1, 5, 10, 30, 50\}$ to estimate the sample mean of Y. For each choice of n generate 100 replications and create a histogram for each. Explain your results in the context of the CLT.

7. Consider Example 5.5 in Chapter 5 about optimizing the purchase of vaccines. Assume that the demand for COVID and flu are lognormally distributed with distributions Lognormal (9.11, 1.0) and Lognormal (9.30, 1.2), and assume the solution to the stochastic program is estimated using 100 sampled scenarios.

a. Compute the mean and variance of the solution to the two-stage stochastic program based on solving 30 instances of the 100 scenario model.

b. Estimate the mean and the 95% CI for the VSS and the EVPI.

c. If a new single-dose multivalent vaccine becomes available that can vaccinate against both flu and COVID with one shot, estimate the maximum cost a decision-maker should be willing to pay if they switch to using the multivalent vaccine. Explain the assumptions behind your analysis and recommendation.

8. Consider a four-state Markov chain with states s_1, s_2, s_3, s_4 and the following transition probability matrix:

$$P = \begin{bmatrix} 0.85 & 0.1 & 0.05 & 0 \\ 0 & 0.6 & 0.3 & 0.1 \\ 0 & 0.3 & 0.65 & 0.05 \\ 0 & 0 & 0 & 1 \end{bmatrix}.$$

Assuming the system starts in state s_1, use Monte Carlo simulation to compute statistical estimates of the following: (a) mean and variance of the number of transitions to reach state s_4; (b) probability of reaching state s_4 within ten transitions; (c) probability of not reaching state s_2. Justify your choice of sample sizes for (a), (b), and (c).

9. Assume a simulation model has outputs $Y_1, Y_2, \ldots, Y_n$ which are independent and identically distributed with mean μ and variance σ^2. Define the sample mean and sample variance after n observations as

$$\mu_n = \frac{1}{n}\sum_{i=1}^{n} Y_i, \qquad \sigma_n^2 = \frac{1}{n-1}\sum_{i=1}^{n} (Y_i - \mu_n)^2.$$

Consider the following sequential stopping rule based on the *half-width*, hw_n (the difference in the upper and lower confidence limit divided by 2, after n samples): Simulate until the half-width of the 95% confidence interval is at most $\epsilon > 0$:

$$N = \min\{n \geq n_{\min} : \text{hw}_n \leq \epsilon\},$$

where $n_{\min}$ is a specified minimum sample size of 30 and ϵ is a maximum tolerance.

a. Explain why choosing the sample size using this stopping rule can be biased, affecting the actual coverage probability of the resulting confidence interval compared to the nominal 95% level.

b. Construct a numerical experiment to illustrate the bias in the sequential stopping rule and experiment with different choices of ϵ.

References

1 Law, A. M. & Kelton, W. D. (2007). *Simulation Modeling and Analysis* 4th. McGraw-Hill, New York, NY.
2 Nelson, B. L. (2010). *Stochastic Modeling: Analysis & Simulation*. Courier Corporation.
3 Ross, S. M. (2022). *Simulation*. Academic Press.
4 Smith, R. L. (1984). Efficient Monte Carlo procedures for generating points uniformly distributed over bounded regions. *Operations Research*, *32* (6), 1296–1308.
5 Zhang, Y., Wu, H., Denton, B. T., Wilson, J. R. & Lobo, J. M. (2019). Probabilistic sensitivity analysis on Markov models with uncertain transition probabilities: An application in evaluating treatment decisions for type 2 diabetes. *Health Care Management Science*, *22*, 34–52.
6 Mak, W. K., Morton, D. P. & Wood, R. K. (1999). Monte Carlo bounding techniques for determining solution quality in stochastic programs. *Operations Research Letters*, *24* (1–2), 47–56.
7 Heitsch, H. & Römisch, W. (2009). Scenario tree modeling for multistage stochastic programs. *Mathematical Programming*, *118*, 371–406.

7

Partially Observable Markov Decision Processes

7.1 Introduction

An important extension of MDPs is *partially observable MDPs* (POMDPs). In this extension, the decision-maker cannot observe the underlying states directly. Rather, alternative observations potentially associated with the unobserved states (and sometimes strongly linked to them) are used to infer the probability distribution over the states the system could be in at each decision epoch. A common example arises in the context of disease diagnosis (e.g., cancer), where the disease is often not directly observable and has no symptoms at early stages. Nevertheless, some alternative observations, such as the presence of a biomarker in the blood, may indicate that the unobserved cancer is present. Indeed, diagnostic tests exist for many common forms of cancer that can be used to acquire information about the *belief* that the patient has cancer. The progression from a noncancerous state to one in which a patient has cancer can happen without detection as the patient ages and transitions through *hidden* states. Thus, a decision maker may wish to determine an optimal policy for when to use biomarker tests as patients age, as part of a health screening program. Generalizing beyond healthcare, Cassandra [1] surveys a wide range of applications of POMDPs in areas such as robotics, machine maintenance, marketing, search and rescue, education, and more, for the interested reader.

This chapter introduces the concept of partial observability in Markov chains, commonly referred to as *hidden Markov models* (HMMs). These models serve as the initial basis for a POMDP. An HMM is to a POMDP, as a Markov chain is to an MDP. In other words, an HMM governs the stochastic (hidden) state transitions and the observable signals that the decision-maker can use to infer the likelihood that the system is in a particular state at a given epoch. HMMs are, by nature, descriptive models appropriate when the states of interest are not directly observable.

After covering the basic concepts relevant to HMMs, POMDP model formulations are discussed that aim to find an optimal policy, which defines a decision for each possible *belief state*, i.e., a probability distribution over the system's states. A reward function is the missing element in an HMM that, in POMDP models, quantifies the objective and enables the computation of optimal policies. POMDP-based policies are defined over the belief state, which represents the system's state and contains all the relevant information needed to make an optimal decision. A belief state is simply a vector with a number of elements equal to the number of hidden states, for which each element is the probability that the system is in the corresponding state. Thus, unlike a standard MDP, a POMDP policy maps decisions to belief states rather than the (unobservable) hidden states of the Markov chain.

The remainder of this chapter introduces mathematical notation and model formulation for HMMs, as well as approaches for estimating their model parameters. Next, POMDPs are discussed, and algorithms for computing optimal or near-optimal policies are described. Motivating examples are discussed and solved using Python implementations to illustrate properties of POMDP solutions.

7.2 Hidden Markov Models

Sequential decision-making situations where the states of the model are not directly observable but can be assumed to be well described by a Markov chain are represented by an HMM. The term *hidden* refers to the fact that the exact state of the system is unknown, but *observations* (sometimes called *emissions* or *messages*) provide information relevant to the decision maker's belief the system is in one of the hidden states, referred to as *core states*. The latter definition is used to avoid confusion with the directly observable states of an MDP. The following is a summary of the key elements of an HMM. The same notation is used as in Chapter 4 where possible.

Core States and Transition Probabilities: The core states are defined on a decision horizon denoted by $t = 1, 2, \ldots, T$. The state at each decision epoch t is denoted by $s_t \in S = \{S_1, S_2, \ldots, S_N\}$. Transition probabilities from state s_t to s_{t+1} are denoted by $p_t(s_{t+1}|s_t)$, which collectively, for all s_t and s_{t+1}, defines the $N \times N$ core state transition probability matrix P_t at epoch t. In situations where decisions affect the transition probabilities, they are denoted as $p_t(s_{t+1}|s_t, a_t)$ and the matrix as $P_t(a_t)$. Since the intent is to use the HMM as the basis for a POMDP, decisions will generally be present in the formulations discussed below.

Observations and the Information Matrix: The observation at epoch t is denoted by $o_t \in O = \{O_1, O_2, \ldots, O_K\}$, where O is a discrete and finite set of K possible observations. The core states, and the observations are tied together probabilistically through an *information matrix* Q_t. Each row of the information matrix corresponds to a core state, and each column corresponds to an observation (note that it is possible to include a null observation indicating no observation is observed). Thus, a particular element $q_t(o_t|s_t)$ denotes the probability of observation o_t given the core state is s_t. As a result, each observation may contain information about the system's core state at each epoch. Similar to a transition probability matrix, P, the rows of the information matrix sum to one, since the columns collectively cover all mutually exclusive outcomes. In situations where decisions may influence the probability distribution of observations, the probabilities are denoted by $q_t(o_t|s_t, a_t)$, and the matrix as $Q(a_t)$.

Belief States: The relationship between the core states and the observations encoded in the information matrix can be used to estimate a *belief vector* sequentially via Bayes' rule, initially discussed in Chapter 4. The belief vector is denoted as $\vec{b}_t = (b_t(S_1), b_t(S_2), \ldots., b_t(S_N))$, where $b_t(s_t)$ is the probability that the system is in core state s_t at epoch t.

The belief state of an HMM contains all the relevant information necessary for a decision-maker to make optimal decisions at each epoch. The belief state is updated sequentially over time using an iterative process called *Bayesian updating* that learns the beliefs at each epoch t from the prior belief at epoch $t-1$ and the most recent observation. Given an initial belief, $\vec{b}_0$, the system transitions probabilistically among core states from s_0 to s_1, followed by an observation o_1. Next, a core state transition occurs from s_1 to s_2 followed by a second observation o_2, and so on. This process continues to the end of the horizon. Given that the core state transition probability matrix, $P_t(a_t)$, and information matrix, $Q_t(a_t)$, are known for all epochs, t, the belief vector at epoch t can be updated at the start of epoch $t+1$ as follows:

$$b_{t+1}(s_{t+1}) = \frac{\sum_{s_t \in S} b_t(s_t) p_t(s_{t+1}|s_t, a_t) q_t(o_t|s_t, a_t)}{\sum_{s_t, s_{t+1} \in S} b_t(s_t) p_t(s_{t+1}|s_t, a_t) q_t(o_t|s_t, a_t)}. \tag{7.1}$$

Recalling the definition of Bayes' rule from Chapter 2, the numerator is the probability of the intersection of two events: (1) transition from core state s_t to s_{t+1}; (2) observation of o_t. The denominator is the probability of observing o_t, averaged over all possible states the system may have transitioned to. Applying the operation in Equation 7.1 for each state yields an updated belief vector $\vec{b}_{t+1} = (b_{t+1}(S_1), b_{t+1}(S_2), \ldots, b_{t+1}(S_N))$.

Example 7.1: *A Coin Flip Game*

Consider the following situation. Suppose two friends are playing a coin game based on coin flips, with one friend picking one side (e.g., heads or tails) immediately before the other friend flips the coin. If the outcome is called correctly, the caller gets a point; otherwise, the flipper gets a point. They continue until one of them reaches 10 points and wins the game. Additionally, the player flipping the coin has two coins to choose from. The first is unbiased, with a probability of 1/2 of landing heads or tails. The second coin is biased toward landing heads with probability 3/4. To make things interesting, the player flipping the coins may switch coins before each flip, with the following transition probability matrix defining the probability of the player switching between core states $S = \{\text{Coin}_1, \text{Coin}_2\}$:

$$P = \begin{bmatrix} 0.90 & 0.10 \\ 0.5 & 0.5 \end{bmatrix}. \tag{7.2}$$

According to the game's rules, the player calling the coins can make the decision to inspect the coin immediately following any coin flip. If the other player is caught flipping the biased coin, they lose, regardless of the score; otherwise, they win. How should the player calling the coin flips decide whether or when to inspect the coin?

This can be viewed as an HMM, with the core states representing which coin is flipped, and transition probabilities representing the probability that the player flipping coins switches coins on any given turn. The set of observations is $O = \{\text{Heads, Tails}\}$ and the information matrix is

$$Q = \begin{bmatrix} 0.5 & 0.5 \\ 0.75 & 0.25 \end{bmatrix}. \tag{7.3}$$

Thus, if the unbiased coin is flipped at a given turn, the next coin flipped will be selected randomly, each with probability 0.5. However, if the biased coin is flipped, then with probability 0.75 the next coin flip will be the unbiased coin, and with probability 0.25 it will remain the biased coin. The game starts with the player choosing a coin. The caller does not know which is chosen, but their initial belief is $\vec{b}_0 = (0.5, 0.5)$. Naturally, the updated beliefs will depend on the observations of the coin flips and the probability of transitions from one coin to another. For example, if the caller observes heads many times, they may suspect that Coin 2 is being flipped and decide to inspect the coin. But how many heads is enough to be confident?

Bayesian updating can be used to decide when to inspect the coin. Consider the following sequence of observations from iteration 1 to 7: (Head, Tail, Tail, Head, Head, Head, Head). Bayesian updating yields the following estimates of the numerator of Equation 7.1.

$$\text{Coin 1: } (0.5 \times 0.9 \times 0.5) + (0.5 \times 0.5 \times 0.5) = 0.225 + 0.125 = 0.35$$

$$\text{Coin 2: } 0.5(\times 0.1 \times 0.75) + (0.5 \times 0.5 \times 0.75) = 0.0375 + 0.1875 = 0.225.$$

The denominator is the same for Coin 1 and Coin 2:

$$0.35 + 0.225 = 0.575.$$

The above calculations yield the belief at epoch 1:

$$\text{Coin 1: } b_1(\text{Coin 1}) = 0.61,$$

$$\text{Coin 2: } b_1(\text{Coin 2}) = 0.39.$$

Continuing with the Bayesian updating for each of the remaining observations in the sequence yields the following:

$$\vec{b}_2 = \begin{bmatrix} 0.85 \\ 0.15 \end{bmatrix}, \vec{b}_3 = \begin{bmatrix} 0.91 \\ 0.09 \end{bmatrix}, \vec{b}_4 = \begin{bmatrix} 0.81 \\ 0.09 \end{bmatrix}, \vec{b}_5 = \begin{bmatrix} 0.76 \\ 0.24 \end{bmatrix}, \vec{b}_6 = \begin{bmatrix} 0.73 \\ 0.27 \end{bmatrix}, \vec{b}_7 = \begin{bmatrix} 0.72 \\ 0.28 \end{bmatrix}.$$

The player who called the coin flips can use this information to decide whether and when to inspect the coin. Based on the belief estimates above, the odds are against inspecting the coin. Of course, other factors might deserve consideration when making a decision, such as the number of points the player has at each epoch. The HMM alone does not prescribe an optimal decision; rather, it is a descriptive model. Shortly, a corresponding POMDP model is discussed that uses the above HMM and Bayesian updating to optimize decisions based on the given reward function.

7.3 HMM Parameter Estimation

An important consideration is how to estimate the parameters that define an HMM. A given model can be denoted as $\mathcal{L} = (P, Q, \vec{b}_0)$, where P and Q denote all relevant core state transitions and observation probabilities (dependencies on t and a_t are suppressed here for simplicity). Model estimation is challenging because the available data typically consist only of observations, yet the model parameters required include the core state transition probability matrix, the information matrix, and the initial beliefs about the starting core state, none of which can be observed directly. For this reason, HMM estimation typically uses longitudinal data from a series of observations to infer the HMM that generated them. In the case of the coin-flipping game, if the initial starting state is unknown (i.e., which coin is flipped first) and the probability transition between coins, and the probabilities of the coin landing heads or tails are unknown, then a decision maker could observe many rounds of play, collecting random sequences of heads and tails to learn the HMM model parameters that are most likely to have generated the observed sequences.

Maximum likelihood estimation is a fundamental approach for estimating the model that is most likely to have generated the observed sequences, denoted as $\mathcal{L}^*$. Consider the case of a set of K observed sequences indexed by $v = 1, \dots, K$ where $O^{(v)} = \{o_1^{(v)}, o_2^{(v)}, \dots o_{T_v}^{(v)}\}$. Denote the set of K observation sequences as $O = \{O^{(1)}, O^{(2)}, \dots, O^{(K)}\}$. Thus, the goal is to find the model $\mathcal{L}^*$ that maximizes the likelihood of observing the observation sequences. The likelihood of the observations is

$$P(O|\mathcal{L})) = \prod_{v=1}^{K} P(O^{(v)}|\mathcal{L}). \tag{7.4}$$

Therefore, the problem of finding $\mathcal{L}^*$ can be expressed as:

$$\mathcal{L}^* = \arg\max_{\mathcal{L}\in\mathcal{M}} \prod_{v=1}^{K} P(O^{(v)}|\mathcal{L}). \tag{7.5}$$

where $\mathcal{M}$ is the set of all possible HMM models.

The optimization model of Equation 7.5 can be challenging to solve, but there is a substantial literature on solution methods. They can be separated into two types. The first type employs optimization algorithms to search the space of the decision variables, i.e., the parameters that define the HMM model. Since the model is non-convex, special considerations are necessary to limit the risk of finding locally optimal solutions and missing the globally optimal solution.

Approaches that use iterative algorithms, known as *expectation-maximization* (EM) methods, have been proposed and are frequently used over direct optimization because they are often more efficient. Moreover, these methods have been refined to handle the common situation in which some data in the observation sequences may be missing. This could happen if the observation process is unreliable for some reason. For example, if the observations are test results from patients being monitored for disease progression, patients may not adhere perfectly to the testing protocol, leaving some epochs without observations. Complete details about these methods are beyond the scope of this book, but the concluding remarks to this chapter provide references for interested readers.

7.4 POMDP Model Formulation

POMDP models incorporate the elements of an HMM and augment them with a reward function that captures the decision-maker's overall goal, as in MDPs. POMDPs assume that the decision-maker does not know the system's exact core state. Instead, the core state is replaced by a *belief state* that defines a probability distribution over the finite set of core states. The POMDP model parameters are as follows:

Decision horizon: $t = 1, 1, 2, \dots, T$,

Core state at epoch $t : s_t \in S = \{S_1, S_s, \dots, S_N\}$,

Decision at epoch $t : a_t \in A = \{A_1, A_2, \dots, A_M\}$,

Observation at epoch $t : o_t \in O = \{O_1, O_2, \dots, O_K\}$,

Reward at epoch t for decision a_t in belief state $\vec{b}_t : r_t(\vec{b}_t, a_t) = \sum_{s_t \in S} b_t(s_t) r_t(s_t, a_t)$,

Discount rate: λ,

Initial state probability distribution: $(b_0(S_1), \dots, b_0(S_N))$,

Transition probabilities from core state s_t to core state s_{t+1} given decision $a_t : p_t(s_{t+1}|s_t, a_t)$.

Observation probabilities conditional on the core state s_t given decision $a_t : q_t(o_t|s_t, a_t)$

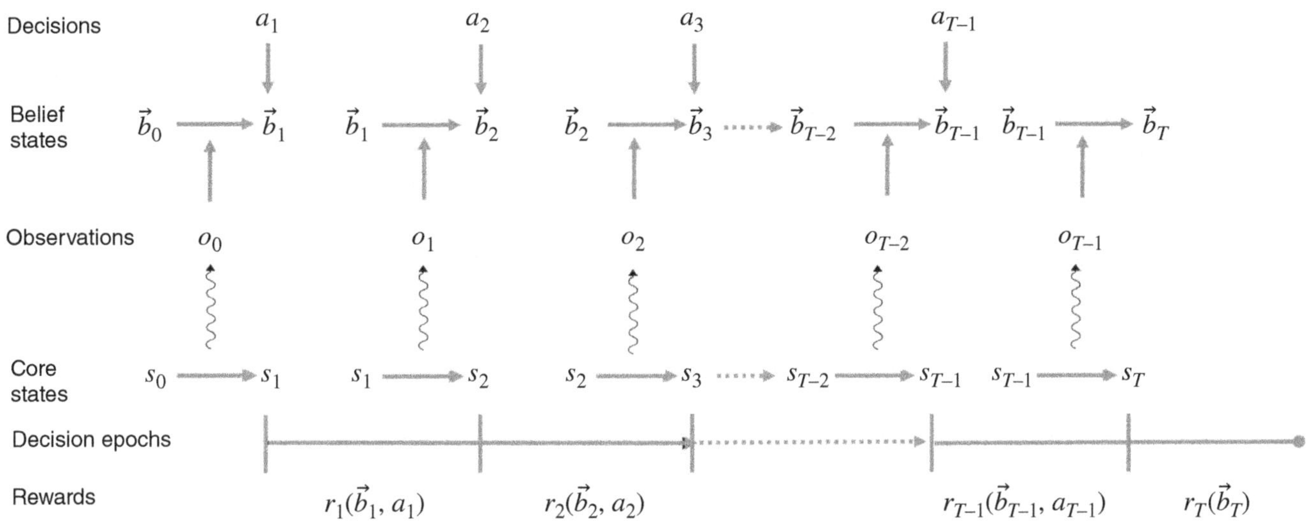

Figure 7.1 Illustration of the sequential decision-making process for a POMDP, including decisions, beliefs, observations, o_t, core state transitions, and rewards, $r_t(b_t, a_t)$, from the start (left) to the finish (right) of the finite time horizon. Observations are signals that depend on the hidden core state at each decision epoch. The observations and prior belief, $\vec{b}_t$, are used to update the belief vector, $\vec{b}_{t+1}$. Decisions are based on the belief vector at each decision epoch with the goal of maximizing expected rewards.

Figure 7.1 illustrates this sequential state transition, observation, and decision process. The core states and their transition probability matrix are denoted as above. The core state transitions at each epoch are determined by the transition probability matrix $P(a_t)$, and the observations are probabilistically linked to the core states through the information matrix, $Q_t(a_t)$. As mentioned earlier, each row of the information matrix corresponds to a core state, and each column entry is the probability of a particular observation conditional on the core state. The probabilistic dependencies among the core state, the observations, and the decision made at each epoch are used to sequentially estimate the belief vector via Bayesian updating. Thus, the system starts at $t = 0$ in state s_0 and emits observation o_0. At the start of the next decision epoch, $t = 1$, the prior observation, prior belief, and decision $(o_0, \vec{b}_0, a_1)$ are used to update the belief, $\vec{b}_1$ at epoch 1. The process continues through the remainder of the decision horizon.

Given the above definitions for a POMDP, the resulting optimality equations are as follows:

$$v_t(\vec{b}_t) = \max_{a_t \in A} \left\{ r_t(\vec{b}_t, a_t) + \lambda \sum_{s_t \in S, s_{t+1} \in S, o_{t+1} \in O} b_t(s_t) p_t(s_{t+1} | s_t, a_t) q_{t+1}(o_{t+1} | s_t, a_t) v_{t+1}(\vec{b}_{t+1}) \right\}, \tag{7.6}$$

for $t = 1, \ldots, T-1$, and a boundary condition $v_T(\vec{b}_T) = r_T(\vec{b}_T)$, where $r_T(\vec{b}_T)$ is the terminal reward, often defined as the expected reward to go beyond the end of the time horizon, where the expectation is with respect to the belief state.

The optimality Equation 7.6 look similar to the standard MDP described in Chapter 4; however, the approaches for solving these problems differ significantly for several reasons, including the fact that $\vec{b}_t$ is continuous. Thus, the system's states are continuous and multidimensional, adding difficulty beyond the simpler finite-state models of Chapters 3 and 4.

In a POMDP model, the decision maker chooses decisions to gather information about the system's state and to achieve a goal defined by the reward function. For example, the problem of optimizing when to perform maintenance on a machine in a factory setting has been modeled as a POMDP, where the decision at each epoch is whether to perform an inspection. Inspections can be costly, but they can also provide information about the machine's underlying condition (e.g., the likelihood of a breakdown in the next time period). Thus, an optimal policy for machine maintenance must balance the cost of inspections with the long-term benefits of preventive maintenance, which could increase overall production and cost efficiency.

7.5 Solution Methods for POMDPs

POMDPs are generally much more challenging to solve than MDPs. Whereas MDPs can be solved exactly in polynomial time, as discussed in Chapter 4, POMDPs have computation time complexity that grows exponentially in the number of decisions, observations, and decision epochs; moreover, the dimension of the belief vector increases with the number of core states. Early methodological studies of POMDPs focused on exact methods that exploit the fact that the optimal value function for a POMDP is convex and, in the finite-horizon case, it is piecewise linear and convex and, therefore, is expressible using a set of lower bounding inequalities known as *supporting hyperplanes*, or in the context of POMDPs, *α-vectors*, which have dimension N corresponding to the number of core states. One of the first exact methods was the *single-pass* method, given in Algorithm 7.1, which, similar to backward induction for MDPs, covered in Chapter 4, uses backward recursion starting with a boundary condition in epoch T. The optimal value function at any epoch before epoch T can be written as a function of the belief vector as follows:

Algorithm 7.1 Algorithm for Computing the Optimal Value Function for a POMDP

Input: POMDP elements: decision epochs, states, decisions, transition probability matrix, information matrix, rewards, discount factor
Boundary Condition: $\Omega_T = \{\vec{r}_T\}$
Backward Induction:
for $t = T - 1$ to 1 **do**
  Generate Ω_t and record the decision a_t associated with each vector generated
  for all $\vec{\alpha}_t \in \Omega_t$ **do**
    if $LP(\vec{\alpha}_t, \Omega_t) > 0$ **then** $\Omega_t \leftarrow \Omega_t \setminus \{\vec{\alpha}_t\}$
    end if
  end for
end for
Return: Minimal α-vector set Ω_t for all $t = 1, \ldots, T$.

$$v_t(\vec{b}_t) = \max_{\vec{\alpha}_t \in \Omega_t} \left\{ \vec{b}_t \cdot \vec{\alpha}_t \right\}, \forall \vec{b}_t, t = 1, \ldots, T - 1 \tag{7.7}$$

where Ω_t, in the context of POMDPs, is a finite set of α-vectors defining the optimal value function at epoch t. Each α-vector is associated with one or more decisions; therefore, Equation (7.7) encodes the optimal decisions at each epoch t and belief point $\vec{b}_t$. At a given epoch t, Ω_t can be recursively generated as follows:

$$v_t(\vec{b}_t) = \max_{a_t \in A} \left\{ \vec{b}_t \cdot \vec{r}_t(a_t) + \lambda \sum_{o_t \in O} \max_{\vec{\alpha}_{t+1} \in \Omega_{t+1}} \left\{ \vec{b}_{t+1} \cdot \vec{\alpha}_{t+1} \right\} \overline{p}_t(o_t | \vec{b}_t, a_t) \right\}. \tag{7.8}$$

The first term in the left-most maximization operation in Equation (7.8) is the dot product of a vector of rewards for decision a_t, an N-dimensional vector $\vec{r}(a_t) = (r(S_1, a_t), \ldots, r(S_N, a_t))$, and the belief vector $\vec{b}_t$. The second term is the expectation over all observations, O, for the optimal value function (with Equation 7.7 substituted for $v_t(\vec{b}_t)$, and $\overline{p}_t(o_t | \vec{b}_t, a_t)$ denotes the probability of observing o_t given the belief state is $\vec{b}_t$, and decision a_t is selected, which can be written as follows:

$$\overline{p}_t(o_t | \vec{b}_t, a_t) = \sum_{s_t \in S} p(o_t | s_t, a_t) b_t(s_t).$$

By substituting Equation (7.1) into Equation 7.8, this can be rewritten in the following form with the belief vector $\vec{b}_t$ conveniently factored out as follows:

$$v_t(\vec{b}_t) = \max_{a_t \in A} \left\{ \vec{b}_t \cdot \left(\vec{r}_t(a_t) + \lambda \sum_{o_t \in O} \max_{\vec{\alpha}_{t+1} \in \Omega_{t+1}} \sum_{s_t \in S} q_t(o_t | s_{t+1}, a_t) p_t(s_{t+1} | s_t, a_t) \vec{\alpha}_{t+1} \right) \right\}, \tag{7.9}$$

for all $\vec{b}_t$ in the *unit simplex*, which is the region defined by the requirements for the belief vector to be a probability distribution, $\{\vec{b}_t \mid \sum_{s_t \in S} b_t(s_t) = 1 \text{ and } 0 \leq b_t(s_t) \leq 1\}$ for all epochs $t = 0, 1, 2, \ldots, T - 1$. The boundary condition at the last epoch, T, is

$$v_T(\vec{b}_T) = \vec{b}_T \cdot \vec{r}_T,$$

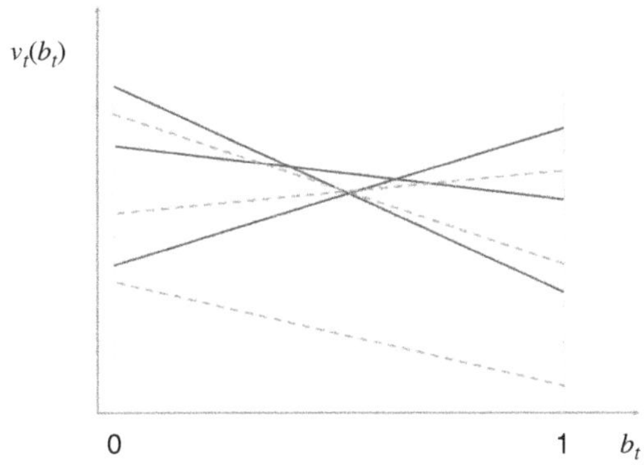

Figure 7.2 Plot of a hypothetical optimal value function for a two-dimensional belief vector at some epoch t where b_t denotes the belief in state 1 and $1 - b_t$ is the belief in state 2 (not pictured). Dashed lines are vectors that are dominated by solid lines.

for all $\vec{b}_T$ in the unit simplex. The vector $\vec{r}_T$ has elements corresponding to the end-of-horizon reward for each state $s_t \in S$. Note that $\vec{r}_T$ is the α-vector at epoch T.

Given the reorganized definition of the optimal value function in Equations 7.9, solving the POMDP becomes equivalent to finding the α-vector set that describes $v_t(\cdot)$ at each epoch t for all belief vectors. The single-pass algorithm, mentioned previously, constructs the set Ω_t from the previous iteration's α-vector set, Ω_{t+1}, using backwards induction, as follows:

$$\Omega_t = \left\{ \vec{\alpha}_t \mid \vec{\alpha}_t = \vec{r}_t(a_t) + \lambda \sum_{o_t \in O} \sum_{s_t \in S} q_t(o_t | s_{t+1}, a_t) p_t(s_{t+1} | s_t, a_t) \vec{\alpha}_{t+1} \mid \forall a_t \in A, \forall \vec{\alpha}_{t+1} \in \Omega_{t+1} \right\}.$$

As defined, Ω_t is the finite set of all possible α-vectors. Iterating backward through decision epochs recursively generates the set of α-vectors at each epoch, which grows exponentially with respect to the number of epochs, T; however, typically, a large number of vectors are unnecessary because one or more other vectors *dominate* them. That is, a vector is *dominated* if no belief point exists at which the given vector is necessary to define the optimal value function (see Figure 7.2 for an illustration of dominated and non-dominated alpha vectors). Thus, such vectors can be discarded. Determining dominated α-vectors at a given epoch t can be done through a process called *pruning*, which can be achieved using linear programming. Let $LP(\vec{\alpha}_t, \Omega_t)$, for a particular α_t denote the following linear program:

$$\begin{aligned} \min \quad & z = y - \sum_{s_t \in S} b_t(s_t)\alpha_t(s_t) \\ \text{Subject to} \quad & \sum_{s_t \in S} b_t(s_t)\alpha'_t(s_t) \le y, \ \forall \vec{\alpha}'_t \in \Omega_t \\ & \sum_{s_t \in S} b_t(s_t) = 1 \\ & 0 \le b_t(s_t) \le 1, \forall s_t \in S. \end{aligned} \tag{7.10}$$

If the optimal solution z^* to $LP(\vec{\alpha}_t, \Omega_t)$, is such that $z^* > 0$, then $\vec{\alpha}_t$ is dominated and can be removed from Ω_t; otherwise, it is non-dominated for at least one belief point and thus is retained in the set of α-vectors. To see why this is the case, consider an optimal solution, $\vec{b}_t^*$ to $LP(\vec{\alpha}_t, \Omega_t)$ for which $z^* > 0$, which implies

$$y^* > \sum_{s_t \in S} b_t^*(s_t)\alpha_t(s_t).$$

The decision variable y is equivalent to the optimal value function for any given $\vec{b}_t$ because

$$y = \max_{\vec{\alpha}_t \in \Omega_t} \left\{ \vec{b}_t \cdot \vec{\alpha}_t \right\}.$$

If no belief point exists for which $z^* > 0$, then $\vec{\alpha}_t$ is unnecessary to define the optimal value function because it is dominated and can be discarded. Thus, solving $LP(\vec{\alpha}_t, \Omega_t)$ can be viewed as a test for $\vec{\alpha}_t$ to see if at least one belief point exists where $\vec{\alpha}_t$

supports the optimal value function. The process can be accelerated by continually removing all dominated α vectors from Ω_t after each test, reducing the size of $LP(\vec{\alpha}_t, \Omega_t)$ by eliminating unnecessary constraints.

Algorithm 7.1 provides pseudo-code for generating the minimal set of non-dominated alpha vectors at each decision epoch from which the optimal policy for any given belief state, $\vec{b}_t$, can be computed using Equation (7.8).

With the minimal α-vector set determined by Algorithm 7.1, the optimal decision at any epoch t and belief vector $\vec{b}_t$ can be determined by first identifying the supporting α-vector at $\vec{b}_t$ as follows:

$$\vec{\alpha}_t^* = \text{argmax}_{\vec{\alpha}_t \in \Omega_t} \{\vec{b}_t \cdot \vec{\alpha}_t\},$$

and the optimal decision is associated with vector $\vec{\alpha}_t^*$.

Example 7.2: *Active surveillance of prostate cancer*

Active surveillance is commonly recommended for patients with low-risk prostate cancer, as defined by a patient's pathology (as determined by prior diagnostic biopsy); it involves routine biopsies for patients to confirm whether their cancer remains a low risk or whether it has progressed to a high risk, in which case it should be treated with options such as surgery or radiation therapy. Biopsies involve sampling to detect a high-grade tumor. Because of the limited number of samples, biopsies are imperfect and subject to sampling error. The HMM has core health states for low- and high-risk cancer. Transition probabilities determine the probability of progression from a low- to a high-grade cancer state, which, if detected, is treated (please see Table 7.1 for a description of the mathematical notation). Annual decision epochs are indexed as $t = 1, \ldots, T$, where $t = 0$ denotes the initial year of diagnosis of a patient with low-risk prostate cancer before the start of the decision process, and T is the age at which surveillance is discontinued. The model core state at epoch t is denoted by $s_t \in S = \{S_L, S_H\}$, where S_L denotes patients with low-risk cancer and S_H denotes patients with high-risk cancer. As patients in the high-risk state do not return to the low-grade state, the transition probability matrix is that of an absorbing Markov chain:

$$P = \begin{bmatrix} p(S_L|S_L) & p(S_H|S_L) \\ 0 & 1 \end{bmatrix}.$$

At $t = 0$, patients begin active surveillance assuming they are in state S_L; however, as mentioned above, they could be in state S_H due to biopsy sampling error (a false negative). Let $\vec{b}_0 = (b_0(S_L),\ b_0(S_H))$ denote the initial belief vector of patients in states S_L and S_H at their first active surveillance biopsy. When the decision is to conduct a biopsy, the model observes $o_t \in O = \{O_-,\ O_+\}$ at epoch t, where O_- denotes a biopsy observation indicating low-risk cancer and O_+ denotes a biopsy observation indicating high-risk cancer. Note that the choice of the word "suggests" for the case of O_- is intentional, because biopsies are imperfect and could miss a high-risk tumor. The following is the information matrix:

$$Q = \begin{bmatrix} q(O_-|S_L) & q(O_+|S_L) \\ q(O_-|S_H) & q(O_+|S_H) \end{bmatrix}.$$

Table 7.1 POMDP model parameter values for active surveillance of prostate cancer.

Model parameter	Value	Description
$\vec{b}_0$	(0.942, 0.058)	Initial belief for the patient immediately following diagnosis.
$P(S_H|S_L)$	0.069	Probability of transition from low to high risk state.
$q(O_+|S_H)$	0.72	Probability of a positive biopsy in the high-risk state.
$q(O_+|S_L)$	0.0	Probability of a positive biopsy in the low-risk state.
θ	−0.5	Cost penalty for a one-year delay in detecting the high-risk state.
η	−0.5	Cost penalty for a biopsy.

The nature of biopsy procedures is that they collect multiple tissue samples for evaluation by a pathologist. If any of the biopsy samples indicate high-risk cancer, then the decision-maker can be assumed to have high-grade cancer with certainty. However, if the biopsy does not indicate high-risk cancer, the decision-maker must be concerned with the non-zero probability that the samples simply missed the high-grade tumor. Therefore, if a biopsy result is O_+, it is assumed the patient exits the system in state S_H and receives treatment such as surgery or radiation therapy. If the biopsy result is O_-, the decision maker is uncertain about the patient's exact state, but the new information can be used to perform a Bayesian update to the patient's belief state. These considerations are important in deciding whether to perform the next biopsy, which carries short-term harms to the patient, or to defer the biopsy, which yields no new information and risks delaying detection of a dangerous, high-risk cancer.

With a model formulation in place, maximum likelihood estimation, as discussed in Section 7.3 can be used to find a model that maximizes the likelihood of the observations, which in this case is the collection of sequences of biopsy observations for each patient.

The reward function is defined based on a weighted sum of two criteria: (1) the number of biopsies a patient receives throughout their time on surveillance and (2) the delay in detecting the patient entering the high-risk state, S_H. Let $r(s_t, a_t, o_t)$ denote the reward function when the system is in state $s_t \in S$, decision $a_t \in A$ is taken, and output $o_t \in O = \{O_-, O_+\}$ is observed if a biopsy is done at epoch t. Thus, the reward function can be written as follows:

$$r(s_t, a_t, o_t) = \begin{cases} 0, & a_t = \text{Defer Biopsy},\ s_t = S_L \\ \theta, & a_t = \text{Defer Biopsy},\ s_t = S_H \\ \eta, & a_t = \text{Biopsy},\ s_t = S_L, o_t = O_- \\ \eta, & a_t = \text{Biopsy},\ s_t = S_H, o_t = O_+ \\ \theta + \eta, & a_t = \text{Biopsy},\ s_t = S_H, o_t = O_-, \end{cases}$$

where θ and η are non-positive scalars that denote the negative reward, i.e., penalty, of one year of delay in detecting high-risk cancer and the burden of a biopsy, respectively. In the optimality equations, therefore, the objective is to minimize a weighted sum of the expected number of biopsies and the number of years of delay in detecting cancer progression. Without loss of generality, the weights are normalized such that $-\theta - \eta = 1$. Now, the expected reward is defined as follows:

$$r(\vec{b}_t, a_t) = \sum_{s_t \in S_t} b_t(s_t) \sum_{o_t \in O_t} Q(o_t|s_t, a_t) r(s_t, a_t, o_t).$$

Consequently, the optimality equations can be written as follows:

$$v_t(\vec{b}_t) = \max_{a_t \in A} \left\{ r(\vec{b}_t, a_t) + \sum_{s_t \in S, s_{t+1} \in S, o_{t+1} \in O} b_t(s_t) p_t(s_{t+1}|s_t, a_t) q_{t+1}(o_{t+1}|s_t, a_t) v_{t+1}(\vec{b}_{t+1}) \right\}, \forall b_t, t = 0, \ldots, T-1,$$

with the following boundary condition at the end of the decision horizon:

$$v_T(\vec{b}_T) = \max_{a_T \in A} \{ r(\vec{b}_T, a_T) \}.$$

Figure 7.3 shows the set of α-vectors obtained from solving the POMDP using Algorithm 7.1 for the optimal value function at $t = 0$. Each α-vector is associated with one of the two decisions: biopsy or defer biopsy. The x-axis represents the element of the belief vector, $b_1(S_H)$, corresponding to the high-risk cancer state, S_H. Each point on the x-axis has a corresponding α-vector that contains the optimal decision for that belief point (the vector with the highest value at the belief point). Thus, Figure 7.3 provides a complete representation of the optimal policy for the POMDP at the first decision epoch. The nature of the optimal policy is such that it is always optimal to conduct a biopsy when $b_1(S_H) \geq 0.28$ and to defer biopsy otherwise.

7.6 Approximation Methods for POMDPs

Algorithm 7.1 conveys a conceptual understanding of an exact approach for POMDPs; however, it is suitable only for very small POMDPs, such as the two-core-state active surveillance example above, due to the exponential growth of the size of such problems in the number of decisions, decision epochs, and observations. Many authors have built on this early

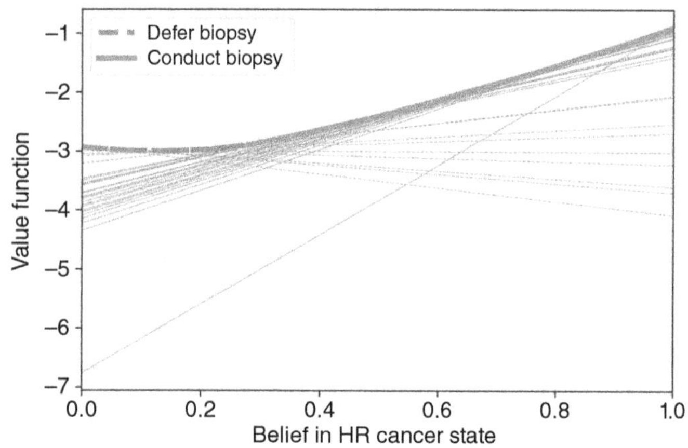

Figure 7.3 Graph of the optimal value function with non-dominated α-vectors for the Partially Observable Markov Decision Process for active surveillance at the first decision epoch, $t = 1$, as a function of $b_1(S_H)$.

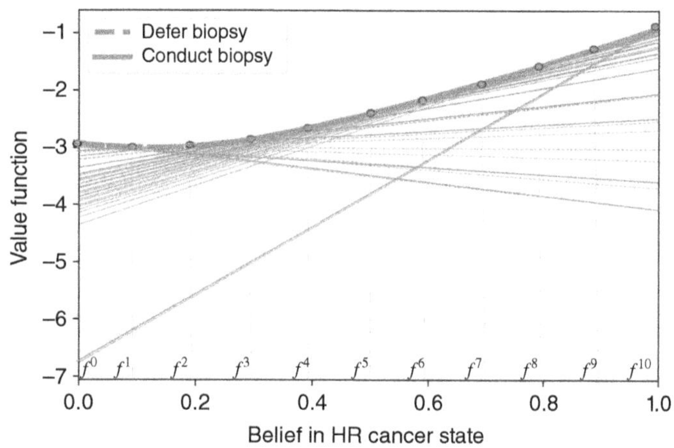

Figure 7.4 Graph of a fixed-finite-grid approximation of the belief space for the partially observable Markov decision process for active surveillance at the first decision epoch, $t = 1$, as a function of $b_1(S_H)$ with 11 equally spaced grid points,f^0, f^1, ..., f^{10}. Only the α-vectors selected at each grid point are shown.

approach to solving POMDPs. As alluded to above, some of these methods seek to more efficiently prune the dominated α-vectors to reduce the computational burden of the single-pass algorithm; however, even these pruning methods are generally limited to small POMDPs, although they have the merit of remaining true to the scenario parameters (in a truncated, or pruned, form). For this reason, they are considered *exact* methods with some efficiency gains over earlier proposed methods.

An alternative approach to addressing the computational burden of large POMDPs is to introduce approximations. Approximation methods have been a focus for solving practical POMDPs since 1966 and subsequent explorations in the 1990s and beyond. One well-known approximation method for POMDPs, originally presented by Eckles, is the *fixed-finite-grid algorithm*. This approach approximates the continuous belief space with a finite set of belief points, yielding a completely observable MDP that approximates the original POMDP. Figure 7.4, using the example of prostate cancer surveillance again, illustrates the basic idea for a small problem with two core states and thus a belief vector with two elements at each epoch, $(b_t^1,\ b_t^2)$. Given that the elements of the belief vector sum to one, the selection of points lies along one dimension. Figure 7.4 illustrates this for the case of an 11-point grid where $\mathcal{B} = \{f^0,\ f^1,\ f^2, ..., f^{10}\}$ defines at equal increments the increasing belief that the patient is in the high-risk state, $b_t(S_H)$ starting from 0, with increments of 0.1. The α-vectors each correspond to a supporting hyperplane for the respective grid point. The total number of α-vectors equals the number of grid points (11 in this case), which is a significant reduction; furthermore, pruning is much more efficient since it only requires determining a dominant α-vector at each grid point.

Python Example 7.3: POMDP for search-and-rescue

The fixed-finite-grid algorithm is illustrated in this example of a search-and-rescue activity, in which a search team navigates a two-dimensional area to find an injured person (target) in one of four regions. The target's exact location at any time is unknown. The search team needs to find the target location within a finite number of decision epochs, T, that defines the decision horizon. The following is the mathematical model formulation and model parameters.

S_1 : Region 1 (search team initial location),

S_2 : Region 2,

S_3 : Region 3,

S_4 : Region 4 (target location).

At each epoch, the search team is in one of the states, indexed by s_t, but the team has imperfect information about their exact location because communications are disrupted due to severe weather. The possible decisions at each epoch, t, indexed by a_t, are as follows:

A_1 : Move Up, $S_3 \to S_1$ or $S_4 \to S_2$,

A_2 : Move Down, $S_1 \to S_3$ or $S_2 \to S_4$,

A_3 : Move Left, $S_2 \to S_1$ or $S_4 \to S_3$,

A_4 : Move Right, $S_1 \to S_2$ or $S_3 \to S_4$,

where the directions of movement (e.g. Move Up, Move Right) are consistent with Figure 7.5. For example, decision A_1 in states S_3 or S_4 tends to result in a transition to (move up to) state S_1 or S_2, respectively. However, while the decision influences the direction of movement, it is subject to uncertainty. The hidden core state transition probability matrix P for each decision $a \in \{A_1, A_2, A_3, A_4\}$ governs the influence of the decision and the core state transitions. The transition probability matrix is denoted as follows:

$$P(a) = \begin{bmatrix} p_{11}(a) & p_{12}(a) & p_{13}(a) & p_{14}(a) \\ p_{21}(a) & p_{22}(a) & p_{23}(a) & p_{24}(a) \\ p_{31}(a) & p_{32}(a) & p_{33}(a) & p_{34}(a) \\ p_{41}(a) & p_{42}(a) & p_{43}(a) & p_{44}(a) \end{bmatrix},$$

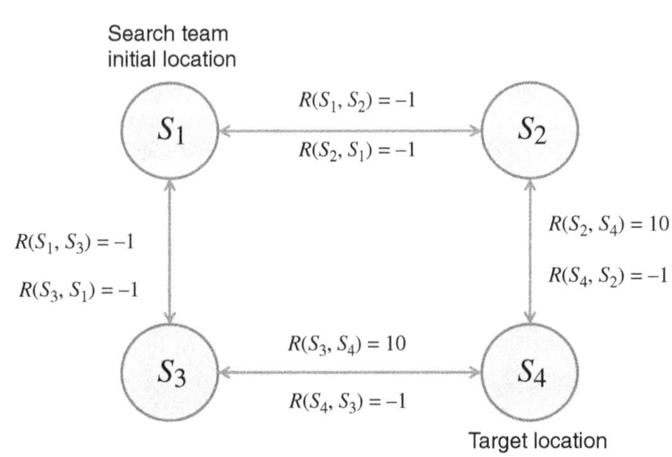

Figure 7.5 Four interconnected regions for search and rescue operations. Rewards denote cost of −1 for moves that do not end in reaching the target location and a reward of 10 for moves that end in the target location.

where $p_{ij}(a_t) = p(S_j|S_i, a)$ denotes the probability of transitioning from state S_i to state S_j given decision a. Note that $P(a)$ has no index for decision epochs because the transition probabilities are stationary, with the following values:

$$P(A_1) = \begin{bmatrix} 0.8 & 0.1 & 0.1 & 0.0 \\ 0.1 & 0.8 & 0.0 & 0.1 \\ 0.8 & 0.0 & 0.1 & 0.1 \\ 0.0 & 0.0 & 0.0 & 1.0 \end{bmatrix}, \quad P(A_2) = \begin{bmatrix} 0.1 & 0.1 & 0.8 & 0.0 \\ 0.1 & 0.1 & 0.0 & 0.8 \\ 0.1 & 0.0 & 0.8 & 0.1 \\ 0.0 & 0.0 & 0.0 & 1.0 \end{bmatrix},$$

$$P(A_3) = \begin{bmatrix} 0.8 & 0.1 & 0.1 & 0.0 \\ 0.8 & 0.1 & 0.0 & 0.1 \\ 0.1 & 0.0 & 0.8 & 0.1 \\ 0.0 & 0.0 & 0.0 & 1.0 \end{bmatrix}, \quad P(A_4) = \begin{bmatrix} 0.1 & 0.8 & 0.1 & 0.0 \\ 0.1 & 0.8 & 0.0 & 0.1 \\ 0.1 & 0.0 & 0.1 & 0.8 \\ 0.0 & 0.0 & 0.0 & 1.0 \end{bmatrix}.$$

Next, the observation probability matrix Q for each core state (rows) and observation (columns) is defined. The elements of the information matrix q_{ij} denote the probability of observation o_j given the search team is in state s_i. The search team periodically receives noisy communications about their location. With probability 0.7, the observation is accurate; otherwise, the true state is equally likely to be one of the other three states, each with probability 0.1. One exception is state S_4, where the search team finds the target and thus accurately confirms the location as S_4, ending the decision-making process. Thus, the information matrix is as follows:

$$Q = \begin{bmatrix} 0.7 & 0.1 & 0.1 & 0.1 \\ 0.1 & 0.7 & 0.1 & 0.1 \\ 0.1 & 0.1 & 0.7 & 0.1 \\ 0 & 0 & 0 & 1.0 \end{bmatrix}.$$

So, for example, if the noisy communication indicates the state is S_2, the probability this is correct is 0.7; the probability the true state is S_1, S_3, S_4, is 0.1 for each. The rewards at each epoch are determined by the core state transitions, which can be expressed in matrix form as follows:

$$R = \begin{bmatrix} r(S_1,S_1) & r(S_1,S_2) & r(S_1,S_3) & r(S_1,S_4) \\ r(S_2,S_1) & r(S_2,S_2) & r(S_2,S_3) & r(S_2,S_4) \\ r(S_3,S_1) & r(S_3,S_2) & r(S_3,S_3) & r(S_3,S_4) \\ r(S_4,S_1) & r(S_4,S_2) & r(S_4,S_3) & r(S_4,S_4) \end{bmatrix} = \begin{bmatrix} 0 & -1 & -\infty & -1 \\ -1 & 0 & 10 & -\infty \\ -\infty & -1 & 0 & -1 \\ -1 & -\infty & 10 & 0 \end{bmatrix},$$

where $r(S_i, S_j)$ represents the reward for transitioning from state S_i to state S_j. Each move that does not result in acquiring the target has a cost of -1. Moves to the target location have a reward of 10, and moves that are not allowed have a reward of $-\infty$. The belief state b_t at epoch t defines the probability distribution over the hidden core states:

$$\vec{b}_t = (b_t(S_1), b_t(S_2), b_t(S_3), b_t(S_4)).$$

The belief vector elements at epoch t are updated based on the decision a_t and the observation o_t, as follows:

$$b_{t+1}(s_{t+1}) \frac{\sum_{s_t \in S} b_t(s_t) p_t(s_{t+1}|s_t, a_t) q_t(o_t|s_t)}{\sum_{s_t, s_{t+1} \in S} b_t(s_t) p_t(s_{t+1}|s_t, a_t) q_t(o_t|s_t)}, \quad \forall s_{t+1} \in S. \tag{7.11}$$

Notice that Equation 7.11 is the same as Equation 7.1 except that the observation probabilities are independent of the decision a_t, as defined in this particular example. Now, for a finite horizon T, the optimal value function is computed using the optimal equations in Equation 7.6, which for this example are the following:

$$v_t(\vec{b}_t) = \max_{a_t \in A} \left\{ r(\vec{b}_t, a_t) + \lambda \sum_{s_t \in S, s_{t+1} \in S, o_{t+1} \in O} b_t(s_t) p(s_{t+1} | s_t, a_t) q(o_{t+1} | s_t) v_{t+1}(\vec{b}_{t+1}) \right\}, \tag{7.12}$$

for decision epochs, $t = 0, \ldots, T-1$, where $r(\vec{b}_t, a_t) = \sum_{s_t \in S} \sum_{s_{t+1} \in S} b_t(s_t) p(s_{t+1} | s_t, a_t) r(s_t, s_{t+1})$ and the boundary condition is

$$v_T(\vec{b}_T) = 0, \quad \forall \vec{b}_T.$$

The following Python code implements this model, using a fixed-finite grid of belief points to approximate the continuous belief.

```
import numpy as np
from itertools import product

# Input data
T = 4
nstates = 4
ndecisions = 4
nobservations = 4
states = ['S1', 'S2', 'S3', 'S4']
decisions = ['Up', 'Down', 'Left', 'Right']
observations = ['Obs_S1', 'Obs_S2', 'Obs_S3', 'Obs_S4']
N = 10 # Number of discrete points for each dimension of the belief vector
P = np.zeros((nstates,ndecisions,nstates)) # Core state transition probabilities
P[:, 0, :] = np.array([[0.8, 0.1, 0.1, 0.0], # d = 'Up'
                         [0.1, 0.8, 0.0, 0.1],
                         [0.8, 0.0, 0.1, 0.1],
                         [0.0, 0.0, 0.0, 1.0]])
P[:, 1, :] = np.array([[0.1, 0.1, 0.8, 0.0], # d = 'Down'
                      [0.1, 0.1, 0.0, 0.8],
                      [0.1, 0.0, 0.8, 0.1],
                      [0.0, 0.0, 0.0, 1.0]])
P[:, 2, :] = np.array([[0.8, 0.1, 0.1, 0.0], # d = 'Left'
                      [0.8, 0.1, 0.0, 0.1],
                      [0.1, 0.0, 0.8, 0.1],
                      [0.0, 0.0, 0.0, 1.0]])
P[:, 3, :] = np.array([[0.1, 0.8, 0.1, 0.0], # d = 'Right'
                       [0.1, 0.8, 0.0, 0.1],
                       [0.1, 0.0, 0.1, 0.8],
                       [0.0, 0.0, 0.0, 1.0]])
# Noisy observation probabilities (rows: core state, columns: observation)
Q = np.array([[0.7, 0.1, 0.1, 0.1],


              [0.1, 0.7, 0.1, 0.1],
              [0.1, 0.1, 0.7, 0.1],
              [0.1, 0.1, 0.1, 0.7]])
# Rewards for each state transition ("-9999" denotes infeasible transitions)
rewards = np.array([[0, -1, -1, -9999],
                    [-1, 0, -9999, 10],
```

```
                        [-1, -9999, 0, 10],
                        [0, 0, 0, 10]])

# Function for belief update given decision and observation
def update_belief(belief, d, o):
    belief = np.dot(P[:, d, :].T, belief)
    belief = belief * Q[:, o]
    belief = belief / np.sum(belief) # Normalize
    return belief

# Function to find the closest grid point to current belief
def find_closest_belief(belief):
    distances = np.linalg.norm(belief_points - belief, axis=1)
    return belief_points[np.argmin(distances)]

# Function to compute the expected reward given belief
def compute_expected_reward(belief, d):
    total_reward = 0
    for s in range(nstates):
        for s_next in range(nstates):
            total_reward += belief[s] * P[s, d, s_next] * rewards[s, s_next]
    return total_reward

# Function to compute expected value to go given belief
def expected_value_to_go(belief, d, v, t):
    expected_value_to_go = 0
    for o in range(nobservations):
        prob_o = 0


        for s in range(nstates):
            for s_next in range(nstates):
                prob_o += belief[s] * P[s, d, s_next] * Q[s_next, o]
        if prob_o == 0:
            continue
        new_belief = update_belief(belief, d, o)
        closest_belief = find_closest_belief(new_belief)
      closest_belief_idx = np.where((belief_points == closest_belief).all(axis=1))[0][0]
        expected_value_to_go += prob_o * v[closest_belief_idx, t + 1]
    return expected_value_to_go

# Create a finite grid, removing combinations deviating from one by > 10^-8
belief_points = np.array([p for p in product(np.linspace(0,1,N),
                                    repeat=nstates) if np.isclose(sum(p), 1.0)])
v = np.zeros((len(belief_points), T + 2))
optimal_policy = [[None] * (T + 2) for _ in range(len(belief_points))]

# Backward induction to compute optimal value function
for t in reversed(range(T + 1)):
    for i, belief in enumerate(belief_points):
        v_values = []
        expected_value = 0
```

```
        for d in range(ndecisions):
            expected_value = compute_expected_reward(belief, d)
            expected_value += expected_value_to_go(belief, d, v, t)
            v_values.append(expected_value)
        v[i, t] = max(v_values)
        optimal_policy[i][t] = decisions[np.argmax(v_values)]

# Print results
print("Optimal Value Function:")
print(v)
print("Approximate Optimal Policy:")
for t in range(T + 1):


    print(f"Time {t}: {[optimal_policy[i][t] for i in range(len(belief_points))]}")
```

Running the above code generates the approximate optimal value function for the POMDP at each decision epoch for each finite belief point used to approximate the continuous belief space. The number of grid points in the above code is 10 per dimension of the belief vector, which is 4-dimensional, with 3 degrees of freedom, because the elements must add to one since they are probabilities. The algorithm starts with all 10^4 belief points and eliminates any whose sum differs by at least 10^{-8}, yielding 240 valid belief points. The resulting approximate policy defines the decision at each of the 240 belief points for each of the 5 decision epochs.

7.7 Concluding Remarks

Some of this chapter's content is drawn from Denton [2], a tutorial on sequential decision-making for chronic diseases that covers model formulation and solution methods for medical decision-making problems. For a general introduction to POMDPs, Cassandra's tutorial is excellent (POMDP.org). Additional technical journal articles that cover early developments in solving POMDPs include Smallwood and Sondik [3], which served as the basis for Algorithm 7.1 in this chapter. Several authors have proposed alternative algorithms to lessen the computational burden, including the *witness method* [4] and *incremental pruning* [5], but even these approaches are limited to small POMDPs. Reviews by Monahan [6] and Lovejoy [7] survey properties and methods for solving POMDPs. Lovejoy, in particular, surveys approximation methods, including grid-based approximations, which have found application to many practical POMDP models (see [8] for early work on fixed-finite grids). Kaelbling et al. [9] comprehensively review the most significant properties and solution methods for POMDPs. A more recent book by Krishnamurthy [10] offers a contemporary treatment of many aspects of POMDPs.

Many resources are available for learning about HMMs and their parameter estimation, including an excellent tutorial by Rabiner [11] and a book by MacDonald and Zuchini [12]. More recently edited volumes cover a wide range of advanced topics and applications of HMMs, including Westhead et al. [13] and Bouguila et al. [14]. Khademi [15] discusses software implementations of HMM methods in R, and Moreno-Pino et al. [16] presents a Python package for estimating HMMs.

Software implementations of POMDP algorithms are also readily found. A few examples are listed here, although this is not a comprehensive list. The Julia POMDP package [17] provides implementations of MDP methods and POMDP algorithms in Julia. Zheng and Tellex [18] describes a Python library that implements methods for model building and solving POMDPs. Cassandra's tutorial on POMDPs (POMDP.org) also provides source code for implementing several popular methods for solving POMDPs.

A common problem in HMM estimation is missing data. This is particularly true when the time of observation is variable. For example, in medical contexts, the observations could be diagnostic test results; however, the frequency of such tests may vary according to clinical recommendations that consider factors such as cost, potential testing-related harm, and patients' behavior. Consequently, some patients may routinely receive recommended tests, while others do not. This results in a missing-data problem that requires careful attention, as HMM parameter estimates may be biased if the data are not missing completely at random. Indeed, missing data may introduce bias because missingness in screening data can be informative (e.g., when "sicker" patients receive more frequent screening). In many cases, missing data points or time intervals between data points are informative and can be included as observation variables in the model (e.g., based on patients' propensity for regular testing, which may be linked to the distance to a testing facility, such as a clinic).

Many methods exist for imputing missing data. Enders' book (2022) covers many such approaches at an introductory level. One particularly relevant method is the EM algorithm. The EM algorithm iteratively generates model estimates that, in theory, converge to a maximum likelihood estimate of the hidden model parameters, which include the core state transition probabilities P, information matrix Q, and initial belief b_0.

The HMM used in the prostate cancer active surveillance problem was estimated using the Baum-Welch algorithm, an EM algorithm. Additional information about this algorithm can be found in Barnett et al. [19]. An article by Li and Denton [20] describes a POMDP model for optimizing active surveillance policies in several medical studies. POMDPs have been used to study many other diseases, such as breast cancer [21, 22], colorectal cancer [23] and heart disease [24]

7.8 Practice Exercises

1. Consider the coin game of Example 7.1. Assume that the coin flipper does not show the opponent the result of the coin flip at each turn; however, the opponent can ask to see the coin. If the coin flipper is found to have been truthful, the opponent loses a point. If the coin flipper lied, the opponent gets two points. If the opponent believes the probability of being truthful is 0.6 compute the probability the opponent will catch the coin flipper in a lie at each turn. What strategy would you recommend?

2. For the search and rescue POMDP problem of Example 7.3, analyze the sensitivity of the optimal value function and the resulting policy to the number of grid points along each dimension of the belief by varying the *num_grid_points*. Summarize your findings regarding dependencies between grid design and the optimal value function and policy.

3. Show that an MDP is a special case of a POMDP.

4. Consider a game show where you have two doors to choose from, one of which reveals a room with a reward and one that reveals an empty room. You believe there is a 50% chance that the reward is behind the right door and a 50% chance that it is behind the left door. If you choose to listen to the audience, you will get a hint, but you are only 85% sure it will be correct. At each decision epoch, the reward location changes 70% of the time (i.e., if the reward is truly behind the left door at time 0, there is a 0.7 probability that it will be behind the right door at the next time period). If you choose to listen to the audience for two decision epochs, and they recommend the left door the first time and the right door the second time. What is your belief vector at decision epochs 1 and 2, respectively? Suppose your policy is to open the door as soon as you are 90% sure of the reward location or the last decision period, whichever comes first. What is your optimal policy under the following observation histories (e.g., left at epoch 1, right at epoch 2; left at epoch 1 and left at epoch 2; right at epoch 1 and left at epoch 2, right at epoch 1 and right at epoch 2)?

5. Consider a POMDP model with two core states, S_1 and S_2, and an absorbing state S_3. The decisions available in states S_1 and S_2 are A_1, A_2, A_3, where A_1 and A_2 lead to a transition to S_3 with probability 1, and A_3 leads to a stochastic state transition between S_1 and S_2 as defined by the transition probability matrices below. There are two observations, O_1, O_2, corresponding to states S_1 and S_2, respectively. The horizon is finite, the discount factor is $\lambda = 0.9$, and below are the rest of the parameters:
Immediate reward:

$$r(S_1, A_1) = -10, r(S_1, A_2) = 10, r(S_1, A_3) = 1, r(S_2, A_1) = 10, r(S_2, A_2) = -5, r(S_2, A_3) = 1$$

Core state transition probability matrix:

$$\begin{aligned} P(S_1 \mid S_1, A_3) &= 0.3, \quad & P(S_1 \mid S_2, A_3) &= 0.7, \\ P(S_2 \mid S_1, A_3) &= 0.7, \quad & P(S_2 \mid S_2, A_3) &= 0.3, \end{aligned}$$

Information matrix:

$$P(O_1 \mid S_1) = 0.8, \quad P(O_2 \mid S_1) = 0.2,$$
$$P(O_1 \mid S_2) = 0.2, \quad P(O_2 \mid S_2) = 0.8,$$

What is the decision rule at decision epoch T? What is the optimal decision rule at decision epoch $T-1$? Create a plot illustrating the set of α-vectors for each decision epoch. Suppose $T = 2$, the initial belief is $b_0(S_1) = 0.5$, $b_0(S_2) = 0.5$, and decision A_3 was made at decision epoch 0, and O_2 was observed at decision epoch 1. What is the optimal decision at epoch 1?

References

1 Cassandra, A. R. (1998). A survey of POMDP applications. In *Working Notes of the AAAI Fall Symposium on Planning with Partially Observable Markov Decision Processes*. AAAI.

2 Denton, B. T. (2018). Optimization of sequential decision making for chronic diseases: From data to decisions. In *Recent Advances in Optimization and Modeling of Contemporary Problems* (pp. 316–348). INFORMS.

3 Smallwood, R. D. & Sondik, E. J. (1973). The optimal control of partially observable Markov processes over a finite horizon. *Operations Research*, *21* (5), 1071–1088.

4 Littman, M. L., Cassandra, A. R. & Kaelbling, L. P. (1995). Efficient dynamic-programming updates in partially observable Markov decision processes.

5 Littman, M. L., Zhang, N. L. & Cassandra, A. (1997). Incremental pruning: Solving MDPs with incomplete state information.

6 Monahan, G. E. (1982). State of the art—A survey of partially observable Markov decision processes: Theory, models, and algorithms. *Management Science*, *28* (1), 1–16.

7 Lovejoy, W. S. (1991). A survey of algorithmic methods for partially observed Markov decision processes. *Annals of Operations Research*, *28* (1), 47–65.

8 Eckles, J. (1966). *Optimum Replacement of Stochastically Failing Systems* (Ph.D. thesis). Department of Engineering-Economic Systems, Stanford University, Stanford, CA.

9 Kaelbling, L. P., Littman, M. L. & Cassandra, A. R. (1998). Planning and acting in partially observable stochastic domains. *Artificial Intelligence*, *101* (1–2), 99–134.

10 Krishnamurthy, V. (2016). *Partially Observed Markov Decision Processes*. Cambridge University Press.

11 Rabiner, L. R. (1989). A tutorial on hidden Markov models and selected applications in speech recognition. *Proceedings of the IEEE*, *77* (2), 257–286.

12 MacDonald, I. L. & Zucchini, W. (1997). *Hidden Markov and Other Models for Discrete-Valued Time Series* (Vol. 110). CRC Press.

13 Westhead, D. R., Vijayabaskar, M. S. & Westhead, M. (2017). *Hidden Markov Models*. Springer, Berlin.

14 Bouguila, N., Fan, W. & Amayri, M. (Eds.). (2022). *Hidden Markov Models and Applications*. Springer.

15 Khademi, A. (2017). Hidden Markov models for time series: An introduction using R. *Journal of Statistical Software*, *80*, 1–4.

16 Moreno-Pino, F., Sükei, E., Olmos, P. M. & Artés-Rodríguez, A. (2022). PyHHMM: A Python library for heterogeneous hidden Markov models. *arXiv* preprint arXiv:2201.06968.

17 Egorov, M., Sunberg, Z. N., Balaban, E.,Wheeler, T. A., Gupta, J.K. & Kochenderfer, M. J. (2017)POMDPs. jl: A framework for sequential decision making under uncertainty. *Journal of Machine Learning Research, 18*(26), 1–5.

18 Zheng, K. & Tellex, S. (2020). pomdp_py: A framework to build and solve POMDP problems. *arXiv* preprint arXiv:2004.10099.

19 Barnett, C., Auffenberg, G., Cheng, Z., Yang, F., Wang, J., Wei, J., Miller, D., Montie, J., Mamawala, M. & Denton, B. T. (2017). Optimizing active surveillance strategies to balance competing goals of early detection of grade progression with minimizing harm from biopsies. *Cancer*, *124* (4), 698–705.

20 Li, W., Denton, B. T. & Morgan, T. M. (2022). Optimizing active surveillance for prostate cancer using partially observable Markov decision processes. *European Journal of Operational Research*, *305* (1), 386–399.

21 Ayer, T., Alagoz, O., Stout, N. K. & Burnside, E. S. (2016). Heterogeneity in women's adherence and its role in optimal breast cancer screening policies. *Management Science*, *62* (5), 1339–1362.

22 Maillart, L. M., Simmons Ivy, J., Ransom, S. & Diehl, K. (2008). Assessing dynamic breast cancer screening policies. *Operations Research*, *56* (6), 1411–1427.

23 Erenay, F. S., Alagoz, O. & Said, A. (2014). Optimizing colonoscopy screening for colorectal cancer prevention and surveillance. *Manufacturing & Service Operations Management*, *16* (3), 381–400.

24 Hauskrecht, M. & Fraser, H. (2000). Planning treatment of ischemic heart disease with partially observable Markov decision processes. *Artificial Intelligence in Medicine*, *18* (3), 221–244.

8

Reinforcement Learning

8.1 Introduction

This chapter introduces the challenging situation in which the model of an underlying system is unknown. Specifically, this challenge involves MDPs for which the transition probability matrix is unknown in advance. This is sometimes referred to as a *model-free* approach to sequential decision-making, and it is important in many contexts where system dynamics cannot be analyzed directly because longitudinal data needed to estimate transition probabilities are unavailable. A common setup for such problems is the following:

1. The state $s_t \in S$ of the system is observable at each of a discrete set of decision epochs, $t = 1, 2, ..., T$, and the sequence of state observations is assumed to come from a Markov chain, albeit one for which the transition probabilities are unknown.
2. A decision-maker selects a decision $a_t \in A$ at each of a discrete set of decision epochs, $t = 1, 2, ..., T$.
3. Based on the observed state and the selected decision, a reward is received at epoch t that depends on the state-decision pair $r_t(s_t, a_t)$.
4. The optimal policy is unclear; the decision-maker seeks to learn it through a sequential process of experimentation whereby decisions are selected, and the subsequent states and rewards are observed, from which the performance, in terms of expected discounted future rewards, can be estimated.

Problems of this type require learning a policy over time as the system's states are observed, and a decision-maker (often referred to as an "agent") experiments with decisions and observes the rewards resulting from state-decision combinations. The experimentation process is used to learn an optimal policy via *reinforcement learning* (RL).

RL is a fundamental algorithmic approach to learning that is analogous to how biological organisms make decisions. Dog owners may find the following quote from the Humane Society helpful in understanding the gist of RL:

> *Positive reinforcement training uses a reward for desired behaviors. Because the reward makes them more likely to repeat the behavior, positive reinforcement is one of your most powerful tools for shaping or changing your dog's behavior.*

The fundamental idea of positive reinforcement for desirable behavior (decisions) is to encourage choices that offer a sufficiently high expected reward, as measured by previous observations of state-decision pairs. This is also related to the general notion of trial and error that is commonly applied intuitively in human learning from an early age.

RL algorithms prescribe approaches for learning the optimal policy systematically over time through sequential interaction with the underlying system. RL approaches find many applications to problems involving sequential decision-making, including automated customer support, autonomous vehicles, fraud detection, games, healthcare, inventory control, and product pricing. The following are notable achievements that illustrate RL's progress over time.

- **1980s**: Renaissance Technologies, a financial investment company, began using RL to automate trading in financial markets.
- **1990s**: TD-Gammon, a backgammon-playing program created by Tesauro (1995) at IBM [1], learned policies by self-playing many instances of the game. It began beating human players and ultimately achieved a skill level comparable to that of the best human players.

- **2000s**: PROS Revenue Management used RL for optimizing automated dynamic ticket pricing in airlines.
- **2010s**: London-based company DeepMind created AlphaGo, a software implementation of RL that defeated the world champion Go player, Lee Sedol, in 2016.
- **2020s**: Today, RL applications are expanding rapidly, driven by companies like Amazon, Google, Meta, OpenAI, and many others to implement autonomous robots and vehicles, automation of manufacturing systems, dynamic marketing and advertising, automated medical decision-making, and many other evolving areas of decision-making.

In this chapter, RL is illustrated through several examples to demonstrate its application. The examples employ RL models and methods that combine aspects of MDPs (Chapter 4) and Monte Carlo simulation (Chapter 6). The literature on RL is vast and frequently updated, as a large research community studies RL due to its extraordinary potential to impact practical applications, particularly with recent advances in computing hardware. This chapter highlights some of the most well-known RL approaches, drawing on material covered in the previous chapters.

8.2 Contexts and Types of Learning

The context in which an RL algorithm is deployed is closely tied to its design and testing. The algorithm is driven by and tailored to the context. The context defines the physical or virtual environment in which the agent operates. Like MDPs, RL uses a system state that contains the information needed to define an agent's policy. The type of system state requires an appropriate learning technique. The approach to learning, therefore, also depends on the context; learning can occur before implementation (e.g., through simulation) or in real time, parallel to implementation, with decisions made in a real-world system. The context in which the learning happens may also vary from one application to another. For instance, some environments may be static, while others vary over time. Some may have a high cost of failure, while others do not. Alternative approaches to learning differentiate RL methods. In the remainder of this section, a general thematic overview of the relationships among contexts, algorithms, and learning approaches is presented, with several useful examples, before moving on to describe the models and methods in detail in subsequent sections.

8.2.1 Offline vs. Online Learning

A common way to categorize RL methods is *offline learning* and *online learning*. In offline learning, an RL agent learns from logged historical data before implementation in the real world. This could involve using sampled data directly or creating a simulation model that can interact with the agent. Often, this approach is applied to situations where a model can be created that describes the underlying system, such as a simulation model, but where creating an MDP is too complex to implement. In some situations, offline learning is safer than online learning because the model can be created and trained offline without interacting with the real decision-making process until the RL-based policy is sufficiently well understood. However, offline learning success depends on the simulation model, which, in many cases, may not capture all the complexities of the real world.

Online learning, conversely, involves the agent interacting directly with the real world, often with some supervision, to avoid the poor outcomes associated with making untested decisions, a process referred to as *exploration* that is critical to the success of RL. For example, the learning process for an autonomous vehicle may involve the vehicle implementing RL algorithms under the oversight of an experienced human driver, who can intervene in the decision-making process to avoid and document potentially unsafe or otherwise undesirable situations.

For many RL applications, a combination of offline and online learning is employed, with offline learning serving as a starting point. The RL algorithm can then learn in a simulated environment separate from the real-world environment it is intended for. With suitable training and without negatively impacting the real-world system, offline learning yields an initial RL algorithm that can continue learning in the online environment, reducing the need for exploration during the initial offline learning stage. This dual approach is sometimes referred to as *hybrid learning*. It offers advantages over purely online and offline learning, from both an efficiency and a safety perspective, for applications in which decisions made during the learning process can cause harm.

8.2.2 Stationary vs. Non-stationary Environments

An important distinction in learning problems is whether the underlying system being observed is stationary. Stationary systems, on the one hand, have a stationary transition probability matrix, i.e., a matrix that remains constant at each decision epoch. For these types of problems, any data observed are useful for training, independent of when they were collected, since the data are always informative about state-decision pair transitions, regardless of when they are observed.

Non-stationary contexts, on the other hand, are those in which the underlying transition probability matrix or the reward function varies over time. This is common in many contexts, such as the monitoring and control of systems that degrade over time. In such cases, failures tend to increase over time. Ignoring the dynamic nature of the environment and the agent's changing status would lead to policies that are poorly justified if all data, regardless of when they were collected, are treated equally. In such cases, the *Robbins–Monro algorithm* discussed in Section 8.5 can be a helpful approach because it enables differentiation of observations based on when they were acquired. Algorithm 8.2 implements the Robbins–Monro algorithm and shows how the recency of observations is reflected by the use of a weight factor that increases the relative effects of recently observed rewards over older observations of rewards that might not be representative of the current status of the underlying system.

8.2.3 Value Iteration vs. Policy Iteration Approaches

Similar to solution methods for MDPs, RL methods can be implemented by sequentially updating approximations of the optimal value function, as in the value iteration algorithm of Chapter 4, or by sequentially evaluating and improving policies, motivated by the policy iteration algorithm of Chapter 4, for solving MDPs. Both approaches are rooted in the use of Bellman's equations within an approximation framework that draws on statistical estimates of the value functions. As explored later in this chapter, these alternative approaches yield powerful algorithms for solving model-free sequential decision-making problems.

8.2.4 On-policy vs. Off-policy Learning

RL algorithms can be categorized based on the source of information used for learning. In *on-policy* RL algorithms, the agent learns by observing outcomes from its current policy and then updates the policy at discrete time steps, selecting a new, improved policy that produces new outcomes to further update and improve the policy. In contrast, with *off-policy* learning, the algorithm uses data from prior policies that were considered or other estimates. Often, off-policy algorithms are considered more efficient than on-policy algorithms in terms of the number of samples needed to train, but they can also be sensitive (and vulnerable) to various potentially harmful factors in the data used for learning and thus may be less predictable and less safe to use in certain contexts.

8.2.5 Exploitation vs. Exploration

The final thematic distinction between learning approaches is *exploitation* vs. *exploration*, which comprises two fundamental characteristics used in combination by all RL methods, as well as by many other algorithms for optimizing stochastic systems. The term exploitation, on the one hand, refers to the use of previously collected data to determine the best policy based on available information about the state-decision pairs and their subsequent rewards. As discussed in later sections of this chapter, the exploitation phase of RL algorithms can be implemented using either value iteration or policy iteration, which use the best-available estimates of the value function to improve the current policy. The term exploration, on the other hand, refers to the selection of decisions via a random process that guarantees each decision has a finite, non-zero probability of being considered during the learning process. Exploration is fundamentally important to ensure that RL algorithms do not become trapped in a local optimum. Without exploration, the exploitation step may become trapped in a suboptimal policy that continually reinforces itself by selecting the same decision for each state.

8.3 Greedy Monte Carlo Policy Iteration

For ease of understanding, the simplest RL algorithm is *Monte Carlo Policy Iteration* (MCPI), as it mirrors standard policy iteration for MDPs, as discussed in Chapter 4. It utilizes concepts from Monte Carlo simulation, covered in Chapter 6, to obtain a statistical estimate of the optimal value function for each state-decision pair. MCPI can be summarized in three steps:

Step 1: Pick an initial policy.
Step 2: Evaluate the policy and estimate expected rewards for using Monte-Carlo simulation.
Step 3: Select a new policy that improves upon the current policy. Return to Step 2.

MCPI is an off-policy and offline learning algorithm in its typical implementation. In the MCPI setting, an agent observes data sequentially over time in the form of state-decision pairs $(s_t,\ a_t)$, which are used to evaluate and improve policies that make decisions intended to achieve the agent's goal. As in the MDPs of previous chapters, the goal is expressed as a reward function that quantifies the agent's overarching objective. In the RL context, the algorithm learns the optimal policy over time by observing the effects of the policy on the resulting state and decision pairs and identifying opportunities to improve the policy.

For MCPI to converge to the optimal policy, each state-decision pair $(s_t,\ a_t)$ must be visited an infinite number of times. However, this asymptotic expectation is clearly not feasible in practice, and, thus, a more practical approach is needed. For this initial introduction, the practical challenges are ignored for now, focusing on conceptual understanding. Afterwards, finite sample approaches, with examples illustrating their applications, are presented as pragmatic alternatives.

The elements of an RL model are the states $s_t \in S$, an initial policy of interest defined by decisions at each state, $\pi(s_t) \in A$, decision epochs $t = 0, \ldots, T$, the discount factor λ, and the reward function $r(s_t, \pi(s_t))$, for all state-decision pairs $(s_t, \pi(s_t))$. The probability distribution of the initial state, denoted by vector $\vec{q}$, starts by randomly sampling according to the distribution defined by q to select an initial state of the system s_0, and the initial policy of interest defines the decision $\pi(s_0)$. At each subsequent epoch t, a decision is selected according to policy π, and then the next state, s_{t+1}, is randomly sampled. The sampling of future states may be done by interacting with an actual system (online learning) or using a simulation model that generates state-decision pairs (offline learning). Regardless of how states are observed, the following rewards are collected at each epoch:

$$r_0(s_0, \pi(s_0)),\ r_1(s_1, \pi(s_1)), \ldots, r_T(s_T, \pi(s_T)).$$

Figure 8.1 illustrates the main steps of this process described above. The sampled rewards are then used to estimate the value function as follows:

$$\tilde{v}_0^{\pi}(s_0) = \left(\sum_{t=0}^{T-1} \lambda^t r_t(s_t, \pi(s_t))\right) + \lambda^T r_T(s_T),$$

where the ~ symbol is used to denote that $\tilde{v}^{\pi}(s_0)$ is a statistical estimate of the true value function. As the number of samples N, tends to ∞, $\tilde{v}_0^{\pi}(s_0)$ asymptotically approaches $v_0^{\pi}(s_0)$. In other words, as the number of samples increases, so does confidence in the statistical estimate of the mean rewards. Thus, by the law of large numbers, the value function estimate converges to the true value function. However, in practice, the number of sample paths N, is finite and must be chosen to trade off between (a) some desired level of confidence and (b) a computational budget, as the following example illustrates.

Example 8.1: *Monte Carlo sampling for decision selection*

Consider a two-player game where player one holds two coins: one is biased towards landing tails up, and the other is biased towards landing heads up. The second player knows there are two different types of coins, but does not know: (a) which coin has which bias and (b) the likelihood of each of them turning up heads. In each turn, the second player selects the coin that player one will flip. If it lands tails, player two gets one point; otherwise, if it lands heads, player one gets a point. What is the best way to play this game?

The game is similar to the example in Chapter 7, which is based on an HMM, with two important differences. First, the system states are directly observable by the decision-maker. Second, the state transition probability matrix is unknown. In the RL context, the decision is the selection of which coin your friend flips, $a_t = \{a_1,\ a_2\}$, and the state is the outcome (heads or tails).

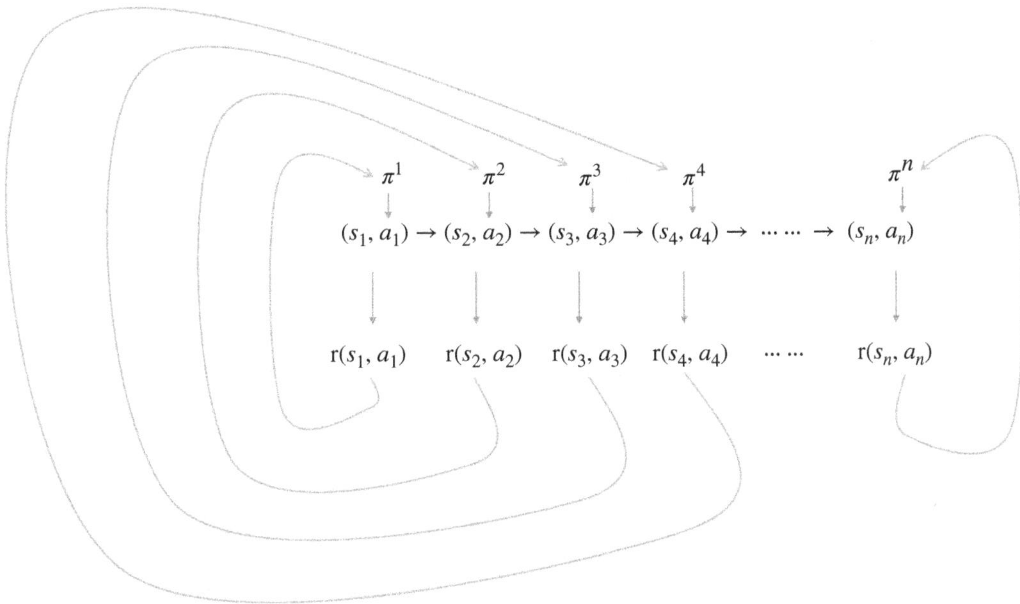

Figure 8.1 Illustration of the RL process, whereby the agent's policy is learned through updating the policy ($\pi^1, \pi^2, ..., \pi^n$) to select decisions given a state of the system based on the state-decision-reward causal relationship.

A natural approach is to experiment to try to infer which coin is which. Suppose player two decides to play the game for 40 turns and uses the first 20 turns to try and identify the better coin and chooses the coin with the highest reward for the remaining 20 turns. In other words, player one tests each coin to explore the two alternatives for a total of 20 turns (10 turns for each coin) and then exploits what was learned for the remaining 20 turns, in the hopes that this will maximize winnings.

Suppose coin 1 has a 0.75 probability of landing tails and coin 2 has a 0.75 probability of landing heads. Player one tests each coin 10 times, receiving the following sampled states:

Coin 1: H, T, T, H, T, T, H, T, T, T; It follows that the mean reward is $4/10 = 0.4$

Coin 2: T, H, H, H, T, H, H, T, H, H; It follows that the mean reward is $(-3)/10 = -0.3$

At this point, player one decides that selecting coin 1 is the optimal decision. Continuing to select coin 1 for the next 20 turns has an expected reward of 0.5 per turn, whereas selecting coin 2 has an expected reward of –0.5 per turn. Overall, player one can expect to win an average of zero points from the first 20 turns as the two options are evaluated and an expected reward of 10 points for the remaining 20 turns.

The decision in the above example worked out satisfactorily, in the sense that the decision-maker selected the optimal coin, but a number of questions nonetheless arise. For example, what if the coin flips, by chance, led to the mistaken selection of coin 2 as the best decision? True, that may seem unlikely in this case, but what if the coins were less biased (e.g., coin 1 lands tails with probability 0.55 and coin 2 lands heads with probability 0.55)? In such cases, will 10 samples be enough to confidently decide? Another important decision is how many samples to collect before committing to a decision: are 10 flips of each coin sufficient or excessive? Remember, the number of turns used for experimentation (exploration) delays the exploitation of the better option. These are all matters that could be decided in this simple two-decision context using statistical hypothesis tests with a selected confidence level (e.g., 95%), but what if the number of decisions grows larger or more complex? Fittingly, the RL framework offers a solution to this problem.

8.3.1 Greedy Policy for Bandit Problems

The previous simple example can be extended to one with more possible decisions, each yielding some reward from an unknown distribution. The generalization is often referred to as a *multi-armed bandit*, a term borrowed from slot machines for gambling, where a player has multiple machines to choose from and does not know the distribution of rewards. A decision-maker with n options wishes to select the best one, but only through experimentation can this be determined. If the expected reward for each decision, $a_t \in A$, is denoted as $Q_t(a_t)$, an estimate of the rewards can be reached by sampling as follows:

$$\tilde{Q}_t(a) = \frac{r_1 + r_2 + \cdots + r_{k_{a_t}}}{k_{a_t}},$$

where k_{a_t} is the number of times decision a_t has been selected. As $k_{a_t} \to \infty$, $\tilde{Q}_t(a_t) \to Q_t(a_t)$, thus, sampling each coin an infinite number of times will identify the optimal decision as follows:

$$a_t^* = \arg\max_{a_t \in A}\{Q_t(a_t)\}.$$

As stated, the above estimates of expected rewards are exact when $k_{a_t} \to \infty$, and the subsequent decisions selected are optimal. However, as noted above, infinite sampling is not a realistic option. Thus, the pragmatic alternative is to settle for finite sample estimates of the decision as follows:

$$\tilde{a}_t = \arg\max_{a_t \in A}\{\tilde{Q}_t(a_t)\}.$$

A natural approach that has been investigated in the context of MCPI is the greedy policy in which an arbitrary estimate of $Q_t(a_t)$ is initialized, and the decision at each epoch is made by selecting the option with the highest estimated reward. In the case of ties, decisions are selected by random sampling. This suggests the following approach to decision-making:

Step 1: Initialize $\tilde{Q}_0(a_0) = 0$ for all $a_0 \in A$. Set $t = 0$.
Step 2: Select the decision with the highest estimated rewards: $\tilde{a}_t = \arg\max\{\tilde{Q}_t(a_t)\}$.
Step 3: If the maximum number of epochs is reached, then stop; otherwise, update $\tilde{Q}_t(a_t)$ with the most recent observation and return to Step 2.

The third step updates the expected rewards for the most recently selected decision, which can be done efficiently by making an incremental update to the previous statistical estimate of $\tilde{Q}_t(a_t)$ using a weighted average as follows:

$$\tilde{Q}_t(a_t) = \frac{1}{k_{a_t} - 1}\tilde{Q}_{t-1}(a_t) + \frac{r_a}{k_{a_t}}.$$

The above steps estimate the expected rewards for each possible decision, $a_t \in A$, each time step 2 is executed and then use the estimate of $Q_t(a_t)$ to pick a new decision. At first glance, this seems like a reasonable approach to eventually make the best decision about which arm to pull. However, this greedy approach, on its own, can perform poorly because it may not adequately explore alternative decisions in the initial stages, leading to a suboptimal policy in early iterations and reinforcing those decisions over time. This is illustrated next.

8.3.2 ϵ-Greedy Approach to the Multi-armed Bandit

The pure greedy approach to the multi-armed bandit problem can fail to converge to an optimum. To address this challenge, it is common to introduce some randomization into the search for which arm to select. This illustrates one of the most fundamental concepts in RL and learning in general: the trade-off between exploration and exploitation. The ϵ-greedy method is a simple yet important modification that explicitly accounts for this trade-off by using a randomized policy. At each step, the ϵ-greedy method chooses one of two options:

1. *Exploitation*: With probability $1 - \epsilon$, select the decision that maximizes rewards based on the current estimate of $\tilde{Q}_t(a_t)$: $\tilde{a}_t = \arg\max_{a_t \in A}\{Q_t(a_t)\}$.
2. *Exploration*: With probability ϵ, randomly select a decision $a_t \in A$.

Thus, the decision is randomized, and it can be expressed as follows:

$$\hat{a}_t = \begin{cases} \tilde{a}_t, & \text{with probability } 1 - \epsilon \\ a, & a \text{ randomly selected decision with probability } \epsilon. \end{cases}$$

The random selection of actions is done using a discrete uniform distribution, or some other choice of distribution suited to exploration. As $k_a \to \infty$, $\tilde{Q}_t(a_t) \to Q_t(a_t)$, and the optimal decision is selected with probability at least $(1 - \epsilon)$, where ϵ is assumed to have a small value, although the exact choice is often problem-dependent and may require some experimentation. As the example below illustrates, this simple modification can significantly enhance the algorithm's performance.

Python Example 8.2: ϵ-Greedy approach to the multi-armed bandit problem

Consider the case of a five-armed bandit where each arm has a different probability distribution of rewards of +1 and −1 with the probabilities provided in Table 8.1. The ϵ-Greedy approach implemented in the Python code below proceeds for 10,000 iterations using an ϵ-greedy policy, with five choices of epsilon ranging from $\epsilon = 0$, which corresponds to the greedy policy with no exploration, and increasing to 0.5, which entails the most exploration of the five choices of ϵ. Thus, the values for ϵ are 0, 0.01, 0.05, 0.1, and 0.5. As ϵ increases, the probability that the agent employs the randomized selection of arms increases. The following Python code implements the algorithm, generating the plot in Figure 8.2. Notice that the $\epsilon = 0$ case performs the worst, because it gets trapped at a suboptimal policy. Increasing ϵ improves the policy up to a point, but a very high value, such as $\epsilon = 0.5$, where the agent randomly selects a decision with probability 0.5, degrades performance. In general, some experimentation with alternative values of ϵ is necessary to find one that performs well.

Table 8.1 Reward distributions for each arm in the 5-armed bandit problem.

Arm	Point distribution (+1, −1)
1	[0.55, 0.45]
2	[0.40, 0.60]
3	[0.60, 0.40]
4	[0.50, 0.50]
5	[0.30, 0.70]

```
import numpy as np
import matplotlib.pyplot as plt

# Input data
reward_probs = [  # Reward probabilities for each arm's outcome (+1,-1)
    [0.55, 0.45],
    [0.4, 0.6],
```

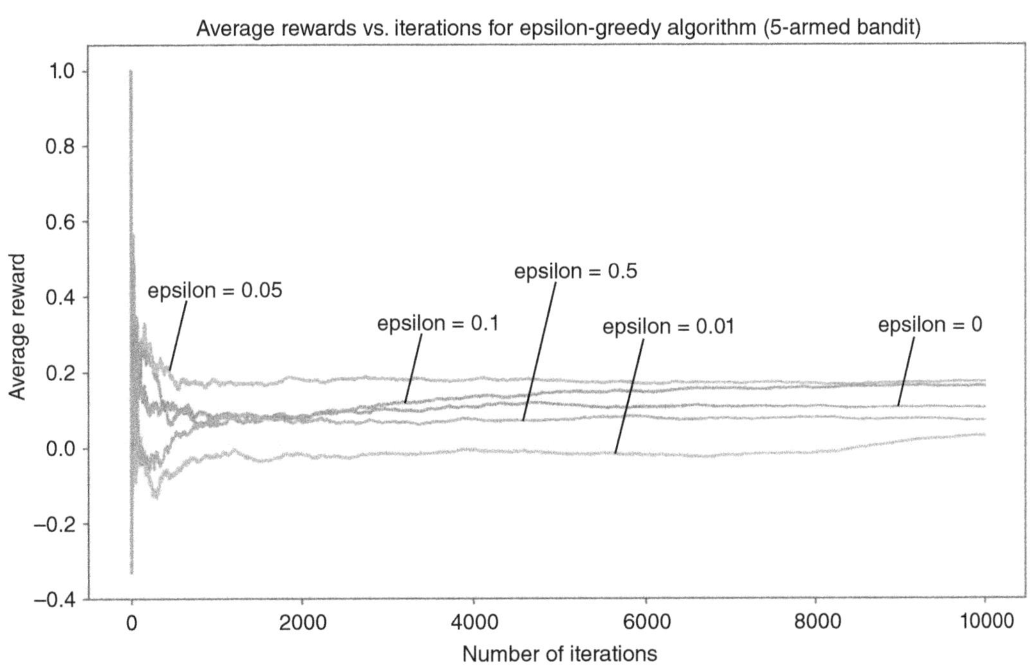

Figure 8.2 Results for the ϵ-greedy algorithm on a five-armed bandit problem for various choices of ϵ, the parameter that determines the probability of exploration through random selection of decisions.

```
    [0.6, 0.4],
    [0.5, 0.5],
    [0.3, 0.7]
]
ndecisions = len(reward_probs) # Number of decisions
nsamples = 10000
epsilon_values = [0, 0.01, 0.05, 0.1, 0.5]
results = {}

for epsilon in epsilon_values:
    rewards = []
    decision_values = np.zeros(ndecisions) # Estimated values for each decision
    decision_counts = np.zeros(ndecisions) # Selection counts for each decision
    for t in range(nsamples):
        # Epsilon-greedy decision selection
        if np.random.rand() < epsilon:
            d = np.random.randint(ndecisions) # Explore
        else:
            d = np.argmax(decision_values) # Exploit
        # Generate reward based on the selected decision's probabilities
        reward = np.random.choice([1, -1], p=reward_probs[d])
        rewards.append(reward)
        # Update estimated value for the selected decision
        decision_counts[d] += 1
        decision_values[d] += (reward - decision_values[d]) / decision_counts[d]
    # Compute average reward over time
    avg_reward = np.cumsum(rewards) / np.arange(1, nsamples + 1)
    results[epsilon] = avg_reward

# Plot results
plt.figure(figsize=(10, 6))
for epsilon, avg_rewards in results.items():
    plt.plot(avg_rewards, label=f"epsilon = {epsilon}")

plt.title("Average Reward vs. Iterations for epsilon-Greedy Alg (5-Armed Bandit)")
plt.xlabel("Number of Iterations")
plt.ylabel("Average Reward")
plt.grid()
plt.show()
```

The above implementation of the ϵ-Greedy approach is the most straightforward conceptually, but it is not very efficient for large problems. An alternative that is straightforward to implement and often leads to better performance is to use a time-dependent exploration probability ϵ_t, to modify exploration over time. For example, a beneficial two-step strategy is to begin with a high degree of exploration (a high initial value of ϵ) when the expected rewards and best decisions are most uncertain; as iterations progress, the initial estimate ϵ_0 is gradually decreased, increasingly favoring exploitation as the statistical estimates of $Q_t(\tilde{a})$ improve.

8.4 Monte Carlo Policy Iteration

The previous examples illustrate how to use Monte Carlo simulation to decide which arm to choose in a multi-armed bandit setting. The underlying assumption for the multi-armed bandit example above is that the arms' reward distributions are static and independent. Further, the formulation implies an absence of stochastic transitions among states. In this subsection, RL, using MCPI, is applied to a stochastic system with two or more states. For problems of this type, the MCPI solution has three standard steps:

Step 1: Create a computer program to simulate the agent's environment (states, decisions, rewards).
Step 2: Use simulation to evaluate the current policy's expected rewards.
Step 3: Try to improve the policy (exploit or explore) and then return to Step 2.

The steps above provide a conceptual framework for explaining how MCPI works, with Step 2 corresponding to policy evaluation and Step 3 to policy improvement. Because this problem involves randomness, an ϵ-randomized policy for the stochastic system is required, as follows:

$$\hat{\pi}(s) = \begin{cases} \tilde{\pi}(s), & \text{with probability } 1-\epsilon \\ a, & \text{randomly selected decision with probability } \epsilon. \end{cases}$$

Next, this policy defines the complete algorithm for MCPI in a finite-horizon problem with T decision epochs, as illustrated in Algorithm 8.1.

Algorithm 8.1 Monte Carlo Policy Iteration

Step 1 (Initialize): Initialize policy, $\tilde{\pi}$; Initialize ϵ to create a randomized policy $\hat{\pi}$

while stopping criterion not satisfied **do**
  Step 2 (Policy Evaluation):
  Randomly select N starting pairs $(s_0,\ a_0)$, and the corresponding sample paths
  for all (s, a) in the sample path **do**
    compute expected rewards:

$$\tilde{Q}^{\hat{\pi}}(s,a) = r_0(s,a) + \sum_{t=1}^{T-1} \lambda^t r_t(s_t, \hat{\pi}(s_t)) + \lambda^T r_T(s_T)$$

  end for

  Step 3 (Policy Improvement):
  for all $s \in S$ **do**
    $\tilde{\pi}(s) \in \arg\max_{a\in A}\{\tilde{Q}^{\hat{\pi}}(s,a)\}$
  end for

end while

Note that the basic idea of MCPI parallels standard PI for MDPs described in Chapter 4 with the following warnings:

1. Warning (Exploration): When following a fixed policy, some state-decision pairs, $(s,\ a)$, may not be visited. Therefore, a randomized policy is necessary to ensure that all state-decision pairs can be visited.
2. Warning (Convergence): An infinite number of sample paths is necessary to guarantee convergence to the optimal policy; however, a finite number, M, must be used in practice.

Warning 1 can be addressed using the above ϵ-randomized policy, $\hat{\pi}$, to guarantee all state-decision pairs can (in theory) be visited. Warning 2 states that a finite number of iterations must be used, and therefore, an approximation of the optimal policy is all that can be hoped for. Despite these cautionary statements, the MCPI has been found effective when the number of samples and the stopping criterion are judiciously selected as follows. First, for the choice of sample size, the decision about how many samples to use typically requires some experimentation with test policies to estimate the influence of sample size on confidence interval estimates on $\tilde{Q}_t^{\hat{\pi}}(s, a)$. Second, for the stopping criterion, various rules can be applied. One such criterion is to test whether the policy has converged by checking whether it is identical across all states from one iteration to the next (as used in the example below). An alternative is to test the norm of the difference $\tilde{Q}_t^{\hat{\pi},n+1}(s, a) - \tilde{Q}_t^{\hat{\pi},n}(s, a)$ from one iteration to the next, stopping if a predefined threshold is satisfied:

$$||\tilde{Q}_t^{\hat{\pi},n+1} - \tilde{Q}_t^{\hat{\pi},n}|| < \delta,$$

where δ is a suitably chosen tolerance on the *norm*, such as the L_∞ norm discussed in Chapter 4, of the difference in the Q-function from one iteration to the next. Both considerations are influenced by sample variation; therefore, the choices of sample size and stopping criterion should be evaluated for stability by running experiments with different random-number generator seeds to assess their sensitivity to sample size N, or tolerance δ.

Python Example 8.3: Monte Carlo policy iteration for grid world

In this example, a robot navigates a grid, shown on the left in Figure 8.3, from a starting point at the top left to a goal state at the bottom right, avoiding boundaries and barriers. The states, S, are defined by the 16 grid locations. The decisions, A, are whether to move up, down, left, right. The robot makes mistakes and executes the selected decision with probability 0.8. With probability 0.2, the robot moves in a perpendicular direction (e.g., if the robot chooses down, then with probability 0.1 it moves left or right). If the robot hits a boundary, it remains at its current location until the next epoch. Each step has a default reward of −0.5; however, if the robot hits the barrier, the reward is −5, and reaching the goal state yields a positive reward of 10. The following Python code implements MCPI for this example.

```
import numpy as np
import random
from collections import defaultdict

# Input data
grid_size = 4  # 4x4 grid
nstates = 16
states = range(nstates)
ndecisions = 4
decisions = ['up', 'down', 'left', 'right']
Lambda = 0.95           # Discount factor
start_state = 0         #top left corner; states indexed left to right, top down
goal_state = 15         # index from 0; 16 states total
penalty_state = 9       # treated as barrier/wall
rewards = {goal_state: 10, penalty_state: -5}  # Goal reward and barrier penalty
```

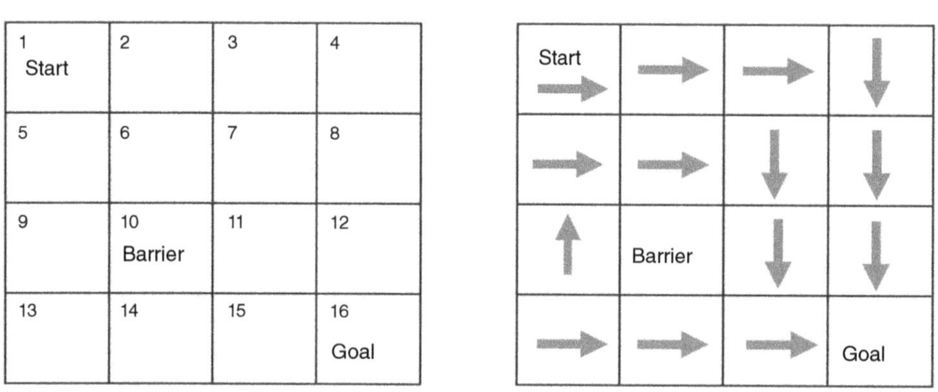

Figure 8.3 Illustration of the solution to the grid world problem, where the agent must learn to navigate the grid via four movement decisions (left, right, up, down) to move from the start state to the goal state, avoiding the barrier. The agent's decisions result in a successful move with probability p, and the agent moves in one of the two perpendicular directions, selected randomly, with equal probabilities, totaling $1 - p$.

```
default_reward = -0.5    # Reward for non-terminal, non-barrier states
max_iterations = 100    # Max iterations of Monte Carlo Policy Iteration.
NUM_EPISODES = 1000
movements = {   #Define decisions
    0: -grid_size,  # 'up'    (row - 1)
    1: grid_size,   # 'down'  (row + 1)
    2: -1,          # 'left'  (col - 1)
    3: 1            # 'right' (col + 1)
}
#Boundary conditions for 4x4 grid
top_row    = {0, 1, 2, 3}
bottom_row = {12, 13, 14, 15}
left_col   = {0, 4, 8, 12}
right_col  = {3, 7, 11, 15}
boundaries = {
    0: top_row,     # 'up'
    1: bottom_row,  # 'down'
    2: left_col,    # 'left'
    3: right_col    # 'right'
}

#State transition and reward
def next_state_and_reward(s, chosen_decision):
    # Check for boundary
    if s in boundaries[chosen_decision]:
        tentative_next = s  # hit edge, stay put
    else:
        tentative_next = s + movements[chosen_decision]
    # Check for barrier
    if tentative_next == penalty_state:
        next_s = s
        reward = rewards.get(penalty_state, default_reward)
    else:
        next_s = tentative_next
        reward = rewards.get(next_s, default_reward)
    return next_s, reward

#Sample movement direction
def step(s, d):
    decision_probs = {d: 0.8}  # move in intended direction with prob 0.8
    if d == 0 or d == 1:       # 'up' or 'down'
        decision_probs[2] = 0.1  # 'left'
        decision_probs[3] = 0.1  # 'right'
    elif d == 2 or d == 3:     # 'left' or 'right'
        decision_probs[0] = 0.1  # 'up'
        decision_probs[1] = 0.1  # 'down'
    chosen_decision = random.choices(
        list(decision_probs.keys()),
        list(decision_probs.values())
    )[0]
    next_s, reward = next_state_and_reward(s, chosen_decision)
    return next_s, reward
```

```
#Generate an episode based on current policy
def generate_episode(policy):
    episode = []
    s = start_state
    steps = 0
    while s != goal_state:
        d = policy[s]
        next_s, reward = step(s, d)
        episode.append((s, d, reward))
        s = next_s
        steps += 1
    return episode

#Evaluate current policy
def policy_evaluation(policy, optimal_value, episodes=NUM_EPISODES):
    returns = defaultdict(list)
    for _ in range(episodes):
        episode = generate_episode(policy)
        G = 0
        visited = set()
        for t in reversed(range(len(episode))):
            s, _, reward = episode[t]
            G = reward + Lambda * G
            if s not in visited:
                returns[s].append(G)
                optimal_value[s] = np.mean(returns[s])
                visited.add(s)

#Estimate Q-function
def Q_value(s, d, V):
    if d == 0 or d == 1:  # up/down --> perpendicular: left, right
        outcomes = [(d, 0.8), (2, 0.1), (3, 0.1)]
    else:                 # left/right --> perpendicular: up, down
        outcomes = [(d, 0.8), (0, 0.1), (1, 0.1)]
    q = 0.0
    for chosen_decision, p in outcomes:
        next_s, r = next_state_and_reward(s, chosen_decision)
        q += p * (r + Lambda * V[next_s])
    return q

#Greedy policy improvement
def policy_improvement(policy, optimal_value):
    policy_stable = True
    for s in states:
        # skip if If goal or barrier state
        if s in [goal_state, penalty_state]:
            continue
        old_decision = policy[s]
        #select a decision based on Q_value
```

```
        decision_values = {
            d: Q_value(s, d, optimal_value) for d in range(ndecisions)
        }
        best_decision = max(decision_values, key=decision_values.get)
        if old_decision != best_decision:
            policy_stable = False
            policy[s] = best_decision
    return policy_stable

#Solve using Monte Carlo Policy Iteration
np.random.seed(42) #Fix seed for repeatability
policy = { # Initialize policy randomly (ignoring goal and barrier states)
    s: random.choice(range(ndecisions))
    for s in states
    if s not in [goal_state, penalty_state]
}
v = np.zeros(nstates)  #initialize value function

for i in range(max_iterations):
    print(f"Iteration {i + 1}")
    policy_evaluation(policy, v, NUM_EPISODES)
    for s in range(nstates):
        print(f"State {s}: {v[s]:.4f}")
    policy_stable = policy_improvement(policy, v)
    if policy_stable:
        print("Policy converged!")
        break

#Print the final policy
print("\nFinal Policy:")
for s in range(nstates):
    if s == goal_state:
        print(f"State {s}: Goal")
    elif s == penalty_state:
        print(f"State {s}: Barrier")
    else:
        print(f"State {s}: {decisions[policy[s]]}")
```

This implementation converged after 9 MCPI iterations, which gives a policy that satisfies the stopping criterion (no change in policy between successive iterations). The policy is illustrated in Figure 8.3, where the arrows indicate the recommended direction for the robot at each grid position. Experimenting with alternative random number generator seeds gives independent estimates of the *Q*-function, which can then be used to estimate the statistical variance across replications (a full run of the Python code) and subsequently confidence intervals on the mean, similar to the approach for stochastic programs discussed in Chapter 6.

8.5 Q-learning

Many RL methods have been developed to enhance the efficiency of MCPI. A common theme of these methods is to learn more quickly by updating the policy more frequently than MCPI (e.g., every time a path is sampled) and using all data collected, including that from sampling policies other than the current one. As described above, these are referred to as off-policy RL algorithms. In this section, *Q-learning* is covered, one of the most well-known off-policy RL methods. The algorithm is named for the Q-value function, denoted by $Q(s, a)$, which is the expected reward assuming decision a is taken, and the optimal policy is subsequently followed for all future decision epochs. This parallels $v(s, a)$ used in Chapter 4 for MDPS. The key difference in the RL context is that $Q(s, a)$ is not known in advance, and it must instead be estimated, as follows:

$$Q^m(s, a) = (1 - \alpha_m)Q^{m-1}(s, a) + \alpha_m \left(r(s, a) + \lambda \max_{a' \in A} Q^{m-1}(s, a') \right).$$

Parameter $\alpha_m \in (0, 1)$ is known as the *learning rate* or sometimes the *step size*. It generates a weighted average that gives the most recent estimate of the Q-value greater weight than the newly updated estimate. Q-learning uses the Robbins–Monro (RM) algorithm, a recency-weighted update rule that incrementally estimates the expectation of random variables. The RM algorithm is explained immediately below, followed by its integration into Q-learning.

Algorithm 8.2 Robbins–Monro Method

1: Let X_i denote the i-th sample of random variable X, with mean μ, and let Y^m denote the estimate of $E[X]$ in the m-th iteration; set $Y^0 = 0$.
2: Update Y as follows:

$$Y^m = (1 - \alpha_m)Y^{m-1} + \alpha_m X_m.$$

3: **if** $|Y^m - Y^{m-1}| < \epsilon$ **then**
4: stop
5: **else**
6: $m = m + 1$ and return to Step 2
7: **end if**

There are many possible choices for the learning rate α_m, and the best choice depends on the context. For instance, in the static context where the random variables sampled do not change over time, a natural choice is $\alpha_m = \frac{1}{m}$. This corresponds to sequentially updating the estimate of the mean one sample at a time. By established proof, Y^m converges to the true mean, μ, asymptotically as $m \to \infty$. In contrast, if the random variable is nonstationary, choosing α_m to place greater weight on the most recent samples, say $\mu = 0.2$, would put a weight of 0.80 on the most recent estimate of the mean and a weight of 0.2 on the most recent sample. The following example illustrates the use of the Robbins–Monro algorithm [2].

Python Example 8.4: Robbins–Monro algorithm

This example assumes a nonstationary random variable that is normally distributed with a mean of $\mu = 5.0 + 0.01t$ and standard deviation $\sigma = 1.0$. Because the mean is changing over time, μ is chosen to give more weight to the most recent sample. Figure 8.4 illustrates the progress of the algorithm and the estimate of the mean for three learning rate parameters, $\alpha = 0.01,\ 0.1,\ 0.5$, showing the influence of α. The first case is slow to adapt to the changes in the mean. The second more closely tracks the mean but still underestimates the true mean, particularly at early iterations. Finally, the third case tracks the changing mean more closely, but it also introduces greater variance because of the high weight given to a single sample at each iteration.

```
import numpy as np
import matplotlib.pyplot as plt

# Input data
nsamples = 300
initial_mean = 5.0
mean_drift_per_step = 0.01   # Drift added to the mean at each step
std_dev = 1.0                # Standard deviation remains fixed
alpha_values = [0.01, 0.1, 0.5]  # Constant stepsizes for nonstationary settings
np.random.seed(42)               # Fix seed for repeatability
```

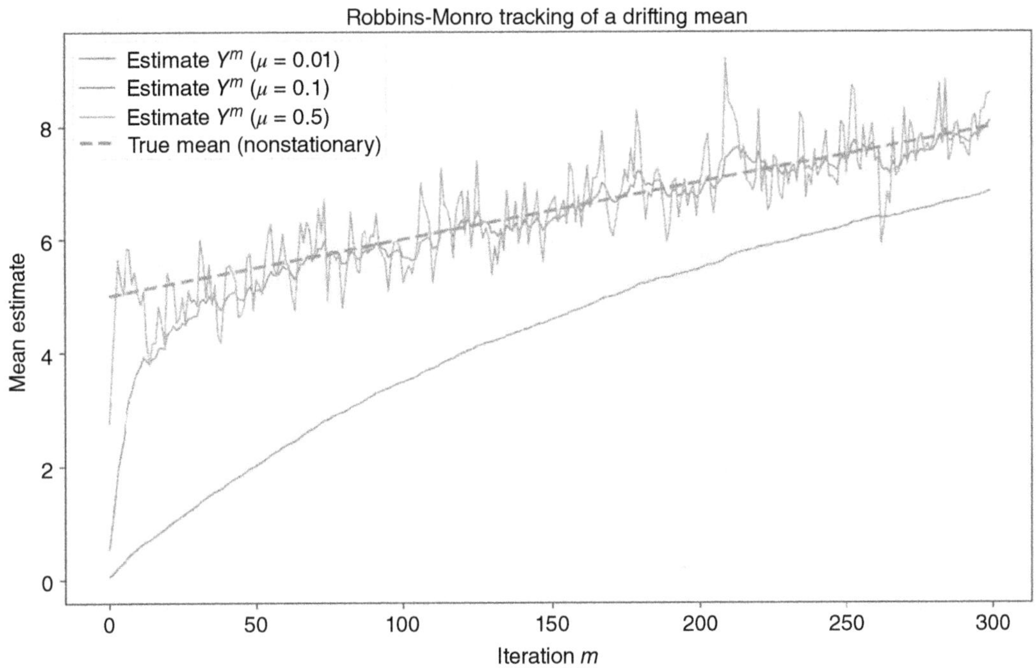

Figure 8.4 Numerical experiments illustrating the application of the Robbins–Monro algorithm to estimate the mean of a nonstationary random variable sampled over 300 iterations, for three choices of weight parameter, μ.

```
# Generate nonstationary samples
np.random.seed(42) #Fix seed for repeatability
mean = initial_mean + mean_drift_per_step * np.arange(nsamples)
sample = np.random.normal(loc=mean, scale=std_dev)

Y = [[0.0] for _ in alpha_values]# One trajectory per mu; start from 0

# Robbins{Monro updates
for i, alpha in enumerate(alpha_values):
    for m in range(nsamples):
        x_m = sample[m]
        y_prev = Y[i][-1]
        y_new = (1 - alpha) * y_prev + alpha * x_m
        Y[i].append(y_new)

# Plot results
colors = ["tab:red", "tab:blue", "tab:gray"]
plt.figure(figsize=(10, 6))
for i, alpha in enumerate(alpha_values):
    plt.plot(Y[i][1:],linewidth=1,color=colors[i],label=fr'$Y^m$ (alpha={alpha})')


plt.plot(mean, label='True mean (nonstationary)', linestyle='--', linewidth=2)
plt.xlabel('Iteration $m$')
plt.ylabel('Mean estimate')
plt.title('Robbins{Monro Tracking of a Drifting Mean')
plt.legend()
plt.show()
```

As stated at the beginning of this section, off-policy RL algorithms, such as Q-learning, typically involve frequent updates to expedite (and subsequently exploit) learning. To this end, the updating policy described just above, using the RM algorithm, integrates seamlessly into Q-learning. Although many variations of Q-learning have been proposed, the one described in Algorithm 8.3 is the most straightforward to implement.

Algorithm 8.3 Q-learning

1: Set $Q^0(s, a) = 0$ for all $s \in S$ and $a \in A$; Set $m = 0$ and maximum iterations M; Choose ϵ.
2: Let the current state be i. With probability ϵ, select decision $a \in A$ randomly; with probability $1 - \epsilon$, choose $\pi(i) = \arg\max\{Q^m(i, a)\}$.
3: Sample a new state, j, given decision $\pi(i)$, and update $Q^m(i, \pi(i))$ as follows:

$$Q^m(i, a) = (1 - \alpha_m)Q^{m-1}(i, a) + \alpha_m \left(r(j, a) + \lambda \max_{a' \in A} Q^{m-1}(j, a') \right).$$

4: **if** $m == M$ OR Stopping Criteria Satisfied **then** Stop
5: **else** Return to Step 2
6: **end if**
7: For all $s \in S$ return $a \in A$ that maximizes $Q^m(s, a)$.

The above Q-learning algorithm updates the value of state-decision pairs incrementally:

$$Q^m(s, a) = (1 - \alpha_m)Q^{m-1}(s, a) + \alpha_m \left(r(s, a) + \lambda \max_{a'} Q^{m-1}(s', a') \right),$$

where α_m controls the update's weighting. The ideal choice of weighting parameters at each iteration is application-specific (e.g., stationary, nonstationary) and typically requires some experimentation to find choices that work well for any particular application. The following example illustrates the application of Q-learning.

Python Example 8.5: Dynamic pricing

Dynamic pricing is used by online retailers to adjust product prices to maximize profits in evolving markets. Components of the problem are these:

- States: market conditions, defined by the competitor's price using predefined price points (e.g., \$10, \$15, \$20, ..., \$50).
- Decisions: agent's product price selection based on a discrete set of predefined price points (e.g., \$10, \$15, \$20, ..., \$50).
- Rewards: Profit = (Unit price − Unit cost) × Demand for each epoch.

The overall objective of the agent is to maximize cumulative profits over a given time horizon. A common assumption based on economic principles is that customers are price-sensitive, meaning that customer demand decreases as the agent's price increases. Q-learning seeks to estimate $Q(s, a)$, where s is the competitor's price and a is the agent's price response.

```
import numpy as np
import random

# Input data
states = range(10, 51, 5) # Possible competitor prices
decisions = range(10, 51, 5) # Possible prices ($10 to $50 in $5 increments)
demand_sensitivity = 0.2 # Sensitivity of demand to price differences
cost_per_unit = 5 # Fixed cost per product
episodes = 1000 # Number of training episodes
max_steps = 12 # Maximum steps per episode
epsilon = 0.2 # Initial exploration rate
epsilon_decay = 0.99 # Rate of decrease in epsilon per epoch
```

```
alpha = 0.05 #recency weight
Lambda = 0.97 # Economic discount factor representing time value of money
base_demand = 10 # Minimum demand independent of price
nstates = len(states)
ndecisions = len(decisions)
np.random.seed(42) #Fix seed for repeatability

# Simulate customer demand as a function of the agent and competitor price
def simulate_demand(agent_price, competitor_price):
    price_difference = competitor_price - agent_price
    demand = max(0,base_demand+demand_sensitivity*price_difference*base_demand)
    return int(demand)

# Compute reward (profit) based on the price decision and the demand generated
def rewards(decision, demand):
    return (decision - cost_per_unit) * demand

# Choose a decision (price) using the epsilon-greedy policy
def choose_decision(state, epsilon):
    if random.uniform(0, 1) < epsilon:
        return random.choice(range(ndecisions)) # Explore
    else:
        return np.argmax(Q_table[state]) # Exploit decision with highest Q-value

# Test the learned policy based on a test scenario
def test_policy():
    s = random.choice(range(nstates))
    total_profit = 0
    for step in range(max_steps):
        d = np.argmax(Q_table[s])
        # Simulate the environment
        demand = simulate_demand(decisions[d], states[s])
        profit = rewards(decisions[d], demand)
        total_profit += profit
        print(f"Step {step + 1}: Competitor Price = ${states[s]}, \
              Agent Price = ${decisions[d]}")
        s = random.choice(range(nstates)) #new competitor price
    print(f"Total Profit: ${total_profit}")

# Initialize Q-table with small random values
Q_table = np.random.uniform(low=1, high=2, size=(nstates, ndecisions))

# Q-Learning Training Loop with epsilon decaying over time
for episode in range(episodes):
    decaying_epsilon = max(0.01,epsilon*(epsilon_decay ** episode))
    for step in range(max_steps):
        s = random.choice(range(nstates))  # Competitor's random price decision
        d = choose_decision(s, decaying_epsilon)
        demand = simulate_demand(decisions[d], states[s])
        reward = rewards(decisions[d], demand)
        next_s = random.choice(range(nstates))
        max_future_q = np.max(Q_table[next_s])
```

```
        # Update Q-value
        current_Q = Q_table[s, d]
        new_q = current_Q + alpha *(reward + Lambda * max_future_q - current_Q)
        Q_table[s, d] = new_q

# Optimal policy
optimal_policy = [decisions[np.argmax(Q_table[s])] for s in range(nstates)]
print("\n Policy:")
for state, decision in zip(states, optimal_policy):
    print(f"Competitor Price: ${state} -> Optimal Agent Price: ${decision}")
```

8.6 Concluding Remarks

RL is a highly active area of sequential decision-making within artificial intelligence and related fields, such as operations research. As opposed to other sequential decision-making models, like the MDPs of Chapter 4, constrained optimization models of Chapter 5, or POMDPs of Chapter 7, decisions in the RL framework are not based on a specific model; rather, the underlying probability model is left unspecified, and the policy is learned through controlled experimentation, trading off the key steps of exploration and exploitation. This can be a powerful approach to learning how to navigate an environment, but as pointed out, directly interacting with a "live" system could be costly, or worse, dangerous. Thus, careful training is often necessary and may involve multiple steps, such as training in a simulated environment, followed by real-world training under the supervision of human decision-makers.

There are many resources for learning more about RL. An excellent introductory resource is the book "Reinforcement Learning: An Introduction" by Sutton and Barto [3], which provides a comprehensive overview of the fundamental concepts of RL and is accompanied by numerous examples. A shorter, focused tutorial by Gosavi [4] offers a concise introduction to key aspects of RL for readers familiar with MDP models. A more recent book by Powell covers a unified framework for reinforcement learning and stochastic optimization [5]. A book by Bertsekas, "A Course in Reinforcement Learning, 2nd Edition," is freely available for educational purposes and provides a comprehensive introduction to the topic [6].

Fundamental work on Q-learning was published in Watkins' dissertation and later in Watkins and Dayan [7], which provides full details of the proof of convergence. Many other approaches were not covered here, which merit attention from readers interested in exploring the methodological aspects of RL in greater depth. These include methods such as *temporal difference learning*, one of the earliest approaches to RL, *policy gradient* methods, and *actor-critic* methods, to name a few. Recent work has increasingly focused on online learning approaches that aim to minimize cumulative regret, an objective that addresses the cost of learning and not just finding a policy that maximizes expected rewards. Regret is the difference between the reward a decision-maker could have received if the optimal decision had been selected and the reward actually received. Cumulative regret is the sum of regrets across decision epochs and tracks the accumulated regret during the learning process. The book by Lattimore and Szepesvári [8] is a good resource to learn more about this topic.

8.7 Practice Exercises

1. Experiment with time-dependent choices of ϵ_t for the 5-armed bandit problem of Section 8.3.1. Start with a constant value $\epsilon_t = \epsilon_0$, for $t = 1, ..., W$, for the first W (warm-up) iterations. At $t = W + 1$, define $\epsilon_t = \frac{\epsilon_0}{t - W}$, for $t = W + 1, W + 2...$ until the stopping criterion is reached. How do the results compare to using $\epsilon_t = \epsilon_0$ for all t? How do the results depend on the choices of ϵ_0 and W? What happens when you vary the rate of decay of ϵ_t?

2. Explain how a greedy algorithm with no exploration ($\epsilon = 0$) can get stuck at a suboptimal solution. Give an example to illustrate this, showing iterations with plausible random outcomes for the rewards from each arm for a two-armed bandit.

3. For the grid world example of section 8.4, answer the following by modifying the Python code to estimate an optimal policy 100 times (replications) with a different seed each time.
   a. Plot a histogram of the result for $\tilde{Q}(s_0)$ where s_0 is the starting point (top left corner of the grid). Compute the mean and variance to assess the accuracy of the solution for the current parameter values in the code.
   b. Experiment with increasing the number of episodes from 100 to 1000 to evaluate the change in the mean and variance.
   c. Implement an alternative stopping criterion based on evaluating the L_2 norm of the difference in the vector of Q-function values. Determine the number of iterations to converge and estimate the variance in the final result for $\tilde{Q}(s_0)$ as a function of δ.

4. Consider a 2-armed bandit problem in which Arm 1 has a reward of +1 with probability 0.55 and a reward of −1 with 0.45. Arm 2 has a reward of +1 with probability 0.4 and −1 with probability 0.6. Both distributions are hidden from the decision-maker. Employ the ϵ-greedy algorithm discussed in Section 8.3.2 for 100 turns of pulling the bandit arms.
   a. Plot the mean reward per turn with respect to the number of turns for $\epsilon = 0, 0.01, 0.05, 0.1, 0.5$.
   b. Rerun your code 30 times with different random number seeds to estimate the 95% confidence intervals on mean rewards. Which choice of ϵ is best?
   c. For your recommended choice in part (b), what is the probability that the accumulated rewards are positive after 100 turns?

5. Consider the multi-armed bandit problem with two arms with the following properties:
   Arm 1: (a_1): this arm yields reward 1 with probability 1/3 and reward −1 with probability 2/3.
   Arm 2: (a_2): this arm yields reward 1 with probability 1/2 and reward −1 with probability 1/2.
   Assume a strictly greedy algorithm ($\epsilon = 0$) has been applied for 11 turns, resulting in pulling Arm 1 eight times and Arm 2 three times, yielding average rewards $Q(a_1) = -1/4$ and $Q(a_2) = -1$.
   a. Assuming a strictly greedy algorithm continues to be followed, what is the probability you will pull Arm 2 at Turns 12, 13, and 14?
   b. Now update $Q(a)$ using Robbins–Monroe updating with step size $\alpha_m = 1/m$ (where m is the number of times that arm has been pulled) to do two iterations of the greedy algorithm, assuming that you observe the following hypothetical rewards based on sampling the distribution for each arm:
      Arm 1: you will observe a reward of −1.0 for each of the next two pulls.
      Arm 2: you will observe 1.0 for the next pull and −1.0 for the second pull.
      Perform two more iterations of the greedy policy (turns 12 and 13) using $Q(a)$ values above as the starting point. Show all your calculations and identify which arm is pulled at each turn.
   c. Explain the deficiency of a strictly greedy algorithm using your results above as evidence.

6. Consider the coin game in which there are 30 coins on a table. A player and their opponent pick up 1, 2, or 3 coins each turn. Play continues until the last coin is picked up. A player who picks up the last coin is the loser. Assume the goal is to find an optimal policy for the player who goes first under the assumption that the opponent randomly selects 1, 2, or 3 coins with equal probability unless limited by the number of available coins (e.g., if the opponent decides to pick up 2 but only 1 remains, then the opponent picks up only 1 coin). Use Monte Carlo policy improvement to solve this problem. Explain the numerical parameters you used to find an optimal policy.

7. Blackjack is a sequential card game with a "player" and a "dealer." The player receives two cards face up; the dealer receives one card face down and one face up. The goal is to get as close to 21 without going over; if the player goes over, the tie goes to the dealer. Face cards are all worth 10, and Aces can be chosen to be 1 or 11. (For the sake of simplicity, assume Aces are always 11.) The player goes first and must decide one at a time whether to take a card ("hit") or not ("stick"). If the player's sum exceeds 21 at any time, they lose. When the player sticks with a sum of 21 or less, the dealer begins taking cards until one of two conditions is met: (a) the sum of the cards exceeds 21, in which case the

dealer loses; (b) the dealer's sum exceeds the player's, in which case the dealer wins. Finally, the dealer is required to take cards until their total exceeds 16. Implement Q-learning to find the optimal policy for the game of blackjack.

8. Consider a robot navigating a grid world. The grid has 3 x 3 cells, numbered 1–9 as follows:

1	2	3
4	5	6
7	8	9

The robot starts at cell 5. The goal is to reach cell 9, where it receives a reward of +10. Moving into cell 7 results in a penalty of −5. Moving to any other cell yields a reward of −0.5. The episode ends when the robot reaches 9 or 7. The robot can move in four directions: up, down, left, and right. If the robot tries to move outside the grid boundaries, it stays in the same cell. The transition probabilities are defined based on the following data:

- The robot moves in the intended direction with a probability of 0.8
- It moves in one of the perpendicular directions (left or right) with a probability of 0.1 each

and the discount factor is $\lambda = 0.95$. Answer the following questions.

a. Policy Evaluation and Iteration:
   i. Define a random initial policy where the robot moves in a randomly chosen direction in every state. Use policy iteration to find the exact optimal policy.
   ii. Report the optimal value function $v(s)$ for all states, $s = 1, 2, \ldots, 9$, under the optimal policy.

b. Simulation of Optimal Policy:
   i. Simulate 500 episodes of the robot navigating the grid world starting from cell 5 and following the optimal policy. For each episode, record the total cumulative discounted reward.
   ii. Estimate the mean and variance of the cumulative rewards across all episodes.

c. Probability Distribution of Rewards:
   i. Create a histogram illustrating the distribution of the cumulative rewards observed during the simulation.
   ii. Analyze the distribution, discussing what it indicates about the variability in outcomes when the robot follows the optimal policy.

9. Prove that the Robbins–Monro algorithm with recency weight $\alpha_m = 1/m$ gives an unbiased estimate of the sample mean at each iteration.

References

1 Tesauro, G. (1994). TD-Gammon, a self-teaching backgammon program, achieves master-level play. *Neural Computation, 6* (2), 215–219.
2 Robbins, H. & Monro, S. (1951). A stochastic approximation method. *The Annals of Mathematical Statistics, 22* (3), 400–407.
3 Sutton, R. S. & Barto, A. G. (2018). *Reinforcement Learning: An Introduction* 2nd. ed. MIT Press.
4 Gosavi, A. (2009). Reinforcement learning: A tutorial survey and recent advances. *INFORMS Journal on Computing, 21* (2), 178–192.
5 Powell, W.B. (Eds.). (2022). *Reinforcement Learning and Stochastic Optimization: A Unified Framework for Sequential Decisions.* Wiley.
6 Bertsekas, D. P. (2025). *A Course in Reinforcement Learning* 2nd ed. Available at: https://www.mit.edu/~dimitrib/RLbook.html. Accessed January 29, 2026.
7 Watkins, C. J. C. H. & Dayan, P. (1992). Q-learning. *Machine Learning, 8*, 279–292.
8 Lattimore, T. & Szepesvári, C. (2020). *Bandit Algorithms.* Cambridge University Press.

9

Multi-agent Markov Decision Processes

9.1 Introduction

Multi-agent models have garnered significant attention due to the wide range of situations in which multiple decision-makers are involved. Applications include search-and-rescue problems in which multiple searchers seek to locate a target with an unknown position, surveillance using multiple sensors to detect system changes, and a team of robots attempting to complete a task. There are many ways to define such models, and they are often distinguished based on assumptions about communications between agents (e.g., perfect, partial, or no communication), the disposition of agents with respect to cooperation or competition, as well as considerations that differentiate the single-agent models of prior chapters, such as model-based versus model-free approaches, or perfect information versus partial observability. This chapter introduces several frameworks that use Markov decision processes (MDPs) as the underlying model for each decision-maker.

This chapter begins with a special case of a two-agent MDP in which a decision-maker seeks an optimal policy while an adversary acts against it. If the decision-maker wishes to maximize rewards, the adversary seeks to mitigate their attempts by selecting some model parameters (typically transition probabilities) under their control that minimize the decision-maker's rewards. This can be viewed as a competitive sequential decision-making process. This type of model is often referred to as a robust Markov decision process (RMDP) because the resulting policies exhibit robustness to changes in model parameters. Therefore, this two-agent adversarial model may be used in situations where a single decision-maker wishes to hedge against uncertainty in model parameters.

Next, the case of multiple decision-makers (agents) who collaborate to achieve an optimal solution to a sequential decision-making problem is introduced. These cooperative multi-agent models comprise multiple underlying Markov chains, one per agent, and aim to coordinate agents' decisions to achieve a mutually agreed-upon objective. Finally, some foundational aspects of multi-agent models are discussed in the context of model-free decision-making, specifically multi-agent reinforcement learning (MARL).

A full treatment of multi-agent models is beyond the scope of this book. Except for RMDPs, the models in this chapter focus on cooperative behavior between agents. Such models largely draw on the work of previous chapters and are a natural extension of the models for sequential decision-making covered thus far. Competitive models, on the other hand, are more complex and require stochastic game theory and other sophisticated approaches to characterize the equilibrium behavior of competing agents. In the closing remarks to this chapter, the interested reader is directed to useful resources for learning more about these topics.

9.2 A Two-agent Model: Decision-maker vs. an Adversary

All decision-making models are subject to uncertainty in their model parameters. In such situations, it is common to employ the notion of an adversary who seeks to undermine the decision-maker's success. Models of this type are RMDPs, which share some similarities with MDPs, their non-robust counterparts. Similar to MDPs of Chapter 4, it is assumed there are a finite and discrete set of decision epochs indexed by $t = 0, \ldots, T$, a discrete set of states $s_t \in S$, and decisions $a_t \in A$, rewards, $r_t(s_t, a_t)$, and transition probability matrices, $P_t(a_t)$, with elements $p_t(s_{t+1}|s_t, a_t)$. The main departure of RMDPs from standard MDPs is the introduction of the adversary's decision process, where the adversary's decisions, denoted by vector $\vec{p}_t(\cdot|s_t, a_t)$, are the selection of transition probabilities given s_t, a_t, subject to constraints on their variation, called an *uncertainty set*, and denoted by $\mathcal{P}_t(s_t, a_t)$ which denotes the set of feasible choices of $p_t(s_{t+1}|s_t, a_t)$, for all $s_{t+1} \in S$. At each decision epoch t, and state s_t, the decision-maker selects $a_t \in A$, and the adversary responds with decision vector $\vec{p}_t(\cdot|s_t, a_t) \in \mathcal{P}_t(s_t, a_t)$. This back-and-forth progression of decisions continues throughout the decision horizon.

With the above definitions, the overall RMDP model can be expressed using the following set of optimality equations:

$$v_t^{\text{RMDP}}(s_t) = \max_{a_t \in A} \min_{\vec{p}_t(\cdot|s_t,a_t) \in \mathcal{P}_t(s_t,a_t)} E_{t,s_t}^{\vec{p}_t(\cdot|s_t,a_t)} \left[\sum_{k=t}^{T-1} \lambda^{k-t} r_k(s_k, a_t) + \lambda^{T-t} r_T(s_T) \right], \forall s_t \in S, t = 0, \ldots, T. \tag{9.1}$$

Models like that of Equation 9.1 are often referred to as *max–min problems*. In this setting, the decision-maker selects a decision $a_t \in A$ at epoch t that maximizes the worst-case expected future rewards with respect to an *inner problem* that defines the adversary's problem. The adversary's decision, represented in Equation 9.1 by $\vec{p}_t(\cdot|s_t, a_t) \in \mathcal{P}_t(s_t, a_t)$, seeks to minimize expected rewards given the decision-maker's choice.

The model in Equation 9.1 provides a general definition of an RMDP that can account for various possible implementations by varying the model parameters under the adversary's control (the adversary's decision variables) or by varying the uncertainty sets that govern the adversary's feasible choices at each epoch. The most common choice for the adversary's decision variables is the transition probabilities, which are the focus of this section (extensions to include reward parameters are straightforward). In this case, the inner (minimization) problem the adversary faces when the decision-maker is in state s_t and takes a decision a_t at decision epoch t can be written as follows:

$$\sigma_t(s_t, a_t) = \min_{\vec{p}_t(\cdot|s_t,a_t) \in \mathscr{P}_t(s_t,a_t)} \sum_{s_{t+1} \in S} p_t(s_{t+1}|s_t) v_{t+1}^{\text{RMDP}}(s_{t+1}, a_t), \forall s_t \in S, \quad t = 0, \ldots, T. \tag{9.2}$$

where $\vec{p}_t(s_t, a_t)$ is a vector of decision variables for the adversary's selected transition probabilities with elements taken from the row of the transition probability matrix corresponding to the current state of the system, s_t. In this inner problem, the decision-maker's choice is fixed, and the decision variables are the elements of $\vec{p}_t$. This formulation explicitly incorporates the adversary's decision-making options, specifically the choice of transition probabilities for the underlying Markov chain at each decision epoch. Combining Equations 9.1 and 9.2 yields the following RMDP:

$$v_t^{\text{RMDP}}(s_t) = \max_{a_t \in A} \{r_t(s_t, a_t) + \lambda \sigma_t(s_t, a_t)\}, \forall s_t \in S, t = 0, \ldots, T, \tag{9.3}$$

with the boundary condition:

$$v_t^{RMDP}(s_T) = r_T(s_T), \forall s_T \in S.$$

In the above RMDP, the adversary's decisions can vary from one epoch to the next. Thus, the sequence of decisions starting at $t = 0$ is decision-maker, adversary, decision-maker, adversary, and so on, until the end of the horizon. In light of this, the decomposable structure of the problem, i.e., the structure that makes MDPs easy to solve due to the independence between the value function for each state at a given decision epoch, is retained. Algorithm 9.1 provides a means to solve RMDPs via backward induction, with some minor modifications relative to Chapter 4, to incorporate an additional step of solving the inner problem for each state at each epoch.

The choice of the uncertainty set plays an important role in determining the computational tractability of solving RMDPs. The simplest choice, known as the *interval model* (IM), uses an uncertainty set defined by the Cartesian product of the uncertainty set for each independent row of the TPM, where each transition probability has a lower and upper bound to express the potential variation.

$$\begin{aligned} \mathscr{P}_t(s_t, a_t) = \Big\{ \vec{p}_t(\cdot|s_t, a_t) \mid & \sum_{s_t \in S} p_t(s_{t+1}|s_t, a_t) = 1, \\ p^l(s_{t+1}|s_t, a_t) \le p(s_{t+1}|s_t, a_t) \le p^u(s_{t+1}|s_t, a_t), & \forall s_{t+1} \in S \Big\}. \end{aligned} \tag{9.4}$$

The quantities $p^l(s_{t+1}|s_t, a_t)$ and $p^u(s_{t+1}|s_t, a_t)$ denote the lower and upper bounds of $p_t(s_{t+1}|s_t, a_t)$, which lie in the interval $[0, 1]$. Moreover, the transition probabilities for each row of a transition probability matrix must sum to 1, which can be viewed as an additional constraint on the adversary's choices.

Algorithm 9.1 Backward Induction for Finite Horizon Robust Markov Decision Process

1: **Input:** RMDP data elements: decision epochs, states, decisions, transition probability matrix, rewards, uncertainty set, discount factor
2: **Boundary Condition:** $v_T^{\text{RMDP}}(s_T) = r_T(s_T), \forall s_T \in S$
3: **Backward Induction:**
4: **for** $t = T - 1$ to 0 **do**
5: **for** all $s_t \in S$ **do**

$$\sigma_t(s_t, a_t) = \min_{\vec{p}_t(\cdot|s_t,a_t)\in\mathcal{P}_t(s_t,a_t)} \sum_{s_{t+1}\in S} p_t(s_{t+1}|s_t, a_t)v_{t+1}^{\text{RMDP}}(s_{t+1}, a_t)$$

$$v_t^{\text{RMDP}}(s_t) = \max_{a_t\in A}\{r_t(s_t, a_t) + \sigma_t(s_t, a_t)\}$$

$$a_t^*(s_t) = \arg\max_{a_t\in A}\{r_t(s_t, a_t) + \lambda \sum_{s_{t+1}\in S} p_t(s_{t+1}|s_t, a_t)v_{t+1}^{\text{RMDP}}(s_{t+1})\}$$

6: **end for**
7: **end for**
8: **Return:** Optimal Policy

Python Example 9.1: Robust medical treatment decisions

Consider an example of an RMDP in the context of medical decision-making, similar to the example in Chapter 4. The problem is that a decision-maker must make a one-time decision (whether and when to begin treatment) at any of a finite number of decision epochs. An adversary controls the probability of adverse events (E) by choosing the transition probabilities for transitions to the E state from the various health states a patient can be in. The MDP elements are defined as follows:

Decision epochs: $t = 0, \ldots, T$.

States: $S = \{L, M, H, V, E\}$, where the health states are defined by four risk categories: Low (L), medium (M), high (H), and very high (V). The E state is an absorbing state that represents the occurrence of an adverse event.

$$\text{Decisions: } a_t = \begin{cases} \{C, Q\}, & s \in \{L, M, H, V\} \\ \emptyset, & s \in \{E\} \end{cases}$$

where C is the decision to defer treatment and Q is the decision to start treatment, resulting in the patient entering the treatment-absorbing state with probability 1, and receiving a reward $g_t(s_t)$, denoting the expected quality-adjusted survival for a patient starting treatment at epoch.
Rewards: Assume that the annual rewards before treatment are

$$r_t(s_t, C) = 1, \ \forall s_t \in S, \ t = 0, \ldots, T - 1,$$

corresponding to a perfect year of health, and the expected quality-adjusted life span for patients starting treatment at epoch t can be estimated using the following equation medical researchers have approximated from a previous study:

$$g_t(s_t) = -0.721 \times (t + 50) + 64.36, \ \forall s_t \in S, \ t = 0, \ldots, T$$

Transition probabilities:

Table 9.1 Transition probabilities among states.

State	L	M	H	V	E
L	$0.74 \times (1 - p_t(E\|L))$	$0.20 \times (1 - p_t(E\|L))$	$0.05 \times (1 - p_t(E\|L))$	$0.01 \times (1 - p_t(E\|L))$	$p_t(E\|L)$
M	$0.10 \times (1 - p_t(E\|M))$	$0.70 \times (1 - p_t(E\|M))$	$0.17 \times (1 - p_t(E\|M))$	$0.03 \times (1 - p_t(E\|M))$	$p_t(E\|M)$
H	$0.01 \times (1 - p_t(E\|H))$	$0.24 \times (1 - p_t(E\|H))$	$0.62 \times (1 - p_t(E\|H))$	$0.13 \times (1 - p_t(E\|H))$	$p_t(E\|H)$
V	$0.01 \times (1 - p_t(E\|V))$	$0.06 \times (1 - p_t(E\|V))$	$0.30 \times (1 - p_t(E\|V))$	$0.63 \times (1 - p_t(E\|V))$	$p_t(E\|V)$
E	0	0	0	0	1

Uncertainty Set: $p_t(E \mid s_t)$ is the probability of transitioning to the E state from health state s_t under decision C, and it satisfies the following:

$$p_t(E \mid s_t) \in [(1-\rho) \times \bar{p}_t(E \mid s_t),\ (1+\rho) \times \bar{p}_t(E \mid s_t)]$$

$$\bar{p}_t(E \mid s_t) = C^R(s_t)0.0003 \times e^{8.91 \times (t+50)}$$

where $C^R(s_t)$ is a parameter determined by the health state of the patient that reflects the risk differences among the states as follows:

$$C^R(s_t) = \begin{cases} 0.1 & \text{for } s_t = L \\ 0.4 & \text{for } s_t = M \\ 2 & \text{for } s_t = H \\ 10 & \text{for } s_t = V \end{cases}$$

Table 9.1 contains the transition probabilities for the model.

The following Python code implements the above model to find an optimal policy.

```
#Chapter 9, Example 9.1: Robust medical treatment decisions
import numpy as np
import gurobipy as gp
from gurobipy import GRB

# Input data
T = 25 # Time horizon (50 (t-0) through 75 (t=25) years
nstates = 5
ndecisions = 2
states = ['L', 'M', 'H', 'V', 'E']  # Health states and adverse event (E)
decisions = ['C', 'Q']              # Decisions: Continue ('C'), Quit/Treat ('Q')
C_R = [0.1, 0.4, 2, 10, 0]          # Risk multipliers by state L, M, H, V, E
D_trt = 0.05                        # Annual disutility of treatment
rho = 0.2                           # Relative deviation for E probability
P = np.array([ # Base transition probabilities for 'C' decision, excluding E
    [0.74, 0.20, 0.05, 0.01],   # From L
    [0.10, 0.70, 0.17, 0.03],   # From M
```

```
    [0.01, 0.24, 0.62, 0.13],  # From H
    [0.01, 0.06, 0.30, 0.63]   # From V
])

rewards = np.zeros((T + 1, nstates, ndecisions)) # Reward function: rewards[t,s,d]
for t in range(T + 1):
    rewards[t, :, 0] = 1  # Decision 'C'
    rewards[t, :, 1] = -0.721*(t + 50) + 64.36 - (25 - t) * D_trt #Decision 'Q'
    rewards[t, -1, :] = 0  # E state reward is 0

# Midpoint of the uncertain P_t(E|s_t) interval
def p_e_mid(t, s): return C_R[s] * 0.0003 * np.exp(0.0891 * (t + 50))

# Uncertainty interval
def p_interval(t, s):
    mid = p_e_mid(t, s)
    lb = float(np.clip(mid * (1 - rho), 0.0, 1.0))
    ub = float(np.clip(mid * (1 + rho), 0.0, 1.0))
    return lb, ub

value = np.zeros((T + 1, nstates, ndecisions))  # Value-to-go function
v = np.zeros((T + 1, nstates))      # Optimal value function
optimal_policy = np.zeros((T + 1, nstates), dtype=int)  # Policy
v[T] = rewards[T, :, 1] # Boundary condition
optimal_policy[T, :-1] = 1  # decision 'Q'
optimal_policy[T, -1] = -1  # E: no decision

# Backward induction
for t in range(T - 1, -1, -1):
    for s in range(nstates):
        if s == nstates - 1:  #E state
            v[t, s] = 0
            optimal_policy[t, s] = -1
            continue
        lb, ub = p_interval(t, s)
        model = gp.Model() # Create a new Gurobi model
        model.Params.OutputFlag = 0
        p = model.addVar(lb=lb, ub=ub, name="p_e") # Adversary decision variables
        obj = (1 - p) * P[s].dot(v[t + 1, :-1]) + \
            p * v[t + 1, -1] # Set objective function
        model.setObjective(obj, GRB.MINIMIZE)
        model.optimize() # Optimize model
        if model.status == GRB.OPTIMAL: min_value = model.ObjVal
        value[t, s, 0] = rewards[t, s, 0] + min_value   # Robust 'C'
        value[t, s, 1] = rewards[t, s, 1] # 'Q'
        optimal_policy[t, s] = np.argmax(value[t, s, :])
        v[t, s] = value[t, s, optimal_policy[t, s]]

# Print results
print("Optimal Policy:")
for t in range(T + 1):
    print(f"Optimal policy at age {t + 50}:")
```

```
    for s in range(nstates):
        if optimal_policy[t, s] == -1:
            decision = '-' # For E state
        else:
            decision = decisions[optimal_policy[t, s]]
        print(f"  State {states[s]}: {decision}")
```

Running the Python code for the selected model parameters yields an optimal policy that recommends treatment for states H and V for patients aged 50–53, and states M, H, V, for ages 54–74, followed by treatment for all patients at age 75 (by assumption). Compared to the non-robust counterpart model in the Python example of Chapter 4, the resulting policy in this case is more aggressive in recommending treatment because it recommends treatment for all but the lowest risk health state L, starting at age 54; whereas the non-robust counterpart delays treatment for the medium risk state M until age 59. This can be viewed as a form of hedging against uncertainty in the (adversary's) choices of transition probabilities.

9.3 Cooperative Multi-agent Markov Decision Processes (MAMDPs)

Multi-agent MDP (MAMDP) models generalize MDPs to settings with multiple agents. Such models can be cooperative or competitive. As mentioned in the introduction, this chapter concentrates on the former because they are the most straightforward, and they can be understood through the lens of MDPs and related models covered in previous chapters. This section provides an introduction to cooperative MAMDPs under the following cases of alternative communication assumptions: (1) independent agents with no communication; (2) perfect communication between agents that allows centralized planning; (3) limited communication requiring decentralized planning of agents' activities. In all cases, agents cooperate and are assumed to share a common objective, such as maximizing or minimizing a common reward. In the first case, conditions under which the problem fully decomposes into independent MDPs are discussed. In the second (centralized) case, complete information is available to each agent, allowing perfect coordination of their decisions, resulting in a large single-agent MDP in which states and decisions are defined by augmenting all the agents' states and decisions, respectively. Finally, in the third (decentralized) case, there are constraints on the amount of information shared, with the most extreme case being the complete absence of information sharing. As shown below, these three cases are in order of increasing computational complexity. In all cases, for simplicity, it is assumed that the agents are identical and thus share the same state space, decision space, transition probabilities, possible observations, and reward functions.

9.3.1 Independent Cooperative Agents with No Information Sharing

The fully decentralized case of independent agents with no information sharing is one that can be easily solved under certain specific assumptions. The first assumption is that the transition probabilities among agents are independent and therefore can be written as:

$$p_t(s_t^1, \dots, s_t^M | s_t^{1'}, \dots, s_t^{M'}, a_t^1, \dots, a_t^M) = \prod_{i=1}^{M} p_t(s_t^i | s_t^{i'}, a_t^i).$$

The second assumption is that the reward for the team of agents is the sum of the individual rewards, which are independent and separable as follows:

$$\bar{r}_t((s_t^1, \dots, s_t^M), \pi(s_t^1, \dots, s_t^M)) = \sum_{i=1}^{M} r(s_t^i, \pi^i(s_t^i)), \forall s_t^i \in S, t = 0, \dots, T.$$

Under these assumptions, the optimal joint policy for all agents is simply the aggregation of the optimal policies for each individual agent's MDP. Thus, each MDP can be solved separately, and the overall model corresponds to a regular MDP, discussed in Chapter 4.

9.3.2 Fully Centralized MAMDPs

An alternative form of the cooperative MAMDP model is one in which the $i = 1, \ldots, M$ agents act cooperatively with perfect information sharing to achieve the shared goal of maximizing total rewards gained. In other words, each agent is aware of all other agents' states s_t^i, and decisions a_t^i, and seeks to act in a way that maximizes total expected rewards. The model parameters for the agents may be the same across all agents (referred to as a homogeneous MAMDP), or they may differ across agents (referred to as a non-homogeneous MAMDP). Without loss of generality for the purposes of this chapter and for simplicity, it is assumed that the MAMDPs are homogeneous, meaning that all agents' MDPs are identical across all decision epochs $t = 0, \ldots, T$. Furthermore, it is assumed that all agents are independent, and thus their state transition dynamics evolve independently of one another.

Assuming each agent i, has a policy π^i, the overall policy for the MAMDP model $\pi = (\pi_1, \ldots, \pi^M)$, is simply the aggregate of the policies for all M agents. Thus, the overarching problem is to solve the following:

$$\max_{\pi \in \Pi} \left\{ E^{\pi} \left[\sum_{t=0}^{T} \sum_{i=1}^{M} \lambda^t r_t((s_t^1, \ldots, s_t^M), \pi^i(s_t^1, \ldots, s_t^M)) + r_T(s_T^1, \ldots, s_T^M) \right] \right\}, \forall s_t^i \in S, i = 1, \ldots, M. \tag{9.5}$$

The MAMDP defined by Equations 9.5 can be expressed as a standard MDP by replacing the sum of agent rewards with a single reward, $\bar{r}_t(s_t^1, \ldots, s_t^M)$ as defined in Equation 9.6 below. Now, simplifying the notation by defining the multidimensional states and decisions to be $s_t = (s_t^1, \ldots, s_t^M) \in \bar{S} = \{S\}^M$ and $a_t = (a_t^1, \ldots, a_t^M) \in \bar{A} = \{A\}^M$, respectively, yields the following optimality equations:

$$v_t(s_t) = \max_{a_t \in \bar{A}} \left\{ \bar{r}_t(s_t, a_t) + \lambda \sum_{s_{t+1} \in \bar{S}} p_t(j \mid s_t, a_t) v_{t+1}(s_{t+1}) \right\}, \quad t = 0, \ldots, T-1, \quad \forall s_t \in \bar{S}. \tag{9.6}$$

$$v_T(s_T) = \bar{r}_T(s_T), \quad \forall s_T \in \bar{S}. \tag{9.7}$$

The MAMDP with fully observable states and decisions and perfect information sharing by agents, can be viewed as the case of a centralized controller. The resulting centralized and cooperative MAMDP can be solved as a standard single-agent MDP with an enlarged state and policy space owing to the need for the central controller to consider all possible combinations of agent states and possible decisions. Problems of this type grow exponentially in size and thus are increasingly computationally difficult to solve as the number of agents, M, increases, because the state space is the Cartesian product of each agent's states, S^M. Thus, much of the work on these types of problems has focused on ways to leverage the special structure of such problems for computational efficiency.

Python Example 9.2: Centralized planning for independent agents in an automated warehouse

Consider the case of two robots tasked with clearing a warehouse's shop floor of debris. Assume the warehouse is laid out as a 5×5 grid, and the robot locations are s^1 and s^2. There are two locations, X and Y, that require debris clearance. The overall state of the system is (s^1, s^2, h_X, h_Y) where s^1 and s^2 can be any of the 25 sectors denoted by the set $S = \{1, 2, \ldots, 25\}$, and h_X, h_Y, are binary indicators for whether either location of the debris has been visited previously or not. The state transitions are assumed to be deterministic and defined by the current state, s^i, of each robot and the robot's decisions to move among the feasible options: left (L), right (R), up (U), down (D), or no movement (N), which are denoted $a^i \in A = \{L, R, U, D, N\}$. Let $s^{i'} = T(s^i, a^i)$ be robot i's next cell if it starts in s^i and makes decision a^i and let h_X and h_Y be 0 if the debris location has never been visited and 1 otherwise. Each move a robot makes has a reward of -1, and if a robot reaches a target location (e.g., $s^i = X$ or Y), it acquires a one-time reward of 10. If the location X or Y has already been reached in a prior epoch, it is recorded as part of the state space as $h_X = 1$ or $h_Y = 1$. Based on this notation, the reward function can be described as follows:

Cost of robot movement:

$$c(a^i) = \begin{cases} -1, & \text{if } a^i \neq N, \\ 0, & \text{otherwise.} \end{cases}$$

Reward for reaching target X or Y:

$$z(s^1, s^2, h_X, h_Y) = \begin{cases} 0, & h_X = 1,\ h_Y = 1, \\ 10, & s^1 = X, h_X = 0 \text{ or } s^1 = Y, h_Y = 0, \\ 10, & s^2 = X, h_X = 0 \text{ or } s^2 = Y, h_Y = 0, \\ 20, & s^1 = X, h_X = 0 \text{ and } s^2 = Y, h_Y = 0, \\ 20, & s^1 = Y, h_Y = 0 \text{ and } s^2 = X, h_X = 0, \\ 0, & \text{otherwise.} \end{cases}$$

Aggregating the two types of rewards yields the following total rewards for movement and target acquisition:

$$r((s^1, s^2, h_X, h_Y), (a^1, a^2)) = c(a^1) + c(a^2) + z(s^1, s^2, h_X, h_Y).$$

Because clearing debris sooner is preferable to later, it is assumed the discount factor is $\lambda = 0.9$. Furthermore, since the end of the horizon is undetermined, the horizon continues until the debris at both locations is removed. Based on this problem description, the joint optimality equations can be written as follows:

$$v(s^1, s^2, h_X, h_Y) = \max_{a^1, a^2 \in A} \left\{ r((s^1, s^2, h_X, h_Y), (a^1, a^2)) + 0.9v(s^{1'}, s^{2'}, h'_X, h'_Y) \right\}, \forall s^1, s^2 \in S, \quad \forall h_X, h_Y \in \{0, 1\}.$$

and the terminal state, when $h_X = 1$ and $h_Y = 1$, is defined as:

$$v(s^1, s^2, 1, 1) = 0, \forall s^1, s^2 \in S.$$

The joint optimality equations can be solved using methods from Chapter 4 (e.g., value iteration, policy iteration) to obtain an optimal joint policy for the robots. The following Python code provides an example assuming the locations of X and Y in the grid are (2, 2) and (3, 1), respectively.

```
import numpy as np

#Input data
GRID_SIZE = 5
DECISIONS = ['L', 'R', 'U', 'D', 'N']
Lambda = 0.9
X_rc = (2, 2) #target X location (row, columns)
Y_rc = (3, 1) #target Y location (row, column)
def rc_to_idx(r, c): #Helper function: convert (row, column) to index
    r = max(1, min(GRID_SIZE, r))
    c = max(1, min(GRID_SIZE, c))
    return (r - 1) * GRID_SIZE + c
def idx_to_rc(idx): #Helper function: convert index to (row, column)
    idx = max(1, min(GRID_SIZE * GRID_SIZE, idx))
    r = (idx - 1) // GRID_SIZE + 1
    c = (idx - 1) % GRID_SIZE + 1
    return r, c
X_idx = rc_to_idx(X_rc[0], X_rc[1])
Y_idx = rc_to_idx(Y_rc[0], Y_rc[1])
S_single = list(range(1, GRID_SIZE*GRID_SIZE + 1)) # Joint states and decisions
STATES = [(s1, s2, hX, hY) for s1 in S_single for s2 \
          in S_single for hX in (0,1) for hY in (0,1)]
JOINT_DECISIONS = [(a1, a2) for a1 in DECISIONS for a2 in DECISIONS]
def is_terminal(s): return s[2] == 1 and s[3] == 1 #Check if target reached
NONTERMINAL_STATES = [s for s in STATES if not is_terminal(s)]
```

```
STATE_TO_INDEX = {s: i for i, s in enumerate(NONTERMINAL_STATES)}
MAX_STEPS=1000 #Upper bound on state transitions

#Robot movement and reward functions
def step(cell_idx, a):
    r, c = idx_to_rc(cell_idx)
    if a == 'L': c -= 1
    elif a == 'R': c += 1
    elif a == 'U': r -= 1
    elif a == 'D': r += 1
    elif a == 'N': pass
    else: raise ValueError(a)
    return rc_to_idx(r, c)

def move_cost(a): return 0 if a == 'N' else -1

def reward_on_arrival(hX, hY, s1p, s2p):
    hitX = int((s1p == X_idx) or (s2p == X_idx))
    hitY = int((s1p == Y_idx) or (s2p == Y_idx))
    return 10 * (1 - hX) * hitX + 10 * (1 - hY) * hitY

def successor(state, joint_decision): #Get next joint state
    s1, s2, hX, hY = state
    d1, d2 = joint_decision
    s1p = step(s1, d1)
    s2p = step(s2, d2)
    r_move = move_cost(d1) + move_cost(d2)
    r_arrival = reward_on_arrival(hX, hY, s1p, s2p)
    r = r_move + r_arrival
    hXp = 1 if (hX or (s1p == X_idx) or (s2p == X_idx)) else 0
    hYp = 1 if (hY or (s1p == Y_idx) or (s2p == Y_idx)) else 0
    return (s1p, s2p, hXp, hYp), r

def policy_evaluation(policy):
    n = len(NONTERMINAL_STATES)
    A = np.eye(n)
    b = np.zeros(n)
    for s in NONTERMINAL_STATES:
        i = STATE_TO_INDEX[s]
        a = policy[s]
        sp, r = successor(s, a)
        b[i] = r
        if not is_terminal(sp): #Deterministic transition under policy
            j = STATE_TO_INDEX[sp]
            A[i, j] -= Lambda  #Move Lambda*v(sp) to LHS
    v_vec = np.linalg.solve(A, b)
    values = {s: 0.0 for s in STATES} #Initialize value to go
    for s in NONTERMINAL_STATES:
        values[s] = float(v_vec[STATE_TO_INDEX[s]])
    return values
```

```
#Policy iteration
def policy_iteration():
    policy = {s: ('N','N') for s in STATES}
    stable = False #Continue iterations until policy does not change
    while not stable:
        v = policy_evaluation(policy)
        stable = True   #Policy improvement steps
        for s in STATES:
            if is_terminal(s): continue
            a_curr = policy[s]   #Current policy value
            sp_curr, r_curr = successor(s, a_curr)
            best_val = r_curr + Lambda * v[sp_curr]
            best_a = a_curr
            for a in JOINT_DECISIONS: #Find the best decision
                sp, r = successor(s, a)
                q = r + Lambda * v[sp]
                if q > best_val + 1e-12: best_val, best_a = q, a
            if best_a != a_curr:
                policy[s] = best_a
                stable = False
    return v, policy

#Output:'.' empty;debris location:'x'/'y';debris cleared:X/Y;'1,2' robots
def print_grid(s1, s2, hX, hY):
    grid = [['.' for _ in range(GRID_SIZE)] for _ in range(GRID_SIZE)]
    rX, cX = X_rc
    rY, cY = Y_rc
    grid[rX-1][cX-1] = 'X' if hX else 'x'
    grid[rY-1][cY-1] = 'Y' if hY else 'y'
    r1, c1 = idx_to_rc(s1)
    r2, c2 = idx_to_rc(s2)
    grid[r1-1][c1-1] = '1'
    grid[r2-1][c2-1] = '2' if (r2, c2) != (r1, c1) else 'B'  #'B' if robots stacked
    header = "    " + " ".join(f"{c:>2}" for c in range(1, GRID_SIZE+1))
    print(header)
    for r in range(GRID_SIZE):
        row_str = " ".join(f"{cell:>2}" for cell in grid[r])
        print(f"{r+1:>2} | {row_str}")

def simulate(policy, start_state, max_steps=MAX_STEPS):
    s = start_state
    sample_path = []
    print_grid(*s)
    for t in range(max_steps):
        if is_terminal(s): break
        a = policy[s]
        sp, r = successor(s, a)
        sample_path.append((t, s, a, r, sp))
        print(f"t={t}, DECISION={a}, reward={r}, next={sp}")
        print_grid(*sp)
```

```
        s = sp
    return sample_path

# Execute policy iteration
print("Running policy iteration...")
print("Legend: '.' empty, x/y sites (uncleared), X/Y sites (cleared),\n \
1/2 robots, B robots in same cell.\n")
print("Start State:")
v, policy = policy_iteration()
start = (rc_to_idx(1, 1), rc_to_idx(5, 5), 0, 0) #Robot start locations
sample_path = simulate(policy, start_state=start, max_steps=MAX_STEPS)
```

Running the above code yields an optimal joint policy. In this particular example, robot 2 remains stationary at all periods, and robot 1 clears the debris at both locations, starting with location X, followed by clearing location Y. Since both locations are close to robot 1, robot 2 is left idle. This can be visualized based on the following output of the Python code:

```
Running policy iteration...
Legend: '.' empty, x/y sites (uncleared), X/Y sites (cleared),
 1/2 robots, B robots in same cell.

Start state
     1  2  3  4  5
 1 |  1  .  .  .  .
 2 |  .  x  .  .  .
 3 |  y  .  .  .  .
 4 |  .  .  .  .  .
 5 |  .  .  .  .  2
t=0, action=('R', 'N'), reward=-1, next=(2, 25, 0, 0)
     1  2  3  4  5
 1 |  .  1  .  .  .
 2 |  .  x  .  .  .
 3 |  y  .  .  .  .
 4 |  .  .  .  .  .
 5 |  .  .  .  .  2
t=1, action=('D', 'N'), reward=9, next=(7, 25, 1, 0)
     1  2  3  4  5
 1 |  .  .  .  .  .
 2 |  .  1  .  .  .
 3 |  y  .  .  .  .
 4 |  .  .  .  .  .
 5 |  .  .  .  .  2
t=2, action=('L', 'N'), reward=-1, next=(6, 25, 1, 0)
     1  2  3  4  5
 1 |  .  .  .  .  .
 2 |  1  X  .  .  .
 3 |  y  .  .  .  .
 4 |  .  .  .  .  .
 5 |  .  .  .  .  2
t=3, action=('D', 'N'), reward=9, next=(11, 25, 1, 1)
```

```
     1  2  3  4  5
1 |  .  .  .  .  .
2 |  .  X  .  .  .
3 |  1  .  .  .  .
4 |  .  .  .  .  .
5 |  .  .  .  .  2
```

Note the contrast between the fully decentralized independent case of two agents with independent and additive rewards and no information sharing, which, as described above, could be solved by solving each MDP separately. In this example, a joint policy is generated by a central planner that optimizes decisions based on full knowledge of the robot locations at each step. The latter problem is significantly more computationally challenging because solving the joint optimality equations becomes exponentially more difficult as the number of agents increases.

In the next example, consider two agents who must search for a target without knowing its location. Unlike the previous example, there is uncertainty about the target's location.

9.3.3 Fully Centralized MAMDPs with Partial Observability

A generalization of the above fully cooperative MAMDP model is the case in which agents do not have perfect information about their environment. If complete information were available to all agents, this would correspond to a standard MDP with a joint state-decision space that can be controlled by a central planner, as described above. However, there are often restrictions on what agents can observe, similar to the POMDP models in Chapter 7. In this case, agents must make decisions based on their *belief* about the environment, without being able to fully observe the system's state. In the case of POMDP models of Chapter 7, it is known that the belief state (the vector of probabilities of the agent being in a particular state) is a summary statistic containing all necessary information to define the state of the system. An alternative formulation, *Decentralized MDPS* (Dec-MDPs), does not necessarily possess this property, and the ideal approach to such problems depends on the constraints on communication between agents. The next example illustrates the case of perfect communication between agents in a partially observable setting.

Python Example 9.3: Searching for a missing hiker in a Canadian national park

Assume two park rangers must try to find a missing hiker. The park is a 3×3 grid with 9 sectors. The hiker is immobile, so their location is fixed but unknown, and the rangers are located at different ranger stations. Since there is no prior information, the rangers assume that the hiker is equally likely to be found in each unsearched sector. Ranger 1 starts at sector $(0,\ 0)$ of the grid, and Ranger 2 starts at sector $(3,\ 3)$. The lost hiker could be in any sector, and the rangers search until they find the missing hiker. The rangers are in constant communication and are aware of each other's positions at all times; therefore, they can keep a log of locations already visited where the hiker was not found to avoid redundant searches of sectors. Each time a sector is eliminated from consideration, the hikers update their belief to be uniformly distributed across all remaining unsearched locations. The search ends when one of the rangers reaches the sector where the hiker is located.

The model formulation is defined as follows. First, the states are defined by combinations of ranger positions, s^1 and s^2. The overall state of the system at epoch t is $s_t = (s^1, s^2)$, where s^1 and s^2 can be any of the 16 sectors ($9^2 = 81$ possibilities), multiplied by the total number of subsets of the 9 sectors that could remain unsearched at a given epoch ($2^9 = 1,024$ possibilities). Thus, the total number of possible states is 82, 944. The hiker's location, s^H, is assumed to be fixed but unknown to the rangers and therefore is not part of the state space. The state transitions are assumed to be deterministic and defined by the current state and the rangers' decisions, a^1 and a^2, which can be to move: up, down, left, right, or stay in place; thus, the set of decisions is $A = \{U,\ D,\ L,\ R,\ S\}$. The reward function is expressed as follows:

$$r(s^1, s^2) = \begin{cases} 10 & \text{if } s^1 = s^H \text{ or } s^2 = s^H \\ -1 & \text{if } s^1 \neq s^H \text{ and } s^2 \neq s^H \text{ and the rangers move in any direction} \end{cases}$$

It includes a cost of movement to adjoining sectors and a reward for finding the hiker.

Because the hiker is at risk and their safety could degrade over time, a discount factor of $\lambda = 0.9$ is used. The rangers' MDP models are identical and stationary, and the search continues until one of the rangers locates the hiker. Thus, the appropriate model is an infinite-horizon MDP that takes the following form in this two-agent context:

$$v\,(s_1, s_2) = \max_{a_1 \in A, a_2 \in A} \left\{ r(s^1, s^2) + \lambda \sum_{s_1' \in S, s_2' \in S} p(s^{1'}, s^{2'} | s^1, s^2, a^1, a^2) v\,(s) \right\}, \forall s^1 \in S, \forall s^2 \in S.$$

Note that the model is stationary based on the model description and per the norm for infinite horizon models as discussed in Chapter 4.

The following Python code solves the model as an infinite-horizon MDP to find the optimal policy for the rangers' coordinated search using the value iteration algorithm.

```
from collections import defaultdict
import numpy as np

# Input Data
N = 3  # dimension of grid with N x N sectors
Lambda = 0.9
FIND_REWARD = 10.0
START_P1 = (0, 0)         #Ranger 1 location
START_P2 = (N-1, N-1)     #Ranger 2 location
TERMINAL = ("T",)         # Terminal state marker
ALL_SECTORS = set(range(N * N))
TOLERANCE, MAX_ITERATIONS =1e-6,1000 #Value iteration tolerance, max iterations
MAX_STEPS = 200
MAX_REACHABLE_STATES = 10**6
np.random.seed(42) #Fix seed for repeatability

#Utility functions
def rc_to_idx(r, c): return r * N + c #convert 2D to 1D index
def idx_to_rc(i):    return (i // N, i % N)# convert 1D to 2D index

#Check feasible decisions given moving off grid is not permitted
def neighbors(idx):
    r, c = idx_to_rc(idx)
    options = []
    if r > 0:          options.append(('U', rc_to_idx(r-1, c)))
    if r < N-1:        options.append(('D', rc_to_idx(r+1, c)))
    if c > 0:          options.append(('L', rc_to_idx(r, c-1)))
    if c < N-1:        options.append(('R', rc_to_idx(r, c+1)))
    options.append(('S', rc_to_idx(r, c)))
    return options  # list of (decision, next_idx)

#Cartesian product of feasible single agent decisions
def joint_decisions(p1, p2):
    for a1, np1 in neighbors(p1):
        for a2, np2 in neighbors(p2):
            yield (a1, a2, np1, np2)

#Rangers search their initial sectors without movement cost
def initial_search():
    p1 = rc_to_idx(*START_P1)
```

```
    p2 = rc_to_idx(*START_P2)
    visited = frozenset({p1, p2})
    return (p1, p2, visited)

#State transition dynamics given current state and join decisions
def step_transition(state, joint_dec):
    p1, p2, visited = state
    a1, a2, np1, np2 = joint_dec
    before_unvisited = set(ALL_SECTORS) - set(visited)
    if not before_unvisited:
        return [(1.0, TERMINAL, 0.0)]
    move_cost = int(np1 != p1) + int(np2 != p2)
    new_sectors = {np1, np2}
    newly_new = [c for c in new_sectors if c in before_unvisited]
    p_success = len(newly_new) / len(before_unvisited)
    r_succ = FIND_REWARD - move_cost
    next_visited = frozenset(set(visited) | new_sectors)
    r_fail = - move_cost
    next_state_fail = (np1, np2, next_visited)
    out = []
    if p_success > 0:
        out.append((p_success, TERMINAL, r_succ))
    out.append((1.0 - p_success, next_state_fail, r_fail))
    return out

#Identify all reachable states
def enumerate_reachable():
    start = initial_search()
    frontier = [start]
    seen = {start}
    while frontier:
        s = frontier.pop()
        p1, p2, visited = s
        for jd in joint_decisions(p1, p2):
            for prob, ns, _ in step_transition(s, jd):
                if ns != TERMINAL and ns not in seen:
                    seen.add(ns)
                    frontier.append(ns)
        if len(seen) == MAX_REACHABLE_STATES:
            print("Hit max reachable states limit")
            break
    return seen

def simulate(policy, MAX_STEPS): #Simulate policy
    start1 = rc_to_idx(*START_P1) #Ranger 1 start sector
    start2 = rc_to_idx(*START_P2) #Ranger 2 start sector
    start_sectors = {start1, start2}
    P0 = len(start_sectors) / (N*N)
    print("\n--- Simulation of policy ---")
    if np.random.random() <= P0:
        print(f"Found at rangers' start points: reward = {FIND_REWARD}")
        return
```

```
    s = initial_search() # Otherwise, continue search
    t, G = 0, 0.0
    while s != TERMINAL and t < MAX_STEPS:
        p1, p2, _ = s
        jd = policy[s]
        a1, a2, _, _ = jd
        print(f"t={t:2d}  pos1={idx_to_rc(p1)} pos2={idx_to_rc(p2)} \
              decision=(R1 {a1}, R2 {a2})")
        outcomes = step_transition(s, jd) # Sample next outcome
        u, acc = np.random.random(), 0.0
        prob, ns, r = outcomes[-1]  # fallback
        for p, nst, rew in outcomes:
            acc += p
            if u <= acc:
                ns, r = nst, rew
                break
        print(f"    reward={r:+.1f}\n")
        G += (Lambda**t) * r
        s, t = ns, t + 1
        if s == TERMINAL:
            print(f"t={t:2d}  FOUND!  discounted return = {G:.3f}")
            break
    if s != TERMINAL:
        print(f"Stopped at t={MAX_STEPS}. discounted return = {G:.3f}")

# Value iteration to solve for near-optimal policy
states = enumerate_reachable()  #Enumerate reachable states
print(f"Reachable non-terminal states: {len(states)}")
v = defaultdict(float) #Optimal value function estimate
policy = {}
for it in range(MAX_ITERATIONS):
    delta = 0.0
    for s in states:
        p1, p2, _ = s
        best_q, best_jd = float('-inf'), None
        for jd in joint_decisions(p1, p2):
            q = 0.0
            for prob, ns, r in step_transition(s, jd):
                q += prob * (r + Lambda * (0.0 if ns == TERMINAL else v[ns]))
            if q > best_q:
                best_q, best_jd  = q, jd
        delta = max(delta, abs(best_q - v[s]))
        v[s], policy[s] = best_q, best_jd
    if delta < TOLERANCE:
        print(f"Converged in {it} iterations (delta={delta:.3e}).")
        break
#Simulate to illustrate application of the optimal policy
simulate(policy, MAX_STEPS)
```

In the above implementation, the rangers search their initial sectors at the start of the horizon. The success probability at a given decision epoch is the ratio of the sectors searched to the unvisited sectors. Running the above Python code yields an optimal cooperative policy to search for a lost hiker with an unknown location. Given the uncertainty about the hiker's location, there is uncertainty about the outcome of the rangers' search at each epoch. The above code generates the following output for an example of a simulated application of the policy.

```
Reachable non-terminal states: 2256
Converged in 4 iterations (delta=0.000e+00).

--- Simulation of policy ---
t= 0  pos1=(0, 0) pos2=(2, 2)                decision=(R1 D, R2 U)
    reward=-2.0

t= 1  pos1=(1, 0) pos2=(1, 2)                decision=(R1 D, R2 U)
    reward=-2.0

t= 2  pos1=(2, 0) pos2=(0, 2)                decision=(R1 R, R2 L)
    reward=+8.0

t= 3  FOUND!  discounted return = 2.680
```

Thus, in this case, the rangers start at positions (0, 0) and (2, 2). They first search the sectors they occupy but do not find the hiker. Next, ranger 1 moves down and ranger 2 moves up, resulting in movement costs of -1 each. In the next epoch, Ranger 1 moves down again, and Ranger 2 moves up again, incurring movement costs again, without finding the hiker. Subsequently, Ranger 1 moves right, then up. Ranger 2 moves left and then stays in place. At decision epoch 4, they locate the lost hiker with probability 1 because only one sector remains unchecked; all others have been ruled out. Running the simulation step multiple times with different random number generator seeds yields different search paths.

9.3.4 Decentralized POMDPs

A more general case of a decentralized model is the *Dec-POMDP*. For such models, observations are available that provide useful information for optimizing decision-making. However, unlike POMDPs, which have optimal policies based on the belief state, Dec-POMDPs do not. Instead, policies must be defined based on each agent's history of observations and decisions, which makes the problem much more challenging. These models are among the most complex and difficult to solve covered in this book.

Consider the following notation for a Dec-POMDP. The local observation space for agent i is denoted by $O^i = \{O_1^i, O_2^i, ..., O_K^i\}$ where O^i is the discrete and finite set of K possible observations for agent i. A joint policy for all agents is $\pi = (\pi^1, \dots, \pi^M)$ where the agents' policies define a mapping from each agent's history of observations and decisions, $h_t^i = (o_0^i, a_0^i, o_1^i, a_1^i, \dots, o_t^i)$ to the next decision at epoch t. The lack of a single summary statistic, such as belief in the case of POMDPs, means that the optimal policy for each agent may be history-dependent, substantially increasing the complexity of solving such problems.

The introduction of policies based on a history of observations and decisions represents a departure from previous models in this book, such as MDPs, where the optimal decision was solely dependent on the current state of the system, thanks to the Markov property, and POMDPs, where the belief is a sufficient statistic that defines the state at each epoch. In the case of Dec-POMDPS, however, the system state is the history of states and decisions, which can be represented as a tree, as shown in Figure 9.1. In this example, there are two decisions (a_1, a_2) and two observations (O_G, O_B). The root node at the top of the tree comprises the history of prior observations and decisions up to decision epoch t. The two branches correspond to the two possible observations, and the subsequent child nodes prescribe the decision to take in each case. Subsequently, new observations occur, followed by further decisions which continue through the end of the horizon T.

Algorithm 9.2 for solving Dec-POMDPs computes optimal joint policies by iteratively expanding and pruning sets of policy trees, such as that in Figure 9.1. The parameter τ denotes the policy-tree depth with a depth-τ tree spanning stages $t = 0$ through $t = \tau - 1$; therefore, $\tau = T + 1$ represents a policy for the full horizon. Q_i^τ denotes the set of feasible

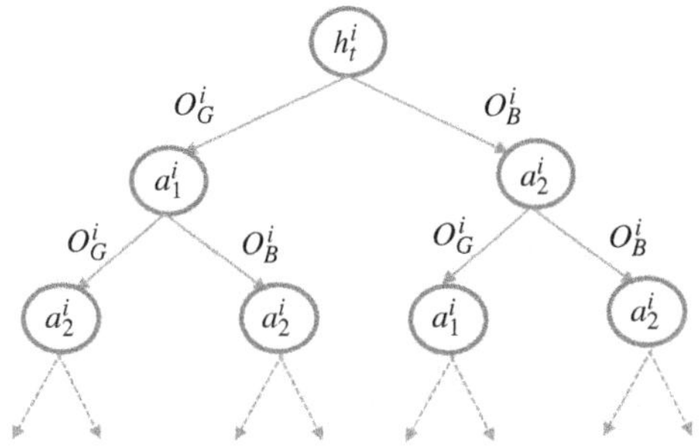

Figure 9.1 Illustration of a policy tree for a Dec-POMDP model starting at an arbitrary decision epoch, t, with agent i with history h_t^i for which the policy depends on the history of observations and decisions, where observations can be good (O_G) or bad (O_B) and there are two possible decisions (a_1 and a_2).

Algorithm 9.2 Algorithm for Solving Dec-POMDPs

Input: Dec-POMDP data elements: decision epochs, states, decisions, transition probability matrix, rewards, observation probabilities, initial belief distribution, discount factor
Initialization: $Q_i^1 \leftarrow A_i,\ i = 1, \ldots, n$
for $\tau = 2$ **to** $T + 1$ **do**
    for $i = 1$ **to** n **do**
        $Q_i^\tau \leftarrow \text{ExhaustiveBackup}(Q_i^{\tau-1})$
    end for
    while some policies have been pruned **do**
        for $i = 1$ **to** n **do**
            $Q_i^\tau \leftarrow \text{Prune}(i, Q_i^\tau, Q_{-i}^\tau, \tau)$
        end for
    end while
end for
$Q^{T+1} \leftarrow Q_1^{T+1} \times \cdots \times Q_n^{T+1}$
for each $\pi \in Q^{T+1}$ **do**
    $v^\pi \leftarrow \sum_{s \in S} b_0(s) v^\pi(s)$
end for
Return $\pi^* \leftarrow \arg\max_{\pi \in Q^{T+1}} v^\pi$

---

depth-τ policy trees for agent i. At each stage, the *exhaustive backup* operation expands the set of policy trees in $Q_i^{\tau-1}$ by adding subtrees for all possible observations, generating new depth-τ candidate trees. Unfortunately, this process creates an exponentially large number of candidate policies. Therefore, *pruning* of suboptimal policies is a necessary step (analogous to pruning α-vectors for POMDPs in Chapter 7). The pruning operations remove dominated policies that cannot be optimal given the other agents' policy trees. Repeating this procedure iteratively until $\tau = T + 1$ yields the final joint policy set for all agents $Q^{T+1} = Q_1^{T+1} \times \cdots \times Q_n^{T+1}$, from which a policy that maximizes expected rewards under the initial belief is selected.

Consider the following example, which illustrates the application of Algorithm 9.2 to solve a problem involving two agents that must coordinate their decisions to complete tasks in a factory setting.

Python Example 9.4: Factory robot coordination problem

Two robots work on an assembly line where new tasks arrive stochastically. At each decision epoch, there is a chance that a new task will arrive, but the robots cannot directly observe whether a task is present. Rather, each robot receives a private, yet noisy, signal indicating the possible presence of a task. To be successful, both robots must choose to process the task at the same decision epoch. If only one robot attempts to complete the task, it fails and incurs a penalty. If neither robot acts, the task is missed, resulting in a penalty. The robots (agents) do not communicate or share their observations, so they must coordinate solely on the basis of their own observation histories and subsequent beliefs. The following is the infinite-horizon decentralized partially observable MDP formulation of this problem.

Decision epochs: $t = 0, \ldots, T$.

States: $s \in S = \{0, 1\}$, where $s = 1$ represents that a task has arrived in the current epoch, and $s = 0$ represents that no task has arrived.

Decisions: The set of possible decisions is $D^i = \{\text{wait, process}\}$, $i = 1, 2$. The joint decision is $d = (d^1, d^2) \in D^1 \times D^2$. That is, each robot independently decides to either process the current task or wait.

Observations: Each agent receives a binary private observation $o_i \in \{0, 1\}$ indicating the probability a task has arrived ($s = 1$) or not ($s = 0$), where:

$$P\{o_i = 1 |\ s = 1\} = 0.9, \quad P\{o_i = 0 |\ s = 1\} = 0.1,$$

$$P\{o_i = 1 |\ s = 0\} = 0.1, \quad P\{o_i = 0 |\ s = 0\} = 0.9.$$

Transition probabilities: The state transition model reflects the dynamics of task availability. If a task is present at epoch t and both agents choose to process it, the task is successfully completed, and a new task is highly likely to arrive in the next step $t + 1$. If the agents fail to process the task (due to miscoordination or choosing to wait), a new task may still arrive, but with lower probability.

$$\text{Transition probabilities:} \quad P\{s_{t+1} = 1 |\ s_t, d_t\} = \begin{cases} 0.9 & \text{if } s_t = 1,\ d_t^1 = d_t^2 = \text{process} \\ 0.9 & \text{if } s_t = 0,\ d_t^1 = d_t^2 = \text{wait} \\ 0.5 & \text{otherwise} \end{cases}$$

$$\text{Rewards: } r(s, d_1, d_2) = \begin{cases} 15 & \text{if } s = 1,\ d_1 = d_2 = \text{process} \\ -12 & \text{if } s = 1,\ \text{exact one of } d_1 \text{ and } d_2 = \text{process} \\ -8 & \text{if } s = 1,\ d_1 = d_2 = \text{wait} \\ 6 & \text{if } s = 0,\ d_1 = d_2 = \text{wait} \\ -10 & \text{if } s = 0,\ \text{exact one of } d_1 \text{ and } d_2 = \text{process} \\ -8 & \text{if } s = 0,\ d_1 = d_2 = \text{process} \end{cases}$$

Discount factor: The discount factor, $\lambda = 0.95$, reflects both the need for timely task completion to maintain high productivity and the increasing cost of prolonged operation due to machine wear.
The following Python code solves this Dec-POMDP using the algorithm described in 9.2.

```
import numpy as np
import itertools
from functools import lru_cache

# Input data
T = 3
states = [0, 1]
nstates = 2
decisions = ['wait', 'process']
ndecisions = 2
nobservations = 2
Lambda = 0.95 # Discount factor
```

```
# Transition probability matrix: TPM[s, d1, d2, s_next]
P = np.zeros((nstates, ndecisions, ndecisions, nstates), dtype=float)
for s in range(nstates):
    for d1 in range(ndecisions):
        for d2 in range(ndecisions):

            # Higher probability of next task if both process at s=1, or wait at s=0
            if (s == 1 and d1 == 1 and d2 == 1) or (s == 0 and d1 == 0 and d2 == 0):
                p_next_task = 0.9

            # Otherwise lower probability
            else: p_next_task = 0.5
            P[s, d1, d2, 1] = p_next_task
            P[s, d1, d2, 0] = 1 - p_next_task

O = np.array([ # Observation probabilities: O[s, o] = P(o|s)
    [0.9, 0.1],  # P(o=0|s=0)=0.9, P(o=1|s=0)=0.1
    [0.1, 0.9]   # P(o=0|s=1)=0.1, P(o=1|s=1)=0.9
], dtype=float)

rewards = np.zeros((nstates,ndecisions,ndecisions),dtype=float) # rewards[s,d1,d2]

# s = 1
rewards[1, 1, 1] = 15.0    # Both process when a task exists
rewards[1, 1, 0] = -12.0   # Exactly one processes
rewards[1, 0, 1] = -12.0   # Exactly one processes
rewards[1, 0, 0] = -8.0    # Both wait
# s = 0
rewards[0, 1, 1] = -10.0   # Both process unnecessarily
rewards[0, 1, 0] = -8.0    # One processes unnecessarily
rewards[0, 0, 1] = -8.0    # One processes unnecessarily
rewards[0, 0, 0] = 6.0     # Both wait

# Initial belief over states
initial_belief = [0.5, 0.5]

# Policy-tree representation
def leaf(d): return ('leaf', d) # take decision d and stop
def node(d, c0, c1):
    return ('node', d, c0, c1) # take decision d then follow subtree c0 or c1

# Generate depth-tau trees from depth-(tau-1) trees
def exhaustive_backup(Q_prev):
    result = []
    for d in range(ndecisions):
        for child_o0, child_o1 in itertools.product(Q_prev, repeat=2):
            result.append(node(d, child_o0, child_o1))
    return result

# Compute the expected value of joint policy starting state s with  depth_h
@lru_cache(maxsize=None) #Forces caching for faster performance
```

```
def value_state(s, pi1, pi2, depth_h):
    if depth_h == 0: return 0
    d1 = pi1[1]
    d2 = pi2[1]
    if depth_h == 1: return rewards[s, d1, d2]
    # Future value
    value = 0
    for s_next in range(nstates):
        for o1 in range(nobservations):
            for o2 in range(nobservations):
                c1 = pi1 if pi1[0] == 'leaf' else (pi1[2] if o1 == 0 else pi1[3])
                c2 = pi2 if pi2[0] == 'leaf' else (pi2[2] if o2 == 0 else pi2[3])
                value += (P[s, d1, d2, s_next] * O[s_next, o1] * O[s_next, o2]
                           * value_state(s_next, c1, c2, depth_h - 1))
    return rewards[s, d1, d2] + Lambda * value

# Compute expected value of a joint policy (pi1, pi2) under initial belief
def value_belief(belief, pi1, pi2, depth_h):
    return float(belief[0] * value_state(0, pi1, pi2, depth_h) +
                 belief[1] * value_state(1, pi1, pi2, depth_h))

# Evaluate pairs (pi1, pi2) on belief points, prune policies never near-optimal
def prune(Q1, Q2, depth_h, grid_pts=41, eps=1e-6, topk=None):
    beliefs = np.linspace(0.0, 1.0, grid_pts)
    keep_pairs = set()
    for p in beliefs:
        b = [1 - p, p]
        vals = [(value_belief(b, pi1, pi2, depth_h), i, j) # Evaluate policies
                for i, pi1 in enumerate(Q1) for j, pi2 in enumerate(Q2)]
        vals.sort(key=lambda x: x[0], reverse=True)
        if topk is None:
            best = vals[0][0]
            winners = [(i, j) for v, i, j in vals if v >= best - eps]
        else:
            cutoff = vals[min(topk - 1, len(vals) - 1)][0]
            winners = [(i, j) for v, i, j in vals if v >= cutoff - eps]
        keep_pairs.update(winners)
    keep_i = {i for (i, _) in keep_pairs}
    keep_j = {j for (_, j) in keep_pairs}
    Q1_new = [pi for i, pi in enumerate(Q1) if i in keep_i]
    Q2_new = [pj for j, pj in enumerate(Q2) if j in keep_j]
    return Q1_new, Q2_new

# Show policy tree
def show_policy(pi, indent=0):
    pad = '   ' * indent
    if pi[0] == 'leaf':
        return f"{pad} d = {decisions[pi[1]]}"
    s = f"{pad}d = {decisions[pi[1]]}\n"
    s += f"{pad} if o = 0:\n{show_policy(pi[2], indent+2)}\n"
    s += f"{pad} if o = 1:\n{show_policy(pi[3], indent+2)}"
    return s
```

```
# Solve Dec-POMDPs
Q1 = [leaf(d) for d in range(ndecisions)] #Initialize policies
Q2 = [leaf(d) for d in range(ndecisions)]

for tau in range(2, T + 2): # Grow tree to depth T + 1 with pruning
    Q1 = exhaustive_backup(Q1)
    Q2 = exhaustive_backup(Q2)
    changed = True  #Iterative pruning until no change
    while changed:
        before = (len(Q1), len(Q2))
        Q1, Q2 = prune(Q1, Q2, tau)
        changed = before != (len(Q1), len(Q2))

# Loop over policy trees, get best value and policy tree
best_pair = None
best_value = -np.inf
for pi1, pi2 in itertools.product(Q1, Q2):
    v = value_belief(initial_belief, pi1, pi2, T + 1)
    if v > best_value:
        best_value = v
        best_pair = (pi1, pi2)

# Print results
print(f"Optimal value (T = {T}, initial belief = {initial_belief}): {best_value:.4f}\n")
print('Robot 1 policy tree:')
print(show_policy(best_pair[0]))
print('Robot 2 policy tree:')
print(show_policy(best_pair[1]))
print(f"\nAfter pruning at depth T + 1 = {T + 1}: |Q1^(T+1)| = {len(Q1)}, \
|Q2^(T+1)| = {len(Q2)}")
```

Running the above Python code yields the following best policy trees of horizon $T + 1$ for each agent, along with the pruned policy counts, which represent the number of policy trees retained for each agent for all initial belief distributions at the final horizon after pruning, indicating the efficiency of the search. The initial output provides the policy tree as a decision rule for Robot 1 and Robot 2. These are the best of the 5 policies remaining after pruning.

```
Optimal value (T = 3, initial belief = [0.5, 0.5]): 28.9927

Robot 1 policy tree:
d = process
 if o = 0:
    d = wait
     if o = 0:
        d = process
         if o = 0:
            d = wait
         if o = 1:
            d = process
     if o = 1:
        d = process
         if o = 0:
            d = process
         if o = 1:
```

```
                d = process
  if o = 1:
     d = process
      if o = 0:
         d = process
          if o = 0:
              d = wait
          if o = 1:
              d = process
      if o = 1:
         d = process
          if o = 0:
              d = process
          if o = 1:
              d = process
Robot 2 policy tree:
d = process
 if o = 0:
     d = wait
      if o = 0:
         d = process
          if o = 0:
              d = wait
          if o = 1:
              d = process
      if o = 1:
         d = process
          if o = 0:
              d = process
          if o = 1:
              d = process
 if o = 1:
     d = process
      if o = 0:
         d = process
          if o = 0:
              d = wait
          if o = 1:
              d = process
      if o = 1:
         d = process
          if o = 0:
              d = process
          if o = 1:
              d = process

After pruning at depth T + 1 = 4: ||Q1^(T+1)|| = 5, ||Q2^(T+1)|| = 5
```

For $T = 3$ (evaluated here using a depth $T + 1 = 4$ policy tree) and initial belief $(0.5,\ 0.5)$, the two robots adopt the same policy tree because the problem data are symmetric. Observation $o = 1$, which strongly suggests $s = 1$ according to the observation probabilities, typically leads both robots to the `process` decision, consistent with the high reward for joint processing when a task exists. In contrast, observation $o = 0$ often leads to `wait`, consistent with the high cost of processing when $s = 0$, while joint waiting yields a positive reward and is also favored by the transition dynamics. After pruning, five candidate trees remain per agent (policies that are optimal for different belief points), and the best remaining pair achieves an approximately optimal value of 28.99 at the initial belief $(0.5,\ 0.5)$.

9.4 Concluding Remarks

The RMDPs presented in a max-min framework have been developed to generate the worst-case optimal policy when MDP parameters, such as TPMs, are uncertain. The addition of an adversary reflects this and emphasizes the worst-case changes to the TPMs resulting from solving an inner optimization problem where the decision variables are the adversary's choice of transition probabilities within an uncertainty set. There are numerous contributions to this body of work, going back decades. Some recent articles that provide a more advanced treatment of this topic include Nilim and El Ghaoui [1] and Iyengar [2], and references therein. These sources extend beyond the notion of an interval model to include more rigorously motivated choices of the uncertainty set, informed by statistical arguments, which lead to elliptical uncertainty sets and others. These provide a more principled representation of the confidence region, but they also lead to a nonlinear formulation of the inner problem (9.2), which can make the RMDP more difficult to solve. Other approaches, such as the likelihood model, the entropy model, and the constrained ellipsoidal model formulations described in Nilim and El Ghaoui [1] and Iyengar [2], include parameters to control the confidence level. A common concern with RMDPs is their focus on the worst-case model parameters, which can lead to policies that are optimal with respect to extremely unlikely scenarios only and thus may not perform well in practice. The medical treatment example of an RMDP in this chapter was motivated by Zhang et al. [3].

Regarding cooperative MAMDPs, early work for the case of perfect information sharing, which corresponds to *factored MDPs*, was published by Boutilier [4]. Many more articles have since fleshed out the details of variations on these problems, such as the Dec-POMDP models covered in this chapter. A useful reference for readers wishing to build their knowledge of this topic is the book by Oliehoek and Amato [5], which covers the topics of this chapter and extends them substantially to include a variety of other environments that influence model formulations and methods, including heuristics that can more realistically handle larger-scale instances than considered here.

The topic of multi-agent decision-making models is an active area of study. This chapter covered the fundamental models that can be understood through the perspectives of MDPs and POMDPs. Multi-agent RL is a natural extension to the above model-based multi-agent problems. In such models, the optimal value function is traded for the Q function. As described in Chapter 8, RL requires a blend of exploration and exploitation to update the Q functions and successively estimate policies. There is a highly active and growing body of research in this area. The interested reader may find Amato [6] a helpful introduction to this advanced topic.

The treatment of multi-agent models in this chapter was entirely based on cooperative models. A rich literature exists on competitive multi-agent models that employ game theory and other approaches to derive equilibrium conditions for understanding multi-agent behavior. Shoham and Leyton-Brown [7] provides a very thorough introduction to this interesting topic.

9.5 Practice Exercises

1. Suppose you can purchase IBM stock at \$50/share today. The stock price is currently \$50, and each day it increases by 5% with a probability of 0.55 and decreases by 10% with a probability of 0.45. You may exercise the option on any of the next five days. What is the optimal exercise policy? How does the policy change under the conditions of an interval model for an RMDP where the probability of increase or decrease could have an error of ± 0.15? How do the results change with respect to the size of the interval?

2. If a task is only considered successful if all agents make the "right" decision, describe a scenario where an agent should not be punished even though the team failed.

3. For the Centralized Planning example with debris-removing robots, solve the model assuming a 10×10 grid with two robots. Evaluate the sensitivity of the results to the number of debris-clearing robots $(3,\ 4,\ 5, \ldots)$. What do you observe as the number of robots increases?

4. Estimate the mean reward for the ranger's search for a missing hiker. How is the mean reward affected by the discount factor, or the cost of movements relative to the reward of finding the missing hiker? Summarize your findings.

5. Consider the example of a Missing Hiker under the assumption that, at each decision epoch with probability 0.2 the hiker moves to an adjacent sector, picking among the feasible options with equal probability. Revise the model to account for the uncertainty in the hiker's location.

6. Explain why agents learning simultaneously makes the environment non-stationary from any one agent's perspective. Give a simple example.

7. For a cooperative K agent-team task, failure occurs if any of the K agents makes a mistake. If all agents choose exploration with probability ϵ, what is the probability that all exploit the current policy at a given decision epoch? Explain why this matters for learning coordination and suggest a way to address this.

8. A hospital deploys two cooperating agents in an emergency room to manage possible sepsis in patients waiting for admission over a 1-hour window divided into four 15-minute decision epochs, $t = 1,\ 2,\ 3,\ 4$. The random variable $X \in \{0, 1\}$ indicates whether the patient has sepsis during this hour. At $t = 1$, $P(X = 1) = 0.25$. If $X = 0$, the patient does not develop sepsis during the hour; if $X = 1$, sepsis is present throughout the hour. Sepsis can only be diagnosed with a test. At each decision epoch, Agent A decides whether to order a diagnostic test or not, $a_A \in \{\text{DT},\ \text{NDT}\}$, and Agent B decides whether to start antibiotics or wait, with decisions $a_B \in \{S, W\}$, which is an irreversible decision that extends for the remaining decision epochs. If DT is chosen at epoch t, then a diagnostic test is performed, and the result, $O_t = X$, is observed immediately by both agents (assume a perfect test). Otherwise, the patient's status is unknown. Antibiotics are delivered on time only if Agent B chooses to start antibiotics for some epoch $t \leq 4$, and there has been a positive test at or before t. The hospital is testing the two-agent system using the following reward function:

$$R = \begin{cases} 10, & \text{if the patient receives antibiotics within one hour of onset of sepsis,} \\ 0, & \text{otherwise.} \end{cases}$$

However, there are also cost penalties of -0.5 for ordering a test, -1 for failing to identify and treat a patient with sepsis, and -0.75 for treating a patient who does not have sepsis. Both agents use the following naive learning rule: if the outcome is positive, increase the probability of repeating their last decision; otherwise, decrease it.

   a. Explain why the naive rule can give misleading feedback to Agent A or Agent B. Give at least two scenarios (joint decision sequences) where an agent took a "reasonable" decision but got negative feedback.
   b. Propose an alternative reward function that reduces misleading feedback without changing the clinical goal, and check whether your solution improves cooperation.

References

1 Nilim, A. & El Ghaoui, L. (2005). Robust control of Markov decision processes with uncertain transition matrices. *Operations Research, 53* (5), 780–798.

2 Iyengar, G. N. (2005). Robust dynamic programming. *Mathematics of Operations Research, 30* (2), 257–280.

3 Zhang, Y., Steimle, L. N. & Denton, B. T. (2017). Robust Markov decision processes for medical treatment decisions. *Optimization Online.*

4 Boutilier, C. (1996). Planning, learning and coordination in multiagent decision processes. In *Proceedings of the International Conference on Theoretical Aspects of Rationality and Knowledge (TARK)* (pp. 195–210).

5 Oliehoek, F. A. & Amato, C. (2016). *A Concise Introduction to Decentralized POMDPs.* Springer.

6 Amato, C. (2025). An initial introduction to cooperative multi-agent reinforcement learning. *arXiv* preprint.

7 Shoham, Y. & Leyton-Brown, K. (2008). *Multiagent Systems: Algorithmic, Game-Theoretic, and Logical Foundations.* Cambridge University Press.

10

Approximation Methods for Large-scale Models

10.1 Introduction

The focus of this chapter is on scaling up the methods from previous chapters to very large-scale problems. As discussed in Chapter 3, the *curse of dimensionality* is a serious problem in many real-world applications when using state-based model formulations based on dynamic programming, such as MDPs, POMDPs, and RL. This raises questions about whether and when the various approaches are practical and when it may be necessary to resort to approximation methods. This chapter introduces *approximate dynamic programming* and several associated methods. It revises the goal from finding optimal policies, the focus of previous chapters, to finding good solutions within a reasonable time. The optimal approximation method for a given application may vary with the nature of the models employed. Some applications may warrant the use of fast heuristics based on decision-makers' intuition, whereas others require more advanced approaches. This chapter aims to provide a brief introduction to the latter case, which is an expansive and rapidly developing research topic. Before discussing methods, however, the following motivating examples illustrate the curse of dimensionality.

- Consider a software company that is developing a decision support system to manage inventory for a retail outlet that sells 100 different types of shoes. Each week, they must decide whether to order from the manufacturer and, if so, how many pairs of each shoe type to order. Assume that the organization's inventory policies limit the number of pairs of shoes per shoe type to 9. Several considerations influence decision-making for different types of shoes, including customer demand, limited inventory, and the substitutability of shoes, meaning that some customers who do not find the specific shoe they want may purchase a similar one. For these reasons, the ordering decisions cannot simply be decomposed into individual products. For an exact treatment of this problem, the state space must include 10 inventory states (0–9 units in stock) for each of the 100 shoe types, yielding 10^{100} possible system states.
- Consider a municipality that is considering implementing a smart traffic system that uses computer vision to detect vehicle traffic and make real-time decisions about traffic light changes at intersections in the road network. If the overall goal is to optimize traffic flow and reduce congestion-related wait times, then decisions at various intersections within the traffic grid are interdependent. Thus, for example, if there are 60 four-way intersections in a municipality, each with as many as 20 cars waiting at any point in time, the number of possible states of the system is $(21^4)^{60}$, a problem size that no methods discussed thus far in this book can handle.
- Consider a team of drones, managed by a central decision-maker, that seeks to complete cooperative tasks, such as searching an area for an injured person who is unconscious and unable to communicate. The area can be defined by a grid of points that the robots can traverse during their search for the injured person, providing regular updates to the central decision-maker. Assuming a drone resolution of 10×10 meters to confirm the presence or absence of an injured person, a 10 km^2 region would have 10 million grid points. The state of the system could be expressed, in part, using a set of binary variables that denote a grid point that has been searched previously (1) or remains to be searched (0), yielding $2^{1000000}$ possible states defining the status of the search, not to mention the location of the drones in the region and other factors that may be relevant to the state definition (e.g., battery charge, weather conditions).

The above examples are just a small sample of the types of problems that arise in a multidimensional context in which the state of the system is defined by multiple factors (dimensions). They illustrate how rapidly the number of states increases as the dimensionality of the factors that influence the state definition increases. For models of this magnitude, finding exact solutions is computationally infeasible. In the last few decades, *approximate dynamic programming* (ADP) methods have been developed. The two main types of ADP are value-function approximations and policy approximations. The former uses an approximate representation of the optimal value function, such as expressing it as a weighted sum of a finite number of *basis functions*. In contrast, policy approximations directly estimate policies, often employing simulation to evaluate candidate policies. Each of these approaches is discussed in the remainder of this section.

The remainder of this chapter provides a brief introduction to approaches for approximating solutions to these large-scale problems, emphasizing the utility of the models and the resulting recommendations with some motivating examples. A complete treatment of the theory and methods of approximate dynamic programming is beyond the scope of this book; however, key points are summarized to provide the reader with a foundation of knowledge. Readers will find more thorough and advanced treatments of this material in recent books by Powell [1, 2] and Bertsekas [3], as well as in the references therein.

10.2 Value Function Approximations

MDPs and their extensions, such as POMDPs, MAMDPs, and model-free (RL) variants, often exhibit exponential state-space growth for complex real-world problems when the state is defined by multiple factors (e.g., inventory levels of multiple products, cars waiting at intersections, grid points in a geographic region). This issue can be addressed in multiple ways, all of which can be viewed as approximation methods that reduce the model's dimensionality while (ideally) providing close approximations to the optimal value function, and hence the resulting policy.

Regardless of the specific approximation method, the goal is to find an approximate value function $\tilde{v}(s)$, that serves as an approximation to the true value function, $\tilde{v}(s) \approx v(s)$ for all states $s \in S$. Any such approximation serves to provide an approximate policy as follows:

$$\tilde{a}(s) = \arg\max_{a \in A}\{r(s, a) + \lambda \sum_{j \in S} p(j \mid s, a)\tilde{v}(j)\}, \forall s \in S.$$

Ideally, the choice of approximation will be efficiently computable and yield a sufficiently close approximation to the true optimal value function. With that caveat in mind, any approach to approximating $\tilde{v}(s)$ could serve to approximate the optimal policy, such as simplified lower-dimensional models that capture the essence of the problem or simulation models that evaluate expected future rewards. In this sense, the breadth of possibilities for developing approximation methods is limited only by creativity and intuition about the factors that influence decisions.

10.2.1 Basis Function Approximations

One of the earliest proposed approximation methods uses the fact that certain sets of functions that form a complete discrete basis can, in theory, be used to estimate more complex functions. A fundamental example is the *Taylor series*, which can be used to express a function as an infinite power series. For instance, the function $\sin(x)$ can be expressed exactly as the following Taylor series expansion:

$$\sin(x) = \sum_{n=0}^{\infty} (-1)^n \frac{x^{2n+1}}{(2n+1)!}.$$

This alone is not very useful because it requires an infinite number of polynomial terms to express $\sin(x)$ exactly. But limiting the number of terms gives an approximation that may be easier to compute for complicated functions. For instance, retaining the first four nonzero terms of the Taylor series expansion yields the following:

$$\sin(x) \approx x - \frac{x^3}{3!} + \frac{x^5}{5!} - \frac{x^7}{7!}.$$

Figure 10.1 illustrates the quality of the approximation compared to the exact function. Increasing the number of terms in the Taylor series yields progressively better approximations that converge to the exact function as the number of terms tends to infinity.

There are many candidates for complete sets of basis functions, including *Legendre polynomials*, *Fourier series*, and *Haar wavelets*. There are also useful basis functions that do not provide exact representations in the limit because they are not complete in the literal infinite-dimensional sense but are nevertheless very effective choices. For example, *radial basis functions*, a class of functions that includes Gaussian functions, have been used successfully for value function approximation in many applications.

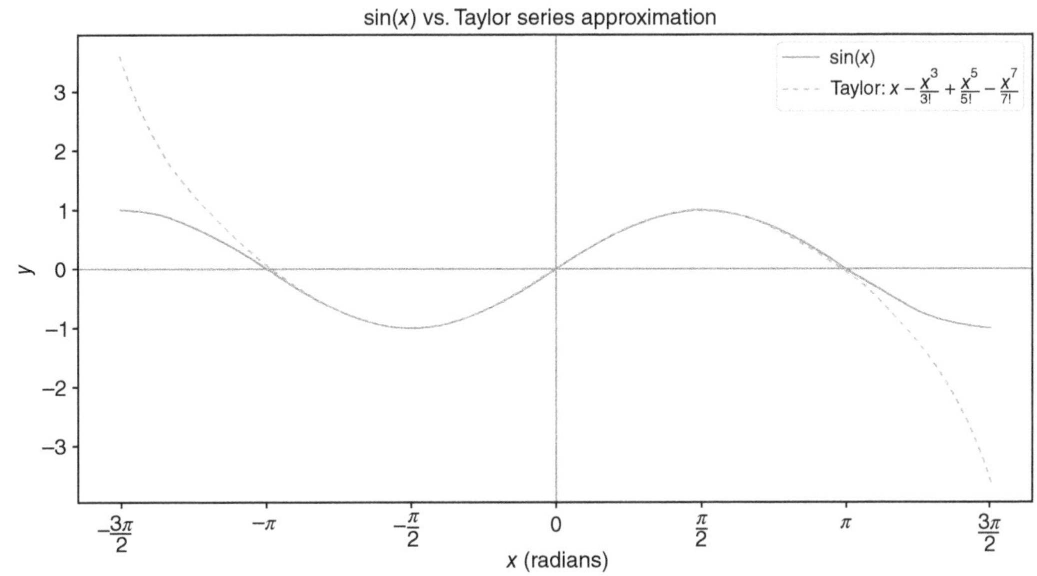

Figure 10.1 Approximation of sin(x) using the first four nonzero terms of the Taylor series expansion.

The general principles of basis function approximations can also be applied to approximating the optimal value function, a fundamental approach in approximate dynamic programming. Regardless of whether an optimal value function $v(s)$ can be expressed exactly as an infinite sum of weighted functions, however, in practice, a finite set of K basis functions $\{\phi_1, \ldots, \phi_K\}$ must be selected, yielding the following approximation (suppressing dependence on decision epochs, as in the stationary infinite horizon context, for now):

$$v(s) \approx \tilde{v}(s) = \sum_{k=1}^{K} \theta_k \phi_k(s) = \vec{\Phi}(s) \cdot \vec{\theta}.$$

where $\vec{\theta} \in R^K$ is a vector of coefficients that must be learned and $\vec{\Phi}(s) \in R^K$ is a vector of basis functions evaluated for a given state s. The parameters that form the vector $\vec{\theta}$ are sometimes called *hyperparameters*; they determine the accuracy of the optimal value function approximation, and their values are typically context-dependent. As such, they define an "outer" optimization model that can be solved to choose hyperparameters that yield good approximations. This itself can be computationally expensive, but once the hyperparameters are selected, the optimal decision for a given state is easy to compute.

One of the first approximation methods motivated by these concepts is approximate value iteration (AVI), proposed by Gordon [4] in the RL context and by Tsitsiklis and Van Roy [5] in the MDP context. The approach uses a combined sample of states at each epoch. It applies backward induction over a finite horizon, as covered in Chapters 3 and 4, to approximate the value function at the sampled states at each epoch, starting from the boundary condition $v_T(s) = R(s)$ for all $s \in S$. Backward induction on the optimality equations yields estimates of $\tilde{v}_{T-1}(s)$, then $\tilde{v}_{T-2}(s)$, and so on, for all $s \in S$, moving backwards recursively until the first decision epoch is reached. These estimates are then used to determine the optimal parameter values θ_t, at each decision epoch via least-squares minimization. The complete AVI algorithm is summarized in Algorithm 10.1 for a finite-horizon MDP with immediate rewards and state transitions, where, at each decision epoch, the policy is influenced by a vector of random variables $\vec{D}_t$.

The quality of the approximation depends on many considerations, including the model and the choice of basis functions. In some cases, it may be advantageous to select basis functions that are specifically related to important factors motivated by the problem structure. Thus, effectively harnessing approximate dynamic programming methods often requires an intuitive understanding of the problem at hand. Another choice that affects the quality of the approximation when implementing AVI is the state sample set at each epoch $\mathcal{X}_t \subseteq S$. The following example illustrates how the various choices influence the approximation.

Algorithm 10.1 Approximate Value Iteration

Require: Decision epochs, $t = 0, \ldots, T$; state space S; decision space A; transfer equation $s_{t+1} = g(s_t, a_t, \vec{D}_t)$ with $\vec{D}_t$ a vector of random variables; immediate reward $r_t(s_t, a_t, \vec{D}_t)$; basis functions $\phi_k(s)$, $k = 1, \ldots, K$; sampled states $\mathcal{X}_t \subseteq S$.

Initialize $v_T(s) = R(s), \forall s \in S$
for $t = T-1, T-2, \ldots, 0$ **do**
  Define the next epoch's approximation $\tilde{v}_{t+1}(s) \leftarrow \vec{\phi}(s)\vec{\theta}_{t+1}$
  for $s \in \mathcal{X}_t$ **do**
    for each $a \in A$ **do**

$$\tilde{v}_t(s, a) \leftarrow E\left[r_t(s, a, \vec{D}_t) + \tilde{v}_{t+1}(g(s, a, \vec{D}_t))\right],$$

    end for
    Set $y_t(s) \leftarrow \max_{a \in A} \tilde{v}_t(s, a)$
  end for
  Fit θ_t by least squares:

$$\theta_t \leftarrow \arg\min_{\theta \in R^K} \sum_{s \in \mathcal{X}_t} \left(y_t(s) - \vec{\phi}(s)\vec{\theta}\right)^2$$

end for
return $\theta_t, t = 0, \ldots, T$ and $\pi_t, t = 0, \ldots, T-1$

Python Example 10.1: Stochastic production lot-sizing

Consider a problem similar to that of Section 4.5 for planning production of a single product over a finite horizon $t = 0, 1, \ldots, T$, with the state defined as the amount of inventory, $s_t \in \{0, 1, 2, \ldots, M\}$, and production decisions, $a_t \in \{0, 1, 2, \ldots, M - s_t\}$, where M denotes the maximum possible inventory that can be held at any point in time. Let demand D_t at each decision epoch be independent and identically distributed according to a Poisson distribution, truncated to have an upper bound M to reflect the maximum demand that can be covered in a single decision epoch (since M is the maximum possible inventory available at each decision epoch). Assume any demand that can not be covered immediately in the current epoch t, is lost. Therefore, the state dynamics are described as follows:

$$s_{t+1} = \max\{0, s_t + a_t - D_t\}, \qquad t = 0, 1, \ldots, T-1.$$

Each unit is sold at a price r, and costs include a fixed (setup) cost c^S, a variable production cost per unit of c^V, and an inventory holding cost per unit c^H. Following the description from Section 4.5, the reward function is

$$r_t(s_t, a_t) = E_{D_t}\left[r \times \min(s_t + a_t, D_t)\right] - c^S \times I(a_t) - c^V(a_t) - c^H \times (s_t + a_t), \forall s_t \in S, \forall a_t \in A,$$

for $t = 0, \ldots, T-1$, where $I(a_t)$ is an indicator function that is 1 if $a_t > 0$ and 0 otherwise, and the terminal reward at epoch T that defines the boundary condition is as follows:

$$r_T(s_T) = 0, \forall s_T \in S.$$

The (exact) optimal value function is defined by the standard form of the optimality equations discussed in Chapter 4. However, to illustrate the use of ADP, apply an approximation of the form:

$$v(s) \approx \tilde{v}_t(s) = \sum_{k=1}^{K} \theta_k \phi_k(s),$$

where $\phi_k(s)$ and θ_k, $k = 1, \ldots, K$, are the basis functions and parameters to be estimated, respectively. For example, for the case of a linear function $\tilde{v}_t(s) = \theta_{0,t} + \theta_{1,t}s$, two parameters define the value-function approximation at each epoch t. Thus, at each epoch, the weights for the constant (intercept) $\theta_{0,t}$, and the slope $\theta_{1,t}$, are estimated using least square minimization. The following Python code implements Algorithm 10.1 to find an approximate value function and its corresponding policy for multiple choices of basis function sets, including a simple single-parameter constant and first through fifth-order polynomials, for a total of six comparisons. The root mean square error (RMSE) and the maximum error over all states for the approximate value function at the first decision epoch are reported as measures of accuracy.

```
import math
import numpy as np
import matplotlib.pyplot as plt

#Input Data
T = 10      # decision epochs
M = 60      # inventory capacity and maximum demand fulfilled
lam = 10    # mean daily demand for Poissone distribution
r = 5.0     # revenue per unit sold
c_S = 20.0  # fixed setup cost
c_V = 1.0   # cost per unit ordered
c_H = 0.50  # holding cost per unit leftover inventory

# Basis function choice:"constant","linear","quadratic","cubic","quartic", "quintic"
BASIS = "cubic"
# Basis functions
def phi(s: int, M: int):
    if BASIS == "constant":
        return np.array([1.0])
    if BASIS == "linear":
        return np.array([1.0, s])
    if BASIS == "quadratic":
        return np.array([1.0, s, s * s])
    if BASIS == "cubic":
        return np.array([1.0, s, s * s, pow(s,3)])
    if BASIS == "quartic":
        return np.array([1.0, s, s * s, pow(s,3), pow(s,4)])
    if BASIS == "quintic":
        return np.array([1.0, s, s * s, pow(s,3), pow(s,4), pow(s,5)])
    raise ValueError(f"Unknown BASIS: {BASIS}")

#Return value function approximation
def vhat_from_w(w: np.ndarray, M: int): return lambda s: float(phi(int(s), M).dot(w))

# Demand model (truncated Poisson): returns probs for d=0,...,M-1 and P(D>=M)
def truncated_poisson_probs(lam: float, M: int):
    probs = np.zeros(M + 1, dtype=float)
    for d in range(M):
        probs[d] = math.exp(-lam) * (lam ** d) / math.factorial(d)
    probs[M] = max(0.0, 1.0 - probs[:M].sum())
    return probs

# Immediate reward: E_D[ r_t(s,a,D) + v_next(s')] using demand truncated at D=M
def expected_future_rewards(s: int,a: int,v_next_fn,demand_probs: np.ndarray,
    M: int,r: float,c_V: float,c_H: float):
    sa = s + a  # inventory after order
    total = 0.0
    def contrib(d: int, prob: float):
        sales = min(sa, d)
```

```
        leftover = max(0, sa - d)
        s_next = max(0, sa - d)  # lost sales => same as leftover
        reward = r * sales - (c_V * a + c_H * leftover)
        return prob * (reward + v_next_fn(s_next))
    for d in range(M): total += contrib(d,demand_probs[d]) # Exact mass for d=0..M-1
    total += contrib(M, demand_probs[M]) # Truncation correction for D=M
    return total

#Compute max value and argmax decision, for state s at epoch t
def bellman_iteration(t: int,s: int,v_next_fn,demand_probs: np.ndarray, M: int,
    r: float,c_V: float,c_H: float):
    best_val = -float("inf")
    best_decision = 0
    for a in range(0, M - s + 1):
        val = expected_future_rewards(s,a,v_next_fn,demand_probs,M,r,c_V,c_H)
        if val > best_val: best_val, best_decision = val, a

    return best_val, best_decision

# Value Function Approximation: compute value function samples, then least squares
def value_function_approx(T: int,M: int,lam: float,r: float,c_V: float,c_H: float):
    demand_probs = truncated_poisson_probs(lam, M)
    m = len(phi(0, M))
    w = [np.zeros(m) for _ in range(T+1)] # Weights at t (w_T zeros because v_T(s)=0)
    pi = np.zeros((T, M + 1), dtype=int)
    for t in range(T - 1, -1, -1):
        v_next = vhat_from_w(w[t + 1], M)
        Phi = np.zeros((M + 1, m)) # Create regression data for all states
        y = np.zeros(M + 1)
        for s in range(M + 1):
            Phi[s, :] = phi(s, M)
            y[s], pi[t,s] = bellman_iteration(t,s,v_next,demand_probs,M,r,c_V,c_H)
        w[t], *_ = np.linalg.lstsq(Phi, y, rcond=None) # Least squares fit
    return w, pi

# Exact DP for comparison to approximation
def solve_exact_dp(T: int,M: int,lam: float,r: float,c_V: float,c_H: float):
    demand_probs = truncated_poisson_probs(lam, M)
    v = np.zeros((T + 1, M + 1))
    pi = np.zeros((T, M + 1), dtype=int)
    v[T, :] = 0.0
    for t in range(T - 1, -1, -1):
        for s in range(M + 1):
          v[t, s], pi[t, s] = bellman_iteration(t,s,v_next_fn=lambda sp: v[t + 1, sp],
                demand_probs=demand_probs,M=M,r=r,c_V=c_V,c_H=c_H)
    return v, pi

# Solve approximate MDP
w_hat, pi_hat = value_function_approx(T, M, lam, r, c_V, c_H)
```

```
v0_hat_fn = vhat_from_w(w_hat[0], M)
s_grid = np.arange(M + 1)

v0_hat = np.array([v0_hat_fn(s) for s in s_grid])

# Solve exact MDP for comparison
v_exact, pi_exact = solve_exact_dp(T, M, lam, r, c_V, c_H)
v0_exact = v_exact[0, :]
rmse = float(np.sqrt(np.mean((v0_hat - v0_exact) ** 2))) #root mean square error
max_err = float(np.max(np.abs(v0_hat - v0_exact)))
print(f"BASIS = {BASIS}")
print(f"RMSE(v_0): {rmse:.4f}    max|err|: {max_err:.4f}")

plt.figure()
plt.plot(s_grid, v0_exact, label="Exact $v_0(s)$")
plt.plot(s_grid, v0_hat, label=f"Approx $\\hat v_0(s)$ ({BASIS} basis)")
plt.xlabel("State $s$ (inventory)")
plt.ylabel("Value $v_0(s)$")
plt.title("Exact vs Approximate Value Function")
plt.legend()
plt.tight_layout()
plt.show()
```

Running the above code for all choices of basis function sets yields Figure 10.2, with six subplots, one for each of the approximations. The improvements in the approximation as the number of basis functions increases are evident from the plots. The larger the number of functions (terms in the polynomial), the better the approximation. With the five-term quintic polynomial (subplot F), the approximate value function has nearly converged to the optimal value function.

In this example, with a relatively small number of states, it was possible to consider all states when generating estimates, rather than resorting to sampling, as in Algorithm 10.1, which would be necessary for problems with very large state spaces, as in the examples described at the beginning of this chapter. In this example, the approximate value function was compared with the exact value function, which is only feasible for problems small enough to be solved exactly, such as this one.

In practice, when approximations are necessary, decisions must be made without knowing the exact value function. Often, these choices are based on the convergence of the approximation, which serves as a surrogate to the true approximation error. It can also be helpful to compare results against a baseline to quantify improvement.

Another early approach that employed a basis-function approximation was proposed by Schweitzer and Seidmann [6], who used the linear programming formulation of MDPs, as discussed in Chapter 4. This was later extended by De Farias and Van Roy [7], who demonstrated the benefits of the approach for computing error bounds on the approximate value function. This linear programming-based approach is particularly well-suited to infinite-horizon MDPs for which the backward induction steps used in the approximate value iteration steps of Algorithm 10.1 are not possible because of the absence of a boundary condition. Recall from Chapter 4 that the linear program formulation of an infinite-horizon MDP can be written as follows:

$$\begin{aligned} \min_{\tilde{v}} \quad & \sum_{s \in \mathcal{S}} \mu(s) v(s) \\ \text{s.t.} \quad & v(s) \geq r(s,a) + \gamma \sum_{s'} p(s' \mid s,a) v(s') \quad \forall s \in S, a \in A, \end{aligned}$$

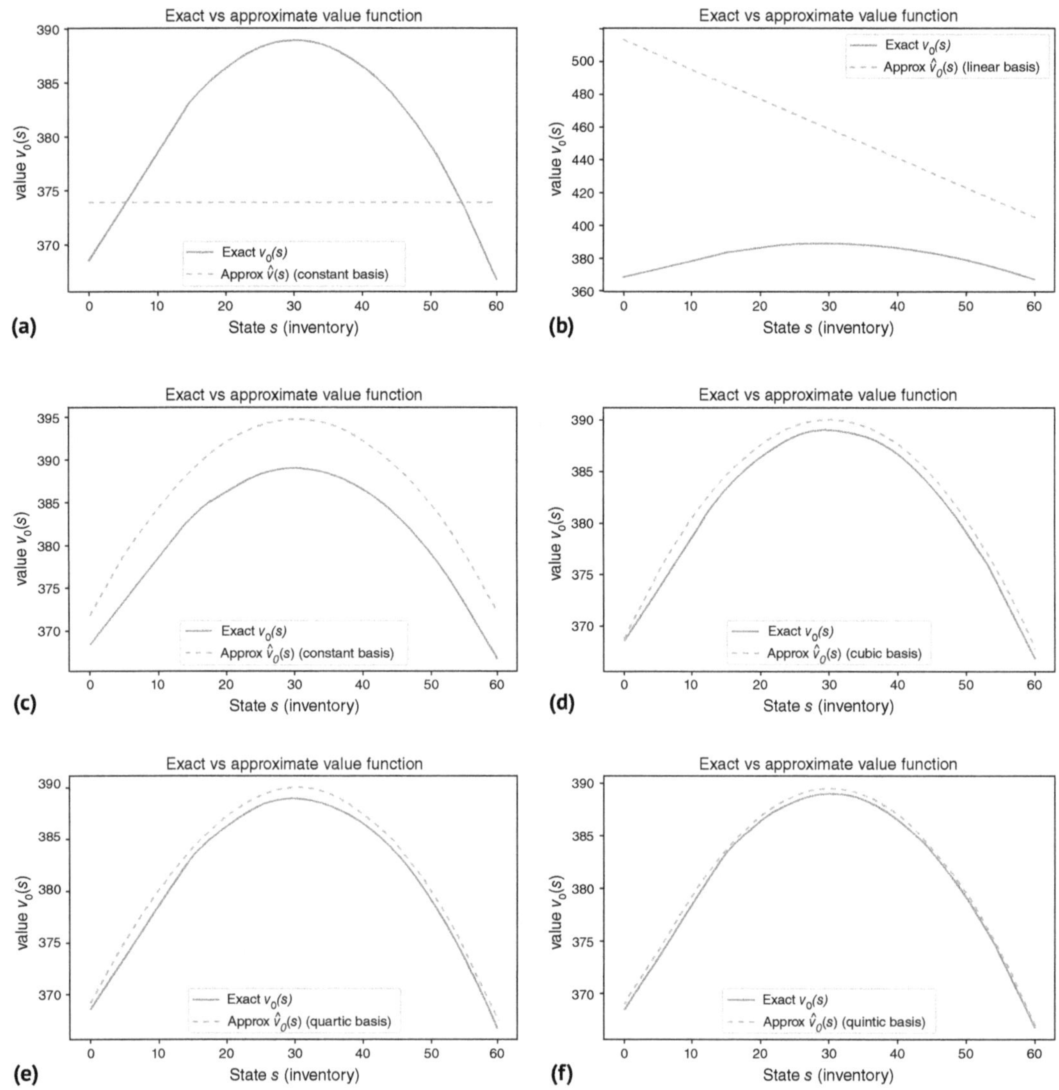

Figure 10.2 Illustration of the approximate value function for the stochastic production lot-sizing example, using six choices of basis function sets, ranging from a single constant (subplot A) to a fifth-order polynomial (subplot F).

where $\mu(s)$ are strictly positive weights for each state, $s \in S$. The approximate linear program replaces $v(s)$ with a linear approximation $\vec{\Phi}(s') \cdot \vec{\theta}$, resulting in:

$$\begin{aligned} \min_{\theta} \quad & \sum_{s \in \mathcal{S}} \mu(s)\vec{\Phi}(s) \cdot \vec{\theta} \\ \text{s.t.} \quad & \vec{\Phi}(s) \cdot \vec{\theta} \geq r(s,a) + \gamma \sum_{s'} p(s' \mid s,a)\vec{\Phi}(s) \cdot \vec{\theta} \quad \forall s \in S, a \in A. \end{aligned}$$

This linear program has K decision variables (the elements of vector $\vec{\theta}$) and $|S| \times |A|$ constraints (one for each state-decision pair). Given this approximate formulation, solution methods aim to identify the optimal parameter values for the vector $\vec{\theta}$, i.e., the vector of basis-function weights. Algorithm 10.2 illustrates how the approach is implemented.

Algorithm 10.2 Approximate Linear Programming Method

Require: Discount factor $\lambda \in (0, 1)$; state space S; decision space A; transfer equation $s' = g(s, a, \vec{D})$ with independent random variables $\vec{D}$; immediate reward $r(s, a, \vec{D})$; basis functions $\vec{\phi}_k(s)$, $k = 1, ..., K$; sampled states $\mathcal{X} \subseteq S$; state-relevance weights $\mu(s) > 0$ for $s \in \mathcal{X}$.

1: Define the value function approximation $\tilde{v}(s) \leftarrow \vec{\phi}(s) \cdot \vec{\theta}$
2: **for** $s \in \mathcal{X}$ **do**
3: **for** each $a \in A$ **do**
4: Compute the expected immediate reward and expected next features:

$$r(s, a) \leftarrow E\left[r(s, a, \vec{D})\right], \qquad \bar{\phi}(s, a) \leftarrow E\left[\phi(g(s, a, \vec{D}))\right]$$

5: **end for**
6: **end for**
7: Fit $\vec{\theta}$ by solving the approximate linear program:

$$\vec{\theta} \leftarrow \arg\min_{\theta \in R^K} \sum_{s \in \mathcal{X}} \mu(s)\, \vec{\phi}(s) \cdot \vec{\theta}$$

subject to, for all $s \in \mathcal{X}$ and $a \in A$,

$$\vec{\phi}(s) \cdot \vec{\theta} \geq r(s, a) + \lambda\, \vec{\phi}(s) \cdot \vec{\theta}.$$

8: Construct a greedy policy $\pi(s)$ using $\tilde{v}$:
9: **for** $s \in \mathcal{X}$ **do**

$$\pi(s) \leftarrow \arg\max_{a \in A} E\left[r(s, a, \vec{D}) + \lambda\, \tilde{v}(g(s, a, \vec{D}))\right]$$

10: **end for**
11: **return** $\vec{\theta}$ and π

Python Example 10.2: Linear programming-based approximation

The stochastic production lot-sizing problem above can be recast as an infinite-horizon MDP by dropping the finite-horizon boundary condition (since t is unbounded) and by adding a discount factor, $\lambda = 0.97$, which is necessary for a well-defined infinite-horizon MDP as discussed in Chapter 4. Otherwise, all model parameters remain the same. Using the linear programming-based approach described in Algorithm 10.2, the following Python code solves the problem as an approximate linear program (ALP).

```
import math
import numpy as np
import matplotlib.pyplot as plt
import gurobipy as gp
from gurobipy import GRB

# Input Data
M = 60        # inventory capacity and maximum demand fulfilled
lam = 10      # mean daily demand for Poisson distribution
r = 5.0       # revenue per unit sold
c_S = 20.0    # fixed setup cost (still unused, as in your original code)
c_V = 1.0     # cost per unit ordered
c_H = 0.50    # holding cost per unit leftover inventory
Lambda = 0.97 # discount factor
```

```
# Basis function choice:"constant","linear","quadratic","cubic","quartic", "quintic"
BASIS = "quintic"

# Basis functions
def phi(s: int, M: int):
    if BASIS == "constant":
        return np.array([1.0])
    if BASIS == "linear":
        return np.array([1.0, s])
    if BASIS == "quadratic":
        return np.array([1.0, s, s * s])
    if BASIS == "cubic":
        return np.array([1.0, s, s * s, pow(s,3)])
    if BASIS == "quartic":
        return np.array([1.0, s, s * s, pow(s,3), pow(s,4)])
    if BASIS == "quintic":

        return np.array([1.0, s, s * s, pow(s,3), pow(s,4), pow(s,5)])
    raise ValueError(f"Unknown BASIS: {BASIS}")

# Return value function approximation
def vhat_from_w(w: np.ndarray, M: int): return lambda s: float(phi(int(s), M).dot(w))

# Demand model (truncated Poisson): returns probs for d=0,...,M-1 and P(D>=M)
def truncated_poisson_probs(lam: float, M: int):
    probs = np.zeros(M + 1, dtype=float)
    for d in range(M):
        probs[d] = math.exp(-lam) * (lam ** d) / math.factorial(d)
    probs[M] = max(0.0, 1.0 - probs[:M].sum())
    return probs

# Expected reward and next-state under demand
def expected_future_rewards(s: int, a: int, v_fn, demand_probs: np.ndarray,
    M: int, r: float, c_V: float, c_H: float, Lambda: float):
    sa = s + a  # inventory after order
    total = 0.0
    def contrib(d: int, prob: float):
        sales = min(sa, d)
        leftover = max(0, sa - d)
        s_next = leftover  # lost sales => leftover inventory
        reward = r * sales - (c_V * a + c_H * leftover)
        return prob * (reward + Lambda * v_fn(s_next))
    for d in range(M):
        total += contrib(d, demand_probs[d])
    total += contrib(M, demand_probs[M])
    return total

# Compute max value and argmax decision, for state s
def bellman_iteration(s: int, v_fn, demand_probs: np.ndarray, M: int,
    r: float, c_V: float, c_H: float, Lambda: float):
    best_val = -float("inf")
```

```
    best_decision = 0
    for a in range(0, M - s + 1):
        val = expected_future_rewards(s,a,v_fn,demand_probs,M,r,c_V,c_H,Lambda)
        if val > best_val: best_val, best_decision = val, a
    return best_val, best_decision

#Compute linear program constraint coefficients for the approximate LP
def expected_future_reward_features(s: int, a: int, demand_probs: np.ndarray,
                                    M: int,r: float, c_V: float, c_H: float):
    sa = s + a
    E_reward = 0.0
    E_phi_next = np.zeros(len(phi(0, M)))
    def contrib(d: int, prob: float):
        sales = min(sa, d)
        leftover = max(0, sa - d)
        s_next = leftover
        reward = r * sales - (c_V * a + c_H * leftover)
        return prob, reward, s_next
    for d in range(M):
        prob, reward, s_next = contrib(d, demand_probs[d])
        E_reward += prob * reward
        E_phi_next += prob * phi(s_next, M)
    prob, reward, s_next = contrib(M, demand_probs[M])
    E_reward += prob * reward
    E_phi_next += prob * phi(s_next, M)
    return E_reward, E_phi_next

#Compute linear program constraint coefficients for exact LP
def expected_reward_and_next_state_probs(s: int, a: int, demand_probs: np.ndarray,
    M: int, r: float, c_V: float, c_H: float):
    sa = s + a
    E_reward = 0.0
    p_next = {}
    def add_mass(d: int, prob: float):
        nonlocal E_reward, p_next
        sales = min(sa, d)
        leftover = max(0, sa - d)
        s_next = leftover
        reward = r * sales - (c_V * a + c_H * leftover)
        E_reward += prob * reward
        p_next[s_next] = p_next.get(s_next, 0.0) + prob
    for d in range(M):
        add_mass(d, demand_probs[d])
    add_mass(M, demand_probs[M])  # truncation mass
    return E_reward, p_next

# Approximate Linear Program solve
def solve_approx_alp(M: int, lam: float, r: float, c_V: float, c_H: float, Lambda: float):
```

```
    demand_probs = truncated_poisson_probs(lam, M)
    m = len(phi(0, M))
    mu = np.ones(M + 1) / (M + 1) #Objective: minimize sum_s mu(s) * vhat(s)
    Phi = np.vstack([phi(s, M) for s in range(M + 1)])
    c = Phi.T @ mu  # c in R^m
    model = gp.Model("ALP") #Gurobi linear program formulation
    w = model.addMVar(shape=m, lb=-GRB.INFINITY, name="w") # Decision vars, w
    model.setObjective(c @ w, GRB.MINIMIZE)  # Objective: min c^T w

    # Constraints: for all (s,a),phi(s)^T w >= E[r] + Lambda * E[phi(s')]^T w
    for s in range(M + 1):
        phi_s = phi(s, M)
        for a in range(0, M - s + 1):
            E_reward, E_phi_next = expected_future_reward_features(
                s, a, demand_probs, M, r, c_V, c_H)
            model.addConstr((phi_s - Lambda * E_phi_next) @ w >= E_reward,
                                name=f"bellman_{s}_{a}")
    model.optimize()

    if model.Status != GRB.OPTIMAL:
        raise RuntimeError(f"ALP (Gurobi) failed. Status={model.Status}")
    w_sol = w.X  # numpy array
    v_fn = vhat_from_w(w_sol, M) # Greedy policy induced by vhat
    pi = np.zeros(M + 1, dtype=int)


    for s in range(M + 1):
      _, best_decision = bellman_iteration(s, v_fn, demand_probs, M, r, c_V, c_H, Lambda)
        pi[s] = best_decision
    return w_sol, pi

def solve_exact_lp(M: int, lam: float, r: float, c_V: float, c_H: float, Lambda: float):
    demand_probs = truncated_poisson_probs(lam, M)
    mu = np.ones(M + 1) / (M + 1)     # Weights for LP objective
    model = gp.Model("Exact_Discounted_MDP_LP")
    model.Params.OutputFlag = 0
    V = model.addVars(M + 1, lb=-GRB.INFINITY, name="V") # Variables: V[s]
    # Objective: minimize sum_s mu[s] * V[s]
    model.setObjective(gp.quicksum(mu[s] * V[s] for s in range(M + 1)), GRB.MINIMIZE)

    # Bellman inequalities as constraints: v(s) >= E[r(s,a)] + Lambda * E[v(s')]
    for s in range(M + 1):
        for a in range(0, M - s + 1):
            E_reward, p_next = expected_reward_and_next_state_probs(
                s, a, demand_probs, M, r, c_V, c_H)
            rhs = E_reward + Lambda*gp.quicksum(prob*V[sp] for
                sp,prob in p_next.items())
            model.addConstr(V[s] >= rhs, name=f"bellman_{s}_{a}")
    model.optimize()
    if model.Status != GRB.OPTIMAL:
        raise RuntimeError(f"Exact LP (Gurobi) failed. Status={model.Status}")
    v = np.array([V[s].X for s in range(M + 1)], dtype=float)
```

```
    pi = np.zeros(M + 1, dtype=int)
    for s in range(M + 1):
        _, best_decision = bellman_iteration(s, v_fn=lambda sp: v[sp],demand_probs=
                          demand_probs, M=M, r=r,c_V=c_V, c_H=c_H, Lambda=Lambda)
        pi[s] = best_decision
    return v, pi

# Solve approximate infinite horizon ALP
w_hat, pi_hat = solve_approx_alp(M, lam, r, c_V, c_H, Lambda)
v_hat_fn = vhat_from_w(w_hat, M)
s_grid = np.arange(M + 1)
v_hat = np.array([v_hat_fn(s) for s in s_grid])

# Solve exact infinite-horizon LP for comparison
v_exact, pi_exact = solve_exact_lp(M, lam, r, c_V, c_H, Lambda)
rmse = float(np.sqrt(np.mean((v_hat - v_exact) ** 2)))
max_err = float(np.max(np.abs(v_hat - v_exact)))

print(f"BASIS = {BASIS}   Lambda = {Lambda}")
print(f"RMSE(v): {rmse:.4f}   max|err|: {max_err:.4f}")

plt.figure()
plt.plot(s_grid, v_exact, label="Exact $v(s)$")
plt.plot(s_grid, v_hat, label=f"Approx $\\hat v(s)$ (ALP, {BASIS} basis)")
plt.xlabel("State $s$ (inventory)")
plt.ylabel("Value $v(s)$")
plt.title("Infinite-Horizon Discounted: Exact vs Approx (ALP)")
plt.legend()
plt.tight_layout()
plt.show()
```

Running the above Python code yields a comparison of the ALP approximation with the exact solution, paralleling the results obtained for the finite-horizon version of the problem using Algorithm 10.1. Figure 10.3 shows the results from the case of a fifth-order polynomial, which provides a good approximation to the optimal value function.

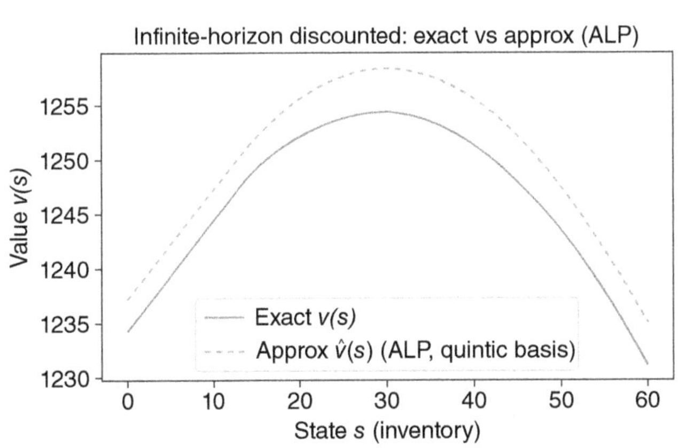

Figure 10.3 Illustration of the approximate value function using the approximate linear programming (ALP) approach for the infinite horizon stochastic production lot-sizing example, using a fifth-order polynomial.

For large-scale problems, ALP may require careful consideration of how to control the size of the resulting linear program. The challenge of exponentially large state spaces is addressed by reformulating the problem as an optimization over basis-function weights, which depend on the number of basis functions. A parsimonious set of basis functions that yields a good approximation will result in a linear program with a manageable number of decision variables. However, the number of constraints may be prohibitive, since the exact representation imposes one for every state-decision pair. For this reason, care must be taken when approximating the resulting linear program. A common approach is to use constraint sampling, in which a subset of constraints is sampled. This approach was motivated by de Farias and Van Roy [8] in the context of dynamic programming and has since been applied to many problems.

10.2.2 Kernel-based Approximations

A related form of value function approximation uses *kernel functions*, which specify a class of functions that provide a measure of similarity between states. For example, selecting a Gaussian kernel function yields the following:

$$k(s, s') = exp\left(\frac{-||s - s'||^2}{2\sigma^2}\right),$$

and the resulting approximate value function is

$$\tilde{v}(s) = \sum_{k=1}^{K} \alpha_k k(s, s_k),$$

where s is the current state and s_k, $k = 1, ..., K$ are neighboring states. The kernel function, $k(s, s')$, is a measure of similarity between states s and s'. In contrast to the basis function approximation described above, where specific functions were selected, with kernel-based approximations, the basis functions $k(\cdot, s_k)$, are determined by the sampled data points s_k, $k = 1, ..., K$. The parameter σ in the kernel function is a hyperparameter that controls the extent to which state proximity influences the value function approximation. If σ is small, the influence of states drops quickly as proximity increases; if σ is large, the kernel function drops off slowly, implying that distant states inform the shape of the local value function approximation.

In some applications, the state s is naturally quantified, and the notion of "distance" as a measure of similarity is natural. For instance, the production planning problems considered above use the inventory level as the state variable, which naturally has a complete ordering from low to high inventory. In other applications, such as those using categorical definitions of states, it may be necessary to introduce a feature vector, $\vec{x}(s) \in R^d$, to quantify d-dimensional states. For instance, in a manufacturing context, the states could be defined by the number of jobs completed, the number waiting in the queue, and the status of machines (e.g., functioning optimally, under repair). In such situations, customized feature vector selection based on expert domain knowledge can be valuable. An example of kernel-based approximations is deferred to the next section to illustrate its use in the context of *policy approximation methods.*

10.3 Policy Iteration-based Approximation

A powerful alternative to value function approximations is algorithms based on the idea of policy iteration, discussed in Chapter 4 and revisited in Chapter 8 in the context of RL in the form of Monte Carlo policy iteration (MCPI), which used a system simulator to avoid constructing a complete MDP model, which may not be feasible for complex systems with a prohibitively large number of states. In fact, MCPI can be viewed as an approximate policy-iteration method that uses statistical sampling to estimate value functions and update the policy. One of the earliest foundational policy approximation algorithms was proposed by Lagoudakis and Parr [9], called *least-squares policy iteration* (LSPI). The basic idea of LSPI is to represent the Q-function as a linear approximation and to estimate the weights that approximately satisfy the optimality equations via least-squares minimization. Subsequently, using the resulting Q-function to find an improved policy completes an iteration. The algorithm proceeds iteratively until a stopping criterion based on a convergence measure is satisfied. Algorithm 10.3 illustrates the main steps of LSPI.

Algorithm 10.3 Least-squares Policy Iteration (LSPI)

Require: Discount factor $\lambda \in (0, 1)$; state space S; decision space A; immediate rewards $r(s, a)$ basis functions $\phi_k(s, a)$, $k = 1, \ldots, K$; batch of sampled transitions $\mathcal{D} = \{(s_i, a_i, r_i, s_i') : i = 1, \ldots, N\}$; tolerance $\epsilon > 0$; maximum iterations M.

Initialize $\vec{\theta}^{(0)}$ and $\pi^{(0)}(s) \in \arg\max_{a \in A} \; \vec{\phi}(s, a)\vec{\theta}^{(0)}$
for $m = 0, 1, \ldots, M - 1$ **do**
    Policy evaluation (least-squares estimation)
    Initialize $L \leftarrow 0^{K \times K}$, $\vec{b} \leftarrow 0^K$
    for each sample $(s_i, a_i, r_i, s_i') \in \mathcal{D}$ **do**
        Set $\bar{a}_i \leftarrow \pi^{(m)}(s_i')$
        Set $\Delta_i \leftarrow \vec{\phi}(s_i, a_i) - \lambda \vec{\phi}(s_i', \bar{a}_i)$
        Update

$$L \leftarrow L + \vec{\phi}(s_i, a_i) \cdot \vec{\Delta}_i, \vec{b} \leftarrow \vec{b} + \vec{\phi}(s_i, a_i) \cdot \vec{r}_i$$

    end for
    Solve $L\,\theta^{(m+1)} = \vec{b}$ for $\vec{\theta}^{m+1}$ to get $\tilde{Q}^{(m+1)}(s, a) \leftarrow \vec{\phi}(s, a) \cdot \vec{\theta}^{(m+1)}$
    Policy improvement:
    Generate improved policy $\pi^{(m+1)}(s) \in \arg\max_{a \in A} \; \tilde{Q}^{(m+1)}(s, a)$
    if $\| \vec{\theta}^{(m+1)} - \vec{\theta}^{(m)} \|_2 \leq \varepsilon$ **then**
        break
    end if
end for
return $\vec{\theta}^{(m+1)}$ and policy $\vec{\pi}^{(m+1)}$

Many examples have been used as a testbed for algorithms like LSPI, such as the following, which dates back to Moore's dissertation [10] and was subsequently popularized as a benchmark problem by Singh and Sutton [11]. Figure 10.4 illustrates the basic idea of the problem, which involves continuous state and decision spaces, where an autonomous vehicle drives on a (one-dimensional) road. The agent controlling the car observes the linear distance to a goal location and determines how much, if any, to apply the accelerator.

Python Example 10.3: Mountain car

This example is based on the original version of the problem, which assumed a 1000 m-long road and a pedal height varying between 0 cm (no acceleration) and 10 cm (maximum acceleration). The road passes through mountains; thus, the car's height varies, and the vehicle dynamics are assumed to be governed by an unobservable physical model. When the car is on an incline, gravity contributes to its deceleration; on a decline, it contributes to its acceleration. This example departs from previous examples because an approximation is necessary due to the infinite number of states and decisions. Furthermore, the exact solution for comparison is not feasible to obtain.

The Mountain Car problem can be solved using radial basis functions as the kernel. To demonstrate, this example uses *Gymnasium* proposed by Towers et al. as a standard interface for RL environments, building upon *OpenAI Gym*, a publicly available software library created by Brockman et al. [12] that includes many benchmark problems, such as the Mountain car. Moreover, this example uses *Scikit-learn*, an open-source Python library that provides utilities for RL, including standard operations such as kernel approximation and least-squares linear regression [13].

```
import time
import numpy as np
import gymnasium as gym
from sklearn.preprocessing import StandardScaler
```

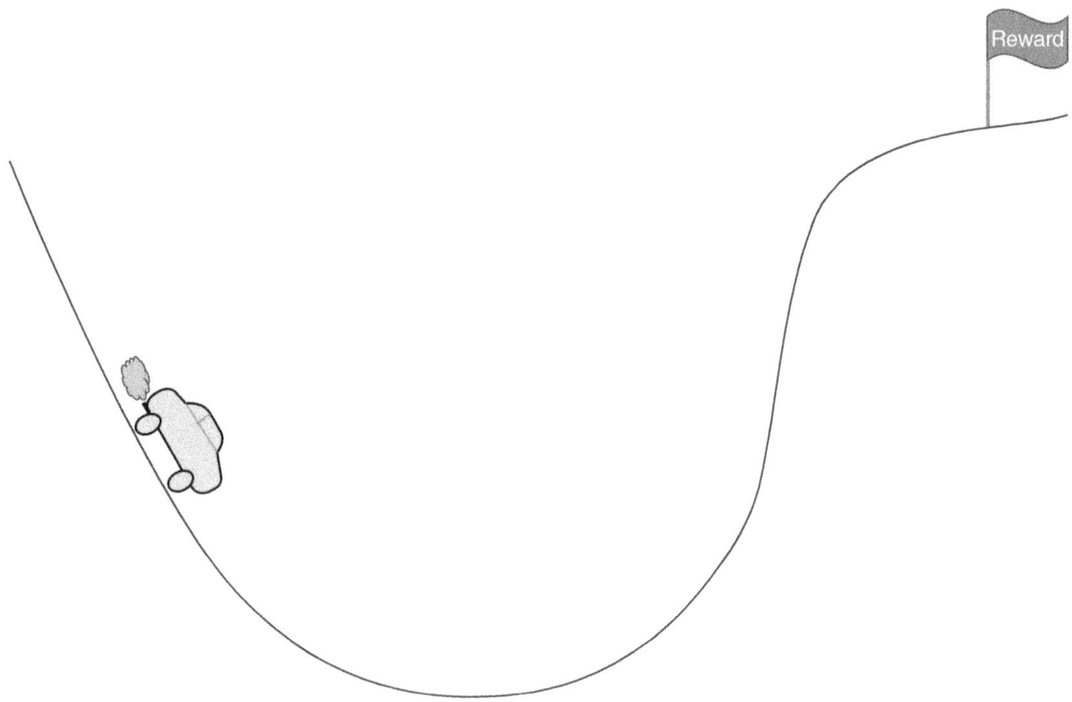

Figure 10.4 The Mountain car benchmark problem with states defined by position and decisions by positive or negative acceleration.

```
from sklearn.kernel_approximation import Nystroem
import imageio
import matplotlib.pyplot as plt
from matplotlib import animation

# Input Data
ENV_ID = "MountainCar-v0"
SEED = 0                      #random number generator seed
DATA_STEPS = 1000             # Total number of state-decision transitions
RBF_COMPONENTS = 30           # Number of radial basis functions
L2_REG = 1e-2                 # keep
RBF_SIGMA = 1.0               # hyperparameter for RBFs
LSPI_MAX_ITERS = 100          # max allowable iterations for LSPI
LSPI_TOL = 1e-6               # stopping criterion for LSPI
DISCOUNT = 0.99               #discount factor
MAX_EPISODE_STEPS = 200
BEHAVIOR_RANDOM_PROB = 0.5  # 50% random, 50% heuristic "push with velocity"
EVAL_EPISODES = 5

# Animation settings
ANIMATE_EPISODE_SEED = 123
ANIMATE_MAX_STEPS = 200       # None uses env default
ANIMATE_INTERVAL_MS = 30       # lower=faster playback
GIF_PATH = "MountainCar_lspi_policy.gif" #animation file

#Define mapping from states to radial basis functions
def fit_feature_map(states: np.ndarray,rbf_components: int, seed: int):
```

```
    scaler = StandardScaler()
    states_scaled = scaler.fit_transform(states)
    rbf_gamma = 1/(2*RBF_SIGMA**2)
    nys = Nystroem(kernel="rbf",gamma=rbf_gamma,n_components=rbf_components,
                   random_state=seed)
    nys.fit(states_scaled)
    def phi_batch(S: np.ndarray) -> np.ndarray:
        S = np.asarray(S, dtype=float)
        S_scaled = scaler.transform(S)
        return nys.transform(S_scaled)  # (N,d)
    def phi_single(s: np.ndarray) -> np.ndarray:
        return phi_batch(np.asarray(s, dtype=float).reshape(1, -1)).ravel()
    d = rbf_components
    return phi_single, phi_batch, d

#Convert state features into state-decision features
def build_decision_features(phi_s: np.ndarray, decisions: np.ndarray,
                            n_decisions: int) -> np.ndarray:
    N, d = phi_s.shape
    D = n_decisions * d
    X = np.zeros((N, D), dtype=float)
    rows = np.arange(N)
    start = decisions * d  # starting column per row
    for k in range(d): # Fill X[rows, start + k] = phi_s[:,k] for k=0..d-1
        X[rows, start + k] = phi_s[:, k]
    return X

#Collect greedy decisions for many states at once
def greedy_decisions_batch(theta: np.ndarray, phi_s: np.ndarray,
                           n_decisions: int) -> np.ndarray:
    N, d = phi_s.shape
    W = theta.reshape(n_decisions, d)        # (A,d)
    Q = phi_s @ W.T                          # (N,A)
    return np.argmax(Q, axis=1).astype(int)

#Collect decision when there is only a single state (during evaluation)
def greedy_decision(theta: np.ndarray,phi_s: np.ndarray,n_decisions: int) -> int:
    d = phi_s.shape[0]
    W = theta.reshape(n_decisions, d)
    q = W @ phi_s
    return int(np.argmax(q))

#Collect state transitions using random or heuristic decisions
def collect_transitions(env_id: str, n_steps: int, seed: int,
                        max_episode_steps:int, behavior_random_prob: float = 1.0):
    env = gym.make(env_id)
    env = gym.wrappers.TimeLimit(env, max_episode_steps)
    rng = np.random.default_rng(seed)
    obs, info = env.reset(seed=seed)
    transitions = []
    for t in range(n_steps):
```

```
        if rng.random() < behavior_random_prob:
            a = int(rng.integers(env.action_space.n))
        else:
            # Heuristic: push with velocity sign (0=left, 1=coast, 2=right)
            a = 2 if obs[1] > 0 else 0
        next_obs, r, terminated, truncated, info = env.step(a)
        done = bool(terminated or truncated)
        transitions.append((obs.copy(), a, float(r), next_obs.copy(), done))
        if done: obs, info = env.reset(seed=seed + t + 1)
        else: obs = next_obs
    env.close()
    return transitions

# Least squares policy iteration (LSPI)
def lspi(transitions, phi_batch_fn, d: int, n_decisions: int,
                    discount: float, l2_reg: float, max_iters: int, tol: float):
    # Unpack transitions
    states   = np.array([tr[0] for tr in transitions], dtype=float)
    decisions = np.array([tr[1] for tr in transitions], dtype=int)
    rewards = np.array([tr[2] for tr in transitions], dtype=float)
    next_s   = np.array([tr[3] for tr in transitions], dtype=float)
    done     = np.array([tr[4] for tr in transitions], dtype=bool)
    # Precompute features
    phi_s  = phi_batch_fn(states)
    phi_sp = phi_batch_fn(next_s)
    X = build_decision_features(phi_s, decisions, n_decisions)
    D = X.shape[1]
    b = X.T @ rewards  #     # Fixed b
    theta = np.zeros(D, dtype=float)
    eye = np.eye(D, dtype=float)

    for it in range(max_iters):
        ap = greedy_decisions_batch(theta, phi_sp, n_decisions)        # (N,)
        Y = build_decision_features(phi_sp, ap, n_decisions)           # (N,D)
        Y[done, :] = 0.0
        A = X.T @ (X - discount * Y)                                 # (D,D)
        A_reg = A + l2_reg * eye
        theta_new = np.linalg.solve(A_reg, b)
        diff = np.linalg.norm(theta_new - theta)
        theta = theta_new
        print(f"LSPI iter {it + 1:2d}: ||Delta|| = {diff:.3e}")
        if diff < tol: break
    return theta

# Policy evaluation + animation of policy application
def evaluate_policy(env_id: str, theta: np.ndarray, phi_single_fn, n_decisions: int,
                    episodes: int, seed: int, max_episode_steps:int):
    env = gym.make(env_id)
    env = gym.wrappers.TimeLimit(env, max_episode_steps)
    returns = []
    for ep in range(episodes):
```

```
        obs, info = env.reset(seed=seed + ep)
        done = False
        total = 0.0
        while not done:
            phi_s = phi_single_fn(obs)
            a = greedy_decision(theta, phi_s, n_decisions)
            obs, r, terminated, truncated, info = env.step(a)
            total += float(r)
            done = bool(terminated or truncated)
        returns.append(total)
    env.close()
    return float(np.mean(returns)), float(np.std(returns))

#Run one episode with render_mode='rgb_array' and collect frames.
def rollout_frames(env_id: str, theta: np.ndarray, phi_single_fn, n_decisions: int,
                   seed: int, max_episode_steps: int):
    env = gym.make(env_id, render_mode="rgb_array")
    env = gym.wrappers.TimeLimit(env, max_episode_steps)
    obs, info = env.reset(seed=seed)
    frames = [env.render()]  # initial frame
    done = False; total = 0.0; steps = 0
    while not done:
        phi_s = phi_single_fn(obs)
        a = greedy_decision(theta, phi_s, n_decisions)
        obs, r, terminated, truncated, info = env.step(a)
        total += float(r)
        done = bool(terminated or truncated)
        frames.append(env.render())
        steps += 1
    env.close()
    return frames, total, steps

def animate_frames(frames, interval_ms: int = 30, title: str = "", gif_path: str = "out.gif"):
    fig, ax = plt.subplots()

    ax.set_axis_off()
    im = ax.imshow(frames[0])
    if title: ax.set_title(title)
    def update(i):
        im.set_data(frames[i])
        return (im,)
   animation.FuncAnimation(fig, update, frames=len(frames),interval=interval_ms, blit=True)
    plt.show()
    imageio.mimsave(gif_path, frames, duration=interval_ms / 1000.0)
    print(f"Saved GIF to: {gif_path}")

# Run LSPI for training, evaluation; generate animation GIF
env0 = gym.make(ENV_ID)
n_decisions = env0.action_space.n
env0.close()
print(f"Collecting {DATA_STEPS} transitions from {ENV_ID}...")
t0 = time.perf_counter()
```

```
transitions = collect_transitions(ENV_ID, DATA_STEPS, SEED,MAX_EPISODE_STEPS,
                                  behavior_random_prob=BEHAVIOR_RANDOM_PROB)
print(f"Collected in {time.perf_counter() - t0:.2f}s")
states = np.array([tr[0] for tr in transitions], dtype=float)
print("Fitting Nyström feature map...")
t0 = time.perf_counter()
phi_single, phi_batch, d = fit_feature_map(states, RBF_COMPONENTS, SEED)
print(f"Fit    in    {time.perf_counter()    -    t0:.2f}s;    d={d},    decision-
aug dim={n_decisions * d}")
print("Running LSPI ...")
t0 = time.perf_counter()
theta = lspi(transitions=transitions,phi_batch_fn=phi_batch,d=d,n_decisions=
   n_decisions,discount=DISCOUNT,l2_reg=L2_REG,max_iters=LSPI_MAX_ITERS,tol=LSPI_TOL)
print(f"LSPI done in {time.perf_counter() - t0:.2f}s")

# Evaluate
mean_r, std_r = evaluate_policy(ENV_ID, theta, phi_single, n_decisions, episodes=
EVAL_EPISODES, seed=SEED, max_episode_steps=MAX_EPISODE_STEPS)
print(f"Evaluation over {EVAL_EPISODES} episodes: mean return = {mean_r:.1f}, std = {std_r:.1f}"

# Animate a single rollout
print("Generating animation...")
frames, total, steps = rollout_frames(ENV_ID, theta, phi_single, n_decisions,seed=
                     ANIMATE_EPISODE_SEED,max_episode_steps=ANIMATE_MAX_STEPS)
title = f"LSPI policy | return={total:.0f}, steps={steps}"
animate_frames(frames,interval_ms=ANIMATE_INTERVAL_MS,title=title,gif_path=GIF_PATH)
```

Running the Python code generates a GIF file named MountainCar_lspi_policy, with an animation of the policy found for controlling the car. Experimenting with algorithmic parameters, including training time, provides insight into their influence.

10.3.1 Policy Approximation Methods

There are many approaches that directly approximate policies. The basic idea is to parameterize the policy function that maps states to decisions and optimize the choice of parameters to find a policy that approximates the optimal solution well. Many approaches deserve mention, several of which are briefly summarized here:

1. *Parametric Policy Approximations*: These approaches pre-specify a functional form for the policy and optimize it by searching over parameter values. For example, with linear decision rules, the form of the approximation can be expressed as

   $$\pi_\theta(s) \in \arg\max_{a \in A} u_\theta(s, a).$$

   where $u_\theta(s, a)$ a *preference function* that serves to rank decisions. The parameters θ are learned from sample paths.
2. *Lookahead Approximations*: These approximations use Monte Carlo simulation to generate sample paths over a subset of the future decision epochs. For example, Bertsekas, Tsitsiklis, and Wu [14] proposed *rollout algorithms* for sequential decision-making in combinatorial optimization problems. Starting with a baseline policy, lookahead approximations simulate sample paths under it to estimate the value function, which is then used to select an improving decision.
3. *Gradient-based Approximations*: These methods seek to find policies using gradient following algorithms based on statistical sampling, which is often credited to the work of Williams [15] in the sequential decision-making context. The basic idea is to use a search algorithm, such as *stochastic gradient ascent*, to optimize the policy parameters θ with respect to an objective measuring policy performance (e.g., expected discounted return under π_θ).

An example of a gradient-based method is *proximal policy optimization* (PPO), which addresses the frequent challenge of high policy variation from one iteration to the next, which can be counterproductive. PPO methods penalize large deviations from the current policy in a single iteration by adding a penalty term to the objective function. The result is to smooth the path of policies generated at each iteration and increase the efficiency of the search. The following example illustrates its use by employing the Stable-Baseline3 (SB3) open-source library, which provides implementations of many *deep reinforcement learning* methods. The package is so named because it employs *deep neural networks* to approximate the policy (or value functions, depending on the selected algorithm) by optimizing the neural network parameters using methods such as PPO to maximize expected return and training on sample paths. While neural networks are beyond the scope of this book, the following example is provided to inspire readers to learn more about the topic.

Python Example 10.4: Training an agent to pilot a lunar lander

This example considers training an agent to pilot a lunar lander using thrusters in a low-gravity environment. The Gymnasium environment (LunarLander-v3) is used in combination with the PPO algorithm implementation in the SB3 library. The problem itself has an eight-dimensional state space. It is defined by the following factors: (1) the two-dimensional lander position coordinates, (2) the two-dimensional velocities, (3) the angle and angular velocity, and (4) the contact sensors on the lander legs. The decisions include: do nothing, apply the left engine, apply the center engine, and apply the right engine. Algorithmic parameters include the training period length, the neural network size, the GIF duration, and the frame rate (FPS). Most of the parameters that can be adjusted by the user are set to their default values in the Python code below; however, for other models, it may be beneficial to adjust parameters such as the number of time steps per episode, regularization parameters, batch size for the stochastic gradient descent algorithms, and many others the reader can learn about from the SB3 library documentation.

```
#Chapter 10, Example 10.4: Training an agent to pilot a lunar lander
import os, imageio
import gymnasium as gym
from stable_baselines3 import PPO
from stable_baselines3.common.monitor import Monitor

#Input data
ENV_ID = "LunarLander-v3"
TRAIN_TIMESTEPS = 20_0000
SEED = 0
GIF_OUTPUT_DIR = "GIFS"
GIF_NAME = "LunarLander_PPO_policy"
GIF_MINUTES = 1
GIF_FPS = 30
NET_SIZE = [128, 128]# PPO network size

# Record a GIF animation of back-to-back episodes until time elapses
def record_gif(model, env_id, out_dir="videos", gif_name="LunaryLander_PPO_policy",
               minutes=1, fps=30, seed=0):
    os.makedirs(out_dir, exist_ok=True)
    target_frames = int(minutes * 60 * fps)
    frames, total_return, episodes = [], 0.0, 0
    env = gym.make(env_id, render_mode="rgb_array")
    obs, info = env.reset(seed=seed)
    terminated = truncated = False
    while len(frames) < target_frames:
        decision, _ = model.predict(obs, deterministic=True)
        obs, reward, terminated, truncated, info = env.step(decision)
        total_return += reward
        frames.append(env.render())
```

```
            if terminated or truncated:
                episodes += 1
                obs, info = env.reset(seed=seed + episodes)
                terminated = truncated = False
        env.close()

        gif_path = os.path.join(out_dir, f"{gif_name}.gif")
        # duration is seconds per frame
        imageio.mimsave(gif_path, frames, duration=1.0 / fps)
        print(f"Saved {gif_path}  (~{minutes} min, {episodes+1} episodes, total return:
        {total_return:.1f})")

# Train proximal policy optimization and record animation
print(f"Training PPO on {ENV_ID} for {TRAIN_TIMESTEPS} timesteps...")
env = Monitor(gym.make(ENV_ID))
model = PPO("MlpPolicy", env, seed=SEED, verbose=0,policy_kwargs=dict(net_arch=NET_SIZE))
model.learn(total_timesteps=TRAIN_TIMESTEPS)
env.close()
print("Recording animation of LunarLandero...")
record_gif(model, ENV_ID, out_dir=GIF_OUTPUT_DIR, gif_name=GIF_NAME,
         minutes=GIF_MINUTES, fps=GIF_FPS, seed=SEED)
```

Running the above Python code yields a PPO policy whose quality varies with the number of time-steps; <10,000 episodes often yield policies that crash the lander, whereas policies with more episodes (e.g., 20,000) often lead to a successful landing. The GIF file generated by running the code, LunarLander_PPO_policy, gives an easy-to-evaluate animation of the policy in action. Many factors can influence the approach's success or failure, and independent runs with different random number seeds exhibit statistical variation in outcomes.

10.4 Concluding Remarks

This chapter briefly covered some fundamental ideas in approximate dynamic programming. Readers interested in learning more are encouraged to explore recent books on approaches to large-scale sequential decision-making, such as Powell [1], Sutton and Barto [16], and Bertsekas [3], all of which provide advanced treatments of the topics discussed in this chapter. Many concepts used in approximation methods are grounded in general optimization. A review by Bottou et al. [17] provides a rigorous overview of stochastic gradient methods for large-scale problems. Other powerful approaches combine ideas from value or policy iteration with neural networks (e.g., PPO) to approximate the optimal value function, thereby giving rise to a body of research known as *deep reinforcement learning*. Lapan [18] provides a practical introduction, and Plaat [19] offers a helpful textbook-style starting point for readers interested in the broader concepts underlying Deep RL.

As is likely apparent to the reader now, approximation methods for large-scale models extend the conceptual ideas from prior chapters to real-world problems. Success with these methods often requires experimentation to identify suitable algorithmic parameters. But this experimentation and the subsequent computational effort often yield policies that can be applied efficiently to the policy at hand. Of course, training algorithms with models like the examples above runs a significant risk of not accounting for failure modes that may occur in the real world. Thus, the practical implementation of policies in the real world typically requires careful investigation and human oversight, with multiple training iterations to ensure a safe and successful launch of a new project, particularly when harm is possible.

10.5 Practice Exercises

1. Consider a discrete-time single-server queuing system with a capacity limit of 5 customers. The number of customers in the system at each decision epoch defines the state of the system $s_t \in S = \{0,\ 1,\ 2,\ 3,\ 4,\ 5\}$. At each epoch, a customer arrives with probability p, and a service completion occurs with probability q if $s_t > 0$. When a customer arrives and $s_t < 5$, a decision-maker chooses to accept (1) or reject (0) the customer, $a_t \in \{0, 1\}$. If no customer arrives or $s_t = 5$ for epoch t, then $a_t = 0$. Indicator functions I_t and J_t keep track of arrivals ($I_t = 1$) and departures ($J_t = 1$) with their associated probabilities. The resulting state transfer equation is:

$$s_{t+1} = \min\{5,\ s_t + a_t I_t\} - J_t.$$

The immediate reward at epoch t is $r_t = R\, a_t I_t - h s_t$, where $R > 0$ is the reward for accepting a customer and $h > 0$ is a cost per epoch per customer in the system. With discount factor $\lambda \in (0, 1)$, the decision-maker wishes to find a stationary policy π that maximizes $E^\pi[\sum_{t=0}^{\infty} \lambda^t r_t]$.

a. Write the transition probabilities $P(s' \mid s, a)$ in terms of p and q.

b. Propose a feature map $\vec{\phi}(s, a) \in R^K$ for a linear approximation $\tilde{Q}_{\vec{\theta}}(s, a) = \vec{\phi}(s, a)\vec{\theta}$ and briefly justify your choice.

c. Use LSPI to approximate an optimal policy and analyze the sensitivity of the method to algorithmic parameters such as the stopping time criterion and the number of basis functions.

2. Battery storage can be modeled as an infinite-horizon MDP. The state is the battery charge $s_t \in \{0,\ 1, \ldots, 10\}$. Decisions are to charge, hold, or discharge, with $a_t \in \{-1,\ 0,\ +1\}$, respectively. Demand at each epoch is uncertain, leading to possible shortfall, no change, or a surplus, defined by $D_t \in \{-1, 0, +1\}$ with probabilities $P(D_t = -1) = 0.25$, $P(D_t = 0) = 0.50$, and $P(D_t = +1) = 0.25$. The transfer equation is

$$s_{t+1} = \min\{10, \max\{0,\ s_t + a_t - D_t\}\}.$$

The immediate reward is avoided grid purchase cost minus a penalty for battery charge or discharge

$$r(s_t, a_t, D_t) = c \max(D_t - a_t, 0) - \kappa |a_t|,$$

with $c = 6$ and $\kappa = 0.4$. Use the approximate linear programming approach, Algorithm 10.2, with basis functions $\phi(s) = [1,\ s,\ s^2, \ldots]$. Experiment with the number of basis functions to evaluate convergence to a near-optimal solution.

3. Describe the key differences between the AVI and LSPI algorithms.

4. Delta Airlines offers callbacks when agents are busy. Let the number of customers in the callback queue be $s_t \in \{0, 1, 2, 3, 4, 5, 6, 7\}$. At each decision epoch, a new callback request arrives with probability $p = 0.40$. If $s_t > 0$, then with probability $q = 0.50$, one callback is completed (and the queue decreases by one). If an arrival occurs and $s_t < 7$, the manager chooses whether to accept the request ($a_t = 1$) or reject it ($a_t = 0$); if $s_t = 7$, requests must be rejected. Accepting a request yields a reward of +3, and a waiting penalty of 0.6 s_t per decision epoch. Assuming a discount factor $\lambda = 0.97$, answer the following:

a. Write a state transition equation for s_{t+1} in terms of s_t, a_t, and Bernoulli random variables representing an arrival and a completion.

b. Propose a feature vector $\phi(s, a)$ with 4–6 components for a linear approximation $\tilde{Q}_\theta(s, a) = \phi(s, a)^\top \theta$.

c. Solve the model using the AVI algorithm.

References

1 Powell, W. B. (2022). *Sequential Decision Analytics and Modeling: Modeling with Python.* now publishers, Boston–Delft. https://doi.org/10.1561/9781638280835.

2 Powell, W. B. (2011). *Approximate Dynamic Programming: Solving the Curses of Dimensionality* 2nd. ed. John Wiley & Sons.

3 Bertsekas, D. P. (2023). *A Course in Reinforcement Learning.* ISBN 978-1-886529-49-6.

4 Gordon, G. J. (1995). Stable function approximation in dynamic programming. In *Machine Learning Proceedings 1995* (pp. 261–268). Morgan Kaufmann.

5 Tsitsiklis, J. N. & Van Roy, B. (1996). Feature-based methods for large scale dynamic programming. *Machine Learning, 22* (1), 59–94.

6 Schweitzer, P. J. & Seidmann, A. (1985). Generalized polynomial approximations in Markovian decision processes. *Journal of Mathematical Analysis and Applications, 110*, 568–582.

7 de Farias, D. P. & Van Roy, B. (2003). The linear programming approach to approximate dynamic programming. *Operations Research, 51* (6), 850–865.

8 de Farias, D. P. & Van Roy, B. (2004). On constraint sampling in the linear programming approach to approximate dynamic programming. *Mathematics of Operations Research, 29* (3), 462–478.

9 Lagoudakis, M. G. & Parr, R. (2003). Least-squares policy iteration. *Journal of Machine Learning Research, 4*, 1107–1149.

10 Moore, A. (1990). *Efficient Memory-Based Learning for Robot Control*. University of Cambridge. Available at: https://www.cl.cam.ac.uk/techreports/UCAM-CL-TR-209.pdf.

11 Singh, S. P. & Sutton, R. S. (1996). Reinforcement learning with replacing eligibility traces. *Machine Learning, 22* (1), 123–158.

12 Brockman, G., Cheung, V., Pettersson, L., Schneider, J., Schulman, J., Tang, J. & Zaremba, W. (2016). OpenAI Gym. *arXiv* preprint arXiv:1606.01540.

13 Pedregosa, F., Varoquaux, G., Gramfort, A., Michel, V., Thirion, B., Grisel, O., Blondel, M., Prettenhofer, P., Weiss, R., Dubourg, V. & VanderPlas, J. (2011). Scikit-learn: Machine learning in Python. *Journal of Machine Learning Research, 12*, 2825–2830.

14 Bertsekas, D. P., Tsitsiklis, J. N. & Wu, C. (1997). Rollout algorithms for combinatorial optimization. *Journal of Heuristics, 3* (3), 245–262.

15 Williams, R. J. (1992). Simple statistical gradient-following algorithms for connectionist reinforcement learning. *Machine Learning, 8* (3), 229–256.

16 Sutton, R. S. & Barto, A. G. (2018). *Reinforcement Learning: An Introduction* 2nd. ed. MIT Press.

17 Bottou, L., Curtis, F. E. & Nocedal, J. (2018). Optimization methods for large-scale machine learning. *SIAM Review, 60* (2), 223–311.

18 Lapan, M. (2024). *Deep Reinforcement Learning Hands-On*. Packt Publishing.

19 Plaat, A. (2022). *Deep Reinforcement Learning*. Springer, Singapore.

Appendix

A.1 Mathematical Notation

Symbol	Meaning
$\forall$	For all
$\exists$	There exists
$\in$	Element of
$\notin$	Not an element of
$\subset$	Proper subset of
$\subseteq$	Subset or equal to
$\cup$	Union
$\cap$	Intersection
$\setminus$	Set difference
$\emptyset$	Empty set
$\mathbb{N}$	Set of natural numbers
$\mathbb{Z}$	Set of integers
$\mathbb{R}$	Set of real numbers
$=$	Equals
$\neq$	Not equal to
$<$	Less than
$>$	Greater than
$\leq$	Less than or equal to
$\geq$	Greater than or equal to
$\pm$	Plus-minus
$\cdot$	Dot product
$\times$	Cartesian product
$\sum$	Summation
$\prod$	Product
$\int$	Integral
∇	Gradient

(*Continued*)

(Continued)

Symbol	Meaning
∞	Infinity
$\perp$	Perpendicular
$\parallel$	Parallel
$\approx$	Approximately equal
$\equiv$	Identically equal
$\propto$	Proportional to
log	Logarithm
ln	Natural logarithm
exp	Exponential function
lim	Limit

A.2 Derivation of Newsvendor Model Optimal Solution

The newsvendor problem arises in the context of a single decision, x, to purchase a supply of a perishable good (e.g., food, vaccines, tickets to a sporting event) to meet random demand. The 2-SLP model was formulated in Chapter 5 as follows. Let $Q(x, z(\omega))$ denote the cost for a particular choice of purchase quantity, x, and scenario ω:

$$Q(x, z(\omega)) = cx + c^u \max\{z(\omega) - x, 0\} + c^o \max\{x - z(\omega), 0\}.$$

The optimization problem is $\min_x\{Q(x) = E_z[Q(x, z)]\}$, where the dependency of z on ω is suppressed for simplicity. The function $Q(x)$ is the *recourse function*. An alternative closed-form solution to this problem is as follows. Using the fact that $x = z + (x-z)^+ - (z-x)^+$ it follows that:

$$\begin{aligned} Q(x, z) &= cz + c(x-z)^+ - c(z-x)^+ - pz + p(z-x)^+ - r(x-z)^+ \\ Q(x) &= E_z[Q(x, z)] = (c-p)\mu + E_z[(c-r)(x-z)^+ + (p-c)(z-x)^+] \end{aligned}.$$

Letting $c^u = p - c$ and $c^o = c - r$ gives:

$$x^* = \arg\min_x\{E_\xi[c^u(z-x)^+ + c^o(x-z)^+]\}.$$

Using Leibniz's rule gives:

$$\frac{\partial}{\partial x}\int_{a(x)}^{b(x)} f(z, x)dz = \int_{a(x)}^{b(x)} \frac{\partial f(z, x)}{\partial x}dz + f(b(x), x)\frac{\partial b}{\partial x} - f(a(x), x)\frac{\partial a}{\partial x},$$

the first term is:

$$c^u\frac{\partial}{\partial x}\int_x^\infty (z-x)f(z)dz = c^u\int_x^\infty (-1)f(z)dz = -c^u(1 - F(x)),$$

and the second term is:

$$c^o\frac{\partial}{\partial x}\int_0^x (x-z)f(z)dz = c^o\int_0^x f(z)dz = c^oF(x).$$

Reorganizing these we find that $x^* = F^{-1}\left(\frac{c^u}{c^u + c^o}\right)$ where $F^{-1}(\cdot)$ denotes the inverse of the cumulative distribution function. It can be interpreted as a function that takes an argument and produces the input to the function $F(\cdot)$ that would generate the argument, which in this case is the optimal purchase quantity, x^*.

A.3 Theorem Proofs

This appendix provides mathematical proofs for any theorems covered in the chapters of this book that are not included in the main body of the text.

Theorem 4.4 *(Bellman, 1957)*: *If there exists a vector $\vec{v}$ that solves $\vec{v} = L\vec{v}$, then $\vec{v}$ is the expected discounted future rewards under the optimal policy, π^*.*

Proof: Let $\vec{v}^*$ denote expected future rewards under the optimal policy π^*. This proof is in two parts: (a) prove that if $\vec{v} \geq L\vec{v}$ then $\vec{v} \geq \vec{v}^*$ and (b) prove if $\vec{v} \leq L\vec{v}$ then $\vec{v} \leq \vec{v}^*$. From which it follows that if $\vec{v} = L\vec{v}$ then $\vec{v}$ is the optimal value function, which we denote here as $\vec{v}^*$.

Part (a): For some arbitrary policy π if $\vec{v} \geq L\vec{v}$ then $\vec{v} \geq \max_\pi\{\vec{r}_\pi + \lambda P_\pi \vec{v}\}$. It follows that:

$$\vec{v} \geq \vec{r}_\pi + \lambda P_\pi \vec{v} = \vec{r}_\pi + \lambda P_\pi(\vec{r}_\pi + \lambda P_\pi \vec{v})$$

$$= \vec{r}_\pi + \lambda P_\pi \vec{r}_\pi + \cdots + \lambda^n P_\pi^n \vec{v}.$$

Therefore it follows from subtracting $\vec{v}^*$ from both sides that:

$$\vec{v} - \vec{v}^* \geq \lambda^n P_\pi^n \vec{v} - \sum_{k=n}^{\infty} \lambda^k P_\pi^k \vec{r}_\pi. \tag{A.1}$$

Since $||\lambda^n P_\pi^n \vec{v}|| \leq \lambda^n ||\vec{v}||$ and $0 < \lambda < 1$ for any ϵ there exists some n sufficiently large:

$$(-\epsilon/2)\vec{e} \leq \lambda^n P_\pi^n \vec{v} \leq (\epsilon/2)\vec{e},$$

where $\vec{e}$ is a vector of ones of dimension equal to that of vector $\vec{v}$. Since the rewards are finite and the norm of a transition probability matrix is 1 it follows that:

$$||\lambda^k P_\pi^k \vec{r}_\pi|| \leq \lambda^k ||P_\pi \vec{r}_\pi|| \leq \lambda^k ||\vec{r}_\pi|| \leq \lambda^k M,$$

where M is a finite upper bound on all elements of the reward vector. Thus, it follows that

$$\sum_{k=n}^{\infty} \lambda^k P_\pi^k \vec{r}_\pi \leq M \sum_{k=n}^{\infty} \lambda^k = \frac{\lambda^n M}{1-\lambda},$$

where the latter equality is obtained by taking the difference between $\sum_{i=0}^{\infty} \lambda^i = \frac{1}{1-\lambda}$ and $\sum_{i=0}^{n} \lambda^i = \frac{1-\lambda^n}{1-\lambda}$. Using this bound, for any given ϵ there exists n sufficiently large that:

$$(-\epsilon/2)\vec{e} \leq \sum_{k=0}^{\infty} \lambda^k P_\pi^k \vec{r}_\pi \leq (\epsilon/2)\vec{e}.$$

Using Equation (1) and the above results, it follows that:

$$\vec{v} \geq \vec{v}^* + \lambda^n P_\pi^n \vec{v} - \sum_{k=n}^{\infty} \lambda^k P_\pi^k \vec{r}_\pi \geq \vec{v}^* - \epsilon.$$

Therefore in the limit $n \to \infty$ and $\epsilon \to 0$ it follows that $\vec{v} \geq \vec{v}^*$. This completes the proof of part (a).

Part (b): If $\vec{v} \leq L\vec{v}$ then there exists some policy π such that $\vec{v} \leq \vec{r}_\pi + \lambda P_\pi \vec{v}$. It follows that $\vec{v} \leq (I - \lambda P_\pi)^{-1} \vec{r}_\pi = \vec{v}^\pi$. Therefore $\vec{v} \leq \max_\pi\{\vec{v}^\pi(s)\} = \vec{v}^*$.

From parts (a) and (b) it follows that if $\vec{v} = L\vec{v}$ then $\vec{v} = \vec{v}^*$.

Theorem 4.5 *(Bellman, 1957)*: *If $0 < \lambda < 1$, then L is a contraction mapping.*

Proof: Let $\vec{v}$ and $\vec{u}$ be arbitrary vectors. Consider an arbitrary state s such that $Lv(s) \geq Lu(s)$ and

$$a^* \in \arg\max_a \left\{ r(s,a) + \lambda \sum_j p(j \mid s,a) v(j) \right\}.$$

Then it follows that

$$0 \leq Lv(s) - Lu(s) \leq r(s,a^*) + \lambda \sum_j p(j \mid s,a^*) v(j) - r(s,a^*) - \lambda \sum_j p(j \mid s,a^*) u(j) \tag{A.2}$$

$$\leq \lambda \sum_j p(j \mid s,a^*)(v(j) - u(j)) \tag{A.3}$$

$$\leq \lambda \sum_j p(j \mid s,a^*) ||v(j) - u(j)|| \tag{A.4}$$

$$= \lambda ||v - u||. \tag{A.5}$$

Thus $Lv(s) - Lu(s) \leq \lambda ||v - u||$ for all s. Repeating this for the case in which $Lv(s) \leq Lu(s)$ yields $||Lv(s) - Lu(s)|| \leq \lambda ||v - u||$ for all s. Therefore, it follows that $||L\vec{v} - L\vec{u}|| \leq \lambda ||\vec{v} - \vec{u}||$ since the property is true for all states, s. Thus, it follows that L is a contraction mapping.

Theorem 4.6 *(Bellman, 1957)*: *For an arbitrary vector $\vec{v}^0$ the sequence $\{\vec{v}^0, \dots, \vec{v}^n\}$ defined by $\vec{v}^{n+1} = L\vec{v}^n$ converges to $\vec{v}^*$ as $n \to \infty$, where $\vec{v}^* = L\vec{v}^*$.*

Proof: First, show that $\{\vec{v}^0, \dots, \vec{v}^n\}$ converges to a limit point. For some n:

$$||\vec{v}^{n+1} - \vec{v}^n|| = ||L^n\vec{v}^1 - L^n\vec{v}^0|| \leq \lambda^n ||\vec{v}^1 - \vec{v}^0||.$$

Therefore, given $\lambda < 1$ it follows that

$$\lim_{n\to\infty} ||\vec{v}^{n+1} - \vec{v}^n|| = 0,$$

and the sequence converges to a limit point, $\vec{v}^*$. Next, show that the limit point $\vec{v}^*$ is the expected future discounted rewards under the optimal policy and a solution to the optimality equations. This follows because

$$0 \leq ||L\vec{v}^* - \vec{v}^*|| = ||L\vec{v}^* - \vec{v}^n + \vec{v}^n - \vec{v}^*|| \tag{A.6}$$

$$\leq ||L\vec{v}^* - \vec{v}^n|| + ||\vec{v}^n - \vec{v}^*|| \tag{A.7}$$

$$\leq \lambda ||\vec{v}^* - \vec{v}^{n-1}|| + ||\vec{v}^n - \vec{v}^*||. \tag{A.8}$$

The first inequality follows from the *triangle inequality*, and the second follows from L being a contraction mapping, which follows from Theorem 4.5 above. From the first part of the proof, it follows that:

$$\lim_{n\to\infty} ||\vec{v}^n - \vec{v}^*|| = 0,$$

from which it follows that in the limit that $n \to \infty$ the right-hand side of the above inequality goes to zero. Therefore

$$0 \leq ||L\vec{v}^* - \vec{v}^*|| \leq 0,$$

and it follows that the limiting point $\vec{v}^*$ is the solution to the optimality equations $L\vec{v} = \vec{v}$.

Theorem 4.7 *(Bellman, 1957)*: *Let the sequence $\{\vec{v}^0, \vec{v}^1, \dots, \vec{v}^n\}$ satisfy $\vec{v}^{n+1} = L\vec{v}^n$, then:*

$$||\vec{v}^{n+1} - \vec{v}^*|| \leq \frac{\epsilon}{2} \quad \text{whenever} \quad ||\vec{v}^{n+1} - \vec{v}^n|| < \frac{\epsilon(1-\lambda)}{2\lambda}.$$

Proof: Drawing on concepts from the proof of Theorem 4.4 and the fact that $\vec{v}\pi^*$ is a limit point of $\vec{v}^{n+1} = L\vec{v}^n$ it follows that:

$$||\vec{v}^{n+1} - \vec{v}\pi^*|| = ||\vec{v}^{n+1} - L\vec{v}^*|| \tag{A.9}$$
$$\leq ||\vec{v}^{n+1} - L\vec{v}^{n+1}|| + ||L\vec{v}^{n+1} - \vec{v}^*|| \tag{A.10}$$
$$\leq ||L\vec{v}^{n} - L\vec{v}^{n+1}|| + ||L\vec{v}^{n+1} - \vec{v}^*|| \tag{A.11}$$
$$\leq \lambda||\vec{v}^{n} - \vec{v}^{n+1}|| + \lambda||\vec{v}^{n+1} - \vec{v}^*||. \tag{A.12}$$

Therefore, it follows that:

$$||\vec{v}^{n+1} - \vec{v}^*|| \leq \frac{\lambda}{1-\lambda}||\vec{v}^{n+1} - \vec{v}^n||.$$

Substituting the condition from the theorem, i.e, that

$$||\vec{v}^{n+1} - \vec{v}^n|| < \frac{\epsilon(1-\lambda)}{2\lambda}$$

yields the desired result, as follows:

$$||\vec{v}^{n+1} - \vec{v}^*|| \leq \frac{\epsilon}{2}.$$

Theorem 4.8 ***(Howard, 1957)*****:** *The policy iteration algorithm terminates in a finite number of iterations with a solution to the optimality equations and an optimal policy π^*.*

Proof: This proof is in two parts. In the first part, it will be shown that if $\vec{v}^n$ and $\vec{v}^{n+1}$ are successive value vectors generated by the policy iteration algorithm. Then,

$$\vec{v}^{n+1} \geq \vec{v}^n.$$

To prove this, let

$$\pi^{n+1} \in \arg\max_{\pi\in\Pi} \{\vec{r}_\pi + \lambda P_\pi \vec{v}^n\}.$$

Then it follows that:

$$\vec{r}_{\pi^{n+1}} + \lambda P_{\pi^{n+1}}\vec{v}^n \;\geq\; \vec{r}_{\pi^n} + \lambda P_{\pi^n}\vec{v}^n = \vec{v}^n,$$

since π^{n+1} maximizes the expression on the left-hand side. Therefore,

$$\vec{r}_{\pi^{n+1}} \geq (I - \lambda P_{\pi^{n+1}})\,\vec{v}^n,$$

since $(I - \lambda P_{\pi^n})\,\vec{v}^n = \vec{r}_{\pi^n}$. Thus, it follows that:

$$(I - \lambda P_{\pi^{n+1}})^{-1}\vec{r}_{\pi^{n+1}} = \vec{v}^{n+1} \geq \vec{v}^n.$$

The matrix I denotes the identity matrix. All entries of $(I - \lambda P_{\pi^{n+1}})^{-1}$ are nonnegative, which can be seen from the Neumann series expansion:

$$(I - \lambda P_{\pi^{n+1}})^{-1} = I + \lambda P_{\pi^{n+1}} + \lambda^2 P^2_{\pi^{n+1}} + \cdots,$$

where each term on the right-hand side has nonnegative elements. Hence, the inverse has nonnegative elements, completing the proof.

In the second part of the proof, it is shown that the policy iteration algorithm converges to the optimal policy π^* in a finite number of iterations. From part one of the proof above, the sequence $\{\vec{v}^n\}$ is nondecreasing. Since there is a finite number of possible policies (the state and decision sets are finite), the algorithm must terminate under the stopping rule:

$$\text{If } \pi^{n+1} = \pi^n, \quad \text{stop and set } \pi^* = \pi^n.$$

At termination, $\pi^n = \pi^{n+1}$ and therefore,

$$\vec{v}^n = \vec{r}_{\pi^{n+1}} + \lambda P_{\pi^{n+1}} \vec{v}^n = \max_{\pi \in \Pi} \{\vec{r}_\pi + \lambda P_\pi \vec{v}^n\}.$$

Thus, $\vec{v}^n$ satisfies the optimality equation

$$\vec{v}^n = L\vec{v}^n,$$

and the policy corresponding to $\vec{v}^n$ is optimal for the MDP.

Index

Note: Page numbers in *italics* and **bold** refers to figures and tables respectively.

www.ingramcontent.com/pod-product-compliance
Lightning Source LLC
LaVergne TN
LVHW080057280926
848956LV00047B/23

* 9 7 8 1 3 9 4 3 4 5 7 8 6 *